Employment of Functionalized Natant Bio-materials as Potential Adsorbents for Industrial Liquid Litters

Scaleup of Plausible Solutions through Greener Approach

N. Muthulakshmi Andal

ELIVA PRESS

ISBN: 978-99949-8-444-2

Cover Design: Eliva Press
Cover Image: Ingimage
Email: info@elivapress.com
Website: www.elivapress.com

This Eliva Press imprint is published by the registered company Eliva Press Global Ltd. part of Eliva Press S.R.L. Publishing Group

The registered company address is: Pope Hennessy Street Level 2, Hennessy Tower Port Louis, Mauritius
Eliva Press S.R.L. Publishing Group address is: Bulevardul Moscova 21, Chisinau, Moldova, Europe

Table of Contents

S.No	Publications	Journal	Page . no
1.	Synthesis, Characterisation and Utilisation of Magnetic Fe_3O_4 – TGT Nanocomposite in the Removal of Pb(II) from Aqueous Solutions	Asian Journal of Water, Environment and Pollution	1
2.	Recovery of Reusable Water and Sludge from Various Wastewater Sources: A Review	Mukt Shabd Journal	16
3.	Sequestration of Cu(II) Ions from Aqueous Mediausing Natant Magnetite-Chitosan Composite	Journal of Materials and Environmental Science	47
4.	Organophosphorus Pesticide Chelation by Modified Pod Shells – Batch Equilibration andIsothermal Studies	Indian Journal of Environmental Pollution	62
5.	Pulp and Paper Industrial Discharges on Soil Environs in Crop Nutrients' Management	Annals of R.S.C.B	75
6.	Sequestration of Zinc Ion using Sugar Industry Residue: An Avenue to Enviro safe	Pollution Research	89
7.	Investigation on Physico – Chemical Characteristics of Pulp and Paper Industrial Discharges into the Environs	Asian Journal of Research in Chemistry	101
8.	A Study on the Bioaccumulation and Screening of Toxic Metal Ions in Effluents, Fodder and Milk Samples Collected from Erode, Tirupur, Coimbatore Districts	Research Journal of Chemistry and Environment	112
9.	A Comparison of Raw and Treated Plant Debris in the Chelation of Anion from Aqueous Media	Indian Journal of	126

	the Chelation of Anion from Aqueous Media	Environmental Protection	
10.	Employment of Surface Modified Biomaterial in Sequestering Divalent Metal Ion	Environment Asia	135
11.	Sorption Capacity of Bivalve Molluscs Shell in the Removal of Divalent Ion from Aqueous Solution	International Journal of Chemtech Research	141
12.	Facile Synthesis, Characterization and Application of MNC's (Magnetic Nanocomposite in Sequestration of Textile Dye	Indian Journal of Environmental Protection	160
13.	Efficacy of Modified Tree Bark in the Screening of Divalent ions at Laboratory Conditions: Equilibrium and Isothermal Studies	International Journal of Current Engineering and Scientific Research	178
14.	Application of *Capra Aegagrus hircus* Dung Activated Carbon for the Removal of Anionic Dyes - Isothermal Studies	International Journal of Current Engineering and Scientific Research	189
15.	Functionalized Novel Nano Composites in Ni(II)Removal from Aqueous/Effluent Media	International Journal of Scientific Engineering and Research	202
16.	Utilization of Magnetic Nano Cobalt Ferrite Doped Capra Aegagrus hircus Dung Activated Carbon Composite for the Adsorption of Anionic Dyes	Journal of Environmental Chemical Engineering	208
17.	Removal of Reactive Dye using Activated Carbon Prepared from *Prosopis juliflora* bark:	Chemical Science Reviews and	237

	Isothermal, Thermodynamics and Kinetic Studies	Letters	
18.	A Comparative Study on the Sorption of Divalent Ions by Bivalves Shells; Equilibrium and Statistical Studies	Chemical Science Reviews and Letters	258
19.	Adsorption Dynamics of Eco friendly Litter Wastes for Zn (II)/Cr (VI) from Electroplating Effluents - Continuous Column Method	SOJ Mater Sci Eng	277
20.	Assessment of Physicochemical Characteristics and the Level of Nutrient Contents as Well as Heavy Metal Ions in Waters of Three Lakes at Coimbatore, Tamil Nadu, India	Journal of Material and Environmental Science	291
21.	Experimental Optimization of an Agricultural Waste in the Uptake of Cu (II) Ions	Chemical Science Reviews and Letters	305
22.	Sorption Kinetics and Equilibrium Studies on the Removal of Toxic Cr (VI) ions Employing Modified Indian Almond Nut Shells	Journal of Energy Technologies and Policy	314
23.	Sorption Ability of Modified Fruit Nut Shells in the Removal of Co(II) ions from Aqueous Solutions	Universal Journal of Environmental Research and Technology	333
24.	Removal of Nickel from Aqueous Solution using *Azaridachta indica* Seed Shell Powder as Adsorbent	Research and Reviews, Journal of Chemistry	351
25.	A Comparative Study on the Sorption Characteristics of Pb(II) and Hg(II) onto Activated Carbon	E- Journal of Chemistry	361
26.	Removal of Pb(II) ions from Aqueous Solutions using *Bombax ceiba* Saw Dust Activated Carbon	Desalination and Water Treatment	374

27.	Adsorption studies of Iron(III) by Chitin	Journal of Chemical Science	396
28.	Adsorption Dynamics and Equilibrium Studies of Zn (II) onto Chitosan	Indian Academic Science (Chemical Science)	413

Synthesis, Characterisation and Utilisation of Magnetic Fe_3O_4– TGT Nanocomposite in the Removal of Pb(II) from Aqueous Solutions

N. Muthulakshmi Andal* and J. Anuradha

Department of Chemistry, PSGR Krishnammal College for Women, Coimbatore – 4,Tamil Nadu, India

*muthulakshmiandal@psgrkcw.ac.in

Received May 8, 2019; revised and accepted July 28, 2022

Abstract: Magnetic Adsorption Separation (MAS) method is an ideal substitute for the environmental clean-up process. Magnetic material embedded with natural litter material is a facile route to achieve maximum adsorptive ability at a lower dosage, contact time and increased chance of metal-laden adsorbents' recovery from aqueous matrices. In the present study, the synthesis of Fe_3O_4–TGT composite (TGT- C2) by the auto combustion method and its employment for Pb(II) removal from aqueous solutions is discussed. TGT- C2 is characterised using VSM (Vibrating Sample Magnetometer), SEM (Scanning Electron Micro- scope), Particle Size Analyzer, EDAX (Energy Dispersive X-ray Spectrometer) and FTIR (Fourier Transform Infra Red Spectrophotometer) techniques. TGT- C2 is found to be magnetic in nature and their saturation magnetization (Ms)/ Co-ercivity (Hc) values are calculated as 1.54 emu g^{-1}/139.83 G, respectively, being less than bare Fe_3O_4. The synthesised nanocomposite registered a maximum of 98% sequestration of Pb(II) ions under the optimised conditions of 100 mg/L initial metal ion concentration, 10 min agitation time, 50 mg dosage and pH 5 environment. Isothermal verification, the kinetics of adsorption and successive de- sorption/ regeneration cycles were performed. The outcomes support the preparation of bio-nanocomposites from animal waste was successful in the efficient trapping of divalent metal ions.

Key words: Nanocomposite, magnetic nature, litter material, characterisation, sequestration.

Introduction

Heavy metals are considered hazardous pollutants due to their toxicity even at low concentrations and non-biodegradable nature. These metallic elements are an intrinsic com- ponent of the environment. A variety of natural processes are responsible for their acute availability even at trace levels in various parts of the biosphere (Mohammed et al., 2011). Rapid industrialisation and their waste discharges have resulted in an unprecedented increase in a heavy metal influx into natural water bodies. The toxic effects of heavy metals while exceeding their tolerance levels affect the normal metabolic process.

Lead is the most abundant element on earth with the properties viz., heavy, soft, highly malleable, bluish– grey colour metal, stable at two oxidation states (+2 and +4). The major sources of lead pollution to the environmentis from the industries of batteries, paint pigments, PVC plastics, X-ray shielding, crystal glass, pesticides, E-waste, smelting operation, ceramics, bangle industry, auto mobile exhaust (Zhang et al., 2015). Excess con- centrations of Pb(II) than the prescribed limits (> 0.01 mg/L) lead to haematological, gastrointestinal and neurological dysfunction. The most insidious characteristic of lead is the ability to replace calcium in the bones and form a semi-permanent reservoir for long term release after its initial absorption

Hence, it has become a necessary and challenging task to treat heavy metal-contaminated wastewater prior to its discharge into the environment. Several methods for the removal of heavy metals, such as chemical precipitation, coagulation-flocculation, ion exchange and membrane filtration have been reported with their inherent advantages and wide- spread limitations in their applications (Madhava Rao et al., 2008).

The adsorption method finds to be an inevitable alternative to remove the toxic sub- stances polluting the environment. Advantages of the adsorption process over other methods include simple design, easy operation, minimum capital investment, operational cost and selective trapping of metals even at lower concentrations (Barakat, 2011). Even though numerous low cost adsorbents from a variety of sources have been explored due to enormous surface areas per unit weight, in recent times, the search for effective/no-cost/ recoverable adsorbents has intensified.

New functional nanocomposite materials with impregnated nanoparticles are at the fore-front of nano- based water treatment methods. A combination of both nano-sized and magnetic properties exhibited by the modified materials seek recent attention as novel sequestrants. The hybrid composites exhibit better efficiencies viz., extremely small size, high surface area/volume ratio, and easy recovery of the adsorbents through an external magnetic field, thus improving the possibilities of reuse than those of conventional mate- rials/products available in the market for heavy metal removal (Horst et al., 2016). More insight is arrived from novel metal oxides and animal litter-based nanocomposites, through auto combustion technique from the present work. Herein, Goat Teeth, a butcher waste are modified and subjected to the preparation of nanocomposite.

Experimental Section

Preparation of Treated Goat Teeth (TGT)

Goat Teeth (GT), causing solid waste disposal problems collected from localities of Coimbatore, Tamil Nadu, India, stockpiled from local carnage, isolated from jaw bones under running water and sun dried for 10 days. Goat Teeth consist of a high percentage of mineral contents, especially exchangeable Ca^{2+} ions and organic/ bio-chemical materials viz lipids, collagen, proteoglycan (chondroitin sulphate, keratin sulfate). Hence, GT was soaked in 0.1N HCl for a period of 4 hrs, after segregating into 0.18 mm particle size using Scientific Test Molecular Sieves. The treated goat teeth (TGT)were washed several times and air dried (Anuradha etal., 2017).

Synthesis of TGT- C2 [TGT- Magnetite (Fe3O4) Nanocomposite

Ferric benzoate nano precursors were prepared through one-pot synthesis as follows:

Ferric nitrate and benzoic acid solutions were prepared as per the literature, mixed and stirred in a magnetic stirrer for 20 min. Yellow crystals of Iron (III) benzoate dihydrate were formed, later filtered, washed thoroughly with hot distilled water and air dried(Ranjithkumar et al., 2014).

A specific dose of TGT was added to ferric benzoate solution after several trials and stirred for 6 hrs, later subjected to auto combustion. The obtained powder material was

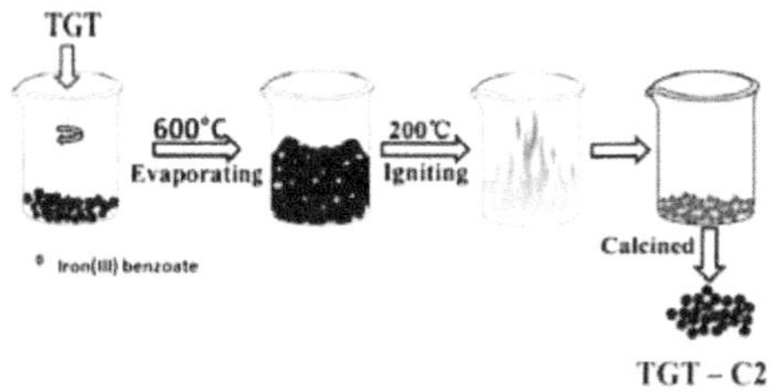

reheated in a muffle furnace for 2 hours which resulted in the in-situ surface formation of magnetic TGT - Fe_3O_4 nanocomposite (TGT- C2). The schematic preparatory procedure of TGT – C2 has been picturised in Figure 1.

Figure 1: Preparation procedure - TGT- C2.

Characterisation Techniques

Magnetic property, nanoparticle nature, surface morphology and the elemental constituents exhibited by synthesised composite are characterised through various analytical techniques. The instruments employedfor analyses include Vibrating Sample Magnetome- ter (Lakeshore 7410), Scanning Electron Microscope(JEOL model JSM-6390 LV), Parti- cle Size Analyzer,

EDAX Spectrometer, FTIR Spectrophotometer, X-ray Powder Diffraction Analyzer.

Adsorbate Preparation

A stock solution of 1000 mg/L Pb(II) was prepared by dissolving the appropriate dos- age of lead nitrate in double distilled water (Distillon 4DQ). Aliquots of adsorbate solutions of varying concentrations were prepared through progressive dilutions of the stock solutions.

Batch Experiments

Removal of Pb using Magnetic Fe3O4-TGT NC **79**

Results and Discussion

The adsorption capacity of TGT- C2 was experimentally verified following Batch mode for 100 mg/L concentration of Pb(II) aqueous solution at pH 5.5. 50 mL of the sorbate species was agitated at a pre-fixed time interval of 10 minutes and pH 5 environment for different doses varying between 25 and 100 mg with increments of 25 mg. The initial and residual Pb(II) concentrations were recorded using Atomic Absorption Spectrophotometer: **Shimadzu (AA 6200)** at 283.3 nm λ_{max} and 0.7 slit width.

Desorption/Regeneration Studies

The economic viability of chosen material was ensured via consecutive desorption/regeneration studies by batch experiment using 0.1 N HCl as the desorbing agent.

Data Analyses

Pb(II) removal in terms of percentage and the amounts adsorbed in terms of sorption capacity, were calculated using the following equations:

Magnetic Sensitivity Test

A magnet was placed externally near bottles containing TGT- C2 composite solution. Figure 2 manifests the attraction of black material towards the magnet within a short span of time, confirming the highly sensitive magnetic flux, in the aqueous matrix.

Figure 2: Magnetic separation – external magnetic field.

VSM Analysis

% adsorption $= C_i * C_e * 100 C_i$

The magnetic property of the nanocomposite was characterised employing VSM (Vibrating Sample Magnetometer) technique with a 15 kOe magnetic field

$$qe = \frac{V(C_i - C_e)}{W}$$

[Ci/Ce - initial/equilibrium metal concentrations (mg/L), V - volume of the solution (L), W - mass of the adsorbent (g)]

The adsorptive nature of the studied system was established using Isothermal (Langmuir, Freundlich) and Kinetic (Pseudo first order, Pseudo second order) parameters. The batch experimental data were applied to the equations (Table 1) and plots were drawn to verify the applicability of above named models.

Table 1: Adsorption isotherm/kinetics factors

Models	*Equation*	*Plots*	*Constants*
Langmuir Isotherm	$(C_e/q_e) = (C_e/q_0) + (K_L/q_0)$	C_e vs (C_e/q_e)	q_m, b, R^2
Freundlich Isotherm	$\log q_e = \log K_F + (1/n) \log C_e$	$\log C_e$ vs $\log q_e$	K_F, 1/n, R^2
Pseudo First Order	$\log(q_e - q_t) = \log q_e - K_1 t / 2.303$	$\log (q_e - q_t)$ vs t	q_{e1}, K_f, R^2
Pseudo Second Order	$t/q_t = 1/(K_s q_e^2) + (1/q_e) t$	t/q_t vs t	q_{e2}, K_s, R^2

at 298K. Figure 3 exhibits the magnetisation (M-H) curve of TGT C2 as characterised by an S-like hysteresis loop, confirming the ferromagnetic nature. Saturation magnetisation strengths (Ms) and coercivity (Hc) valuesderived from the hysteresis loop are listed in Ta- ble 2. Ms value of TGT-C2 (1.54 emu/g) is markedly lessin comparison with bare mag- netite (92 emu/g) (Ristić et al., 2013), which could be due to the encapsulation of TGT within iron oxide particles. Also, a greater Hc value (139 Oe) supports the property of a specific magnetic field.

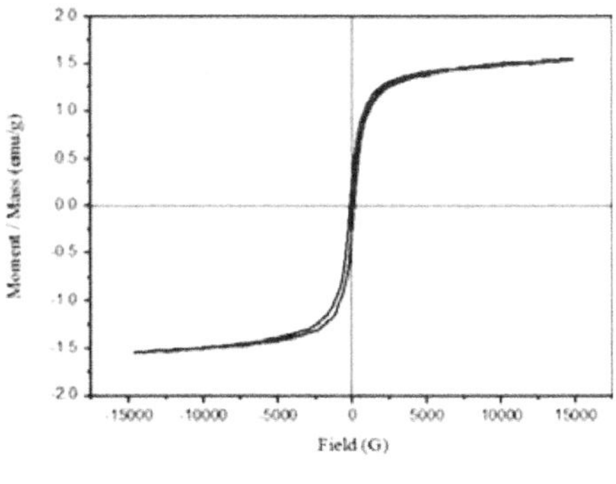

Figure 3: Magnetisation curve – TGT C2.

Table 2: VSM measurements

Magnetic Parameters	*TGT- C2*

Magnetisation (Ms) (emu g^{-1})	1.54
Coercivity (Hc) (Oe)	139.83

SEM/Particle Size Analyser Techniques

SEM micrograph (Figure 4a) reveals a homogeneous non-sphere/plate-shaped rough sur- face of iron oxide particles dispersed onto the TGT surface. The particle size analyser data evidenced the nano size of TGT- C2 as 57.35 nm (Figure 4b) which could be possible dueto auto combustion synthetic route.

EDAX Study

EDAX spectra (Figure 5b), show the presence of Fe and C obviously, in TGT-C2 against its counterpart (TGT) in Figure 5a. Also, a decline in the % weight of

Figure 4a: SEM image – TGT C2.

elementals viz., Ca, P, O inFigure 5b, confirms their participation in composite formation.

FTIR Technique

FTIR spectra of TGT, TGT- C2 are shown in Figure 6. Peak shifts from 556 cm^{-1} to 581 cm^{-1} (characteristics of PO^{3-} group), and 1515 cm^{-1} to 1456 cm^{-1} (characteristics of CO^{2-} group) indicate the successful doping of iron

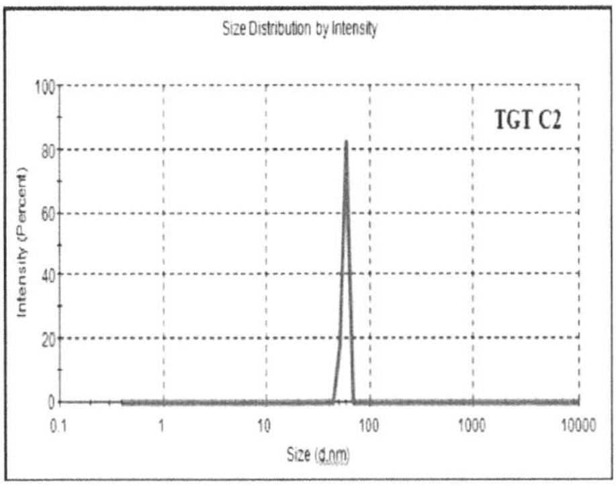

Figure 4b: Particle size Analyser data – TGT C2.

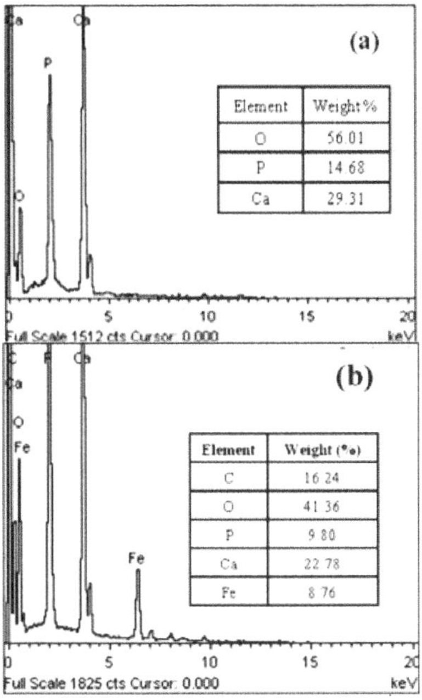

Figure 5: EDAX spectra: (a)TGT (b)TGT - C2.

oxides with the hydroxyl apatite of TGT. Asymmetric P-O stretching and peak occurrencecorresponding to the

Figure 6: FTIR spectra.

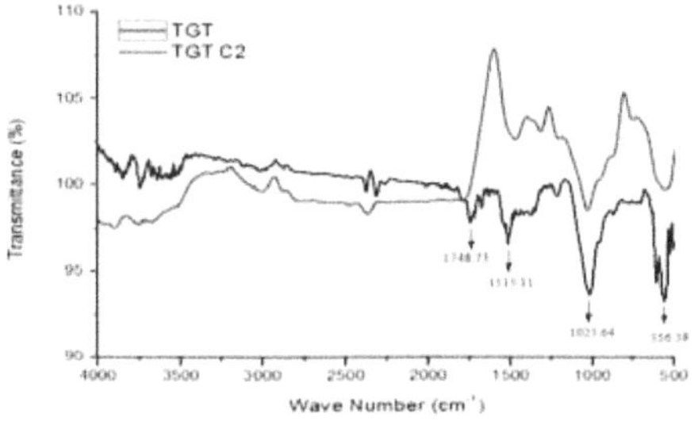

–OH group at 1025 cm^{-1} and 1748 cm^{-1} are identical tothat of TGT, emphasising the func-tional group identityof the parent precursor.

Batch Studies

Pb(II) sorption characteristics of TGT- C2 verified through batch experiments are listed in Table 3, for different doses (25-100 mg: 25 mg interval) of TGT-C2, where other pa- rameters were pre-fixed (pH: 5.5, Adsorbate Volume/Concentration: 50 mL/100 mg/L, Contact Time: 10 min). The percentage removal of Pb(II) is observed to be maximum at 50 mg TGT- C2 dosage itself, beyond which the least increase in the removal process had occurred. This exhibits the better efficiency of TGT-C2 at lower doses under optimised conditions.

Table 3: Batch study

TGT C2 dos-age (mg)	*Percentage re-moval (%)*
25	86.38
50	98.40
75	98.26
100	98.28

Desorption and Regeneration Studies

Desorbing ability of loaded metals and the regeneratingcapacity of TGT – C2 in consecu- tive cycles, determinedby batch mode studies are shown in Table 4. A very low desorption rate is recorded invariably at successive processes without major variations in adsorption capacities. This could be attributed to the immobilisingnature of TGT- C2, i.e., a high ra- tio of exchangeable calcium ions. Results evidenced that the parental TGT's (Anuradha et al., 2017) property has not been alteredby the composite preparation. As far as the regen- eration capability of TGT is concerned, the adsorption values observed in the three con- secutive cycles, have less difference in the amounts adsorbed, (97.56: 97.10: 96.77) dis- closing the excellent reusability of material.

Table 4: Desorption and regeneration study

[Pb(II) - TGT C2]	*Amount adsorbed (mg/g)*	
	Adsorption	*Desorption*
Cycle 1	97.56	0.23
Cycle 2	97.10	1.46
Cycle 3	96.77	2.39

Isothermal and Kinetic Verifications

(a) Isothermal Models

Langmuir, Freundlich isotherms were applied to the experimental data and the iso- thermal constants, and correlation coefficients (R^2) were calculated from linear plots (Figures 7 and 8) using isothermal equations (Table 1) which are summarised in Table 5. Langmuir constant (q_m) calculated as 98.03 mg/g for Pb(II) from the graphical represen- tations exhibit good agreement with the adsorption capacities of TGT- C2 as per batch results (97.56 mg/g). Langmuir constant [b (L/g)] relating to the free energy of sorption being smaller (0.39) value implies greater affinity of adsorbent towards the selected metal ion through saturated monolayer sorption. Correlation coefficient R^2 values about 0.99 exhibiting linearity of the plots favour the fit in of the Langmuir model to Pb(II) – TGT C2 system. A smaller KF (sorption capacity) value indicates the high adsorption process

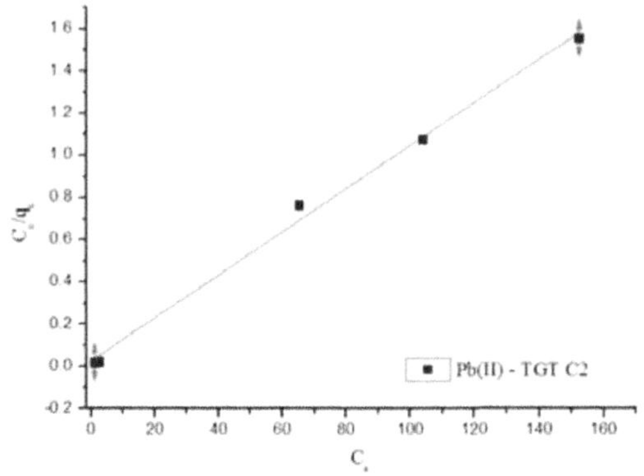

while a 1/n (sorption intensity)

Figure 7: Langmuir plot.

value lying between 0.1 and 1.0 favour physisorption. Approximity of the R^2 values to unity and less linearitythan the Langmuir model, favour lesser multilayer

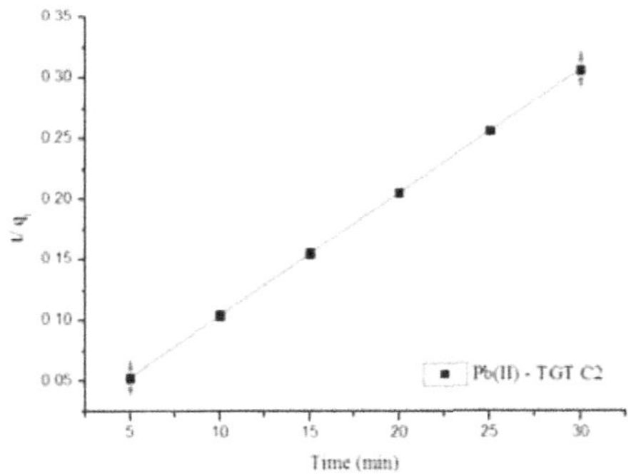

adsorption.

Figure 8: Freundlich Plot

Table 5: Isothermal constants [Pb(II)- TGT C2]

Langmuir			*Freundlich*		
q_m (mg/g)	b (L/g)	R^2	K_f (L/mg)	1/n	R^2
98.0392	0.3968	0.9959	76.5597	0.0472	0.9772

Pseudo Models

Table 6 shows data for pseudo-first/second order kinetic models at optimised concentra- tions using kinetic equations (Table 1) were employed to sketch out the respective plots (Figures 9 and 10). The resulting data reveal R^2 values nearness to unity, in the case of pseudo second order model than its counterpart being more suitable to describe the ad- sorption kinetics.

Table 6: Kinetic constants

Pb(II)- TGT C2	
Pseudo First Order	
q_e (mg/g)	8.43
K_f (min^{-1})	14.90
R^2	0.9370
Pseudo Second Order	
q_e (mg/g)	98.42
K_s (g/mg min)	3.62
R^2	0.9998

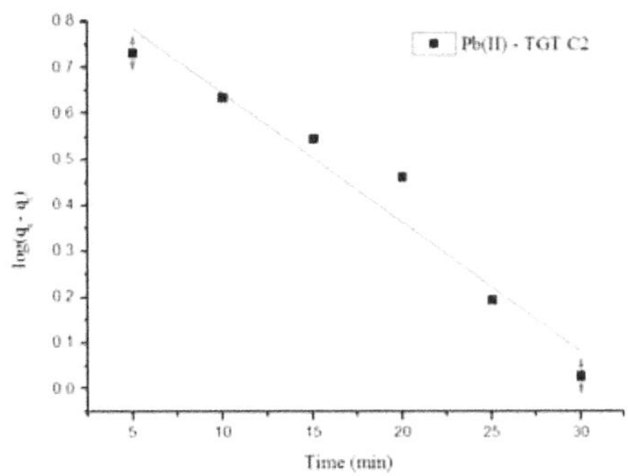

Figure 9: Pseudo first order kinetics.

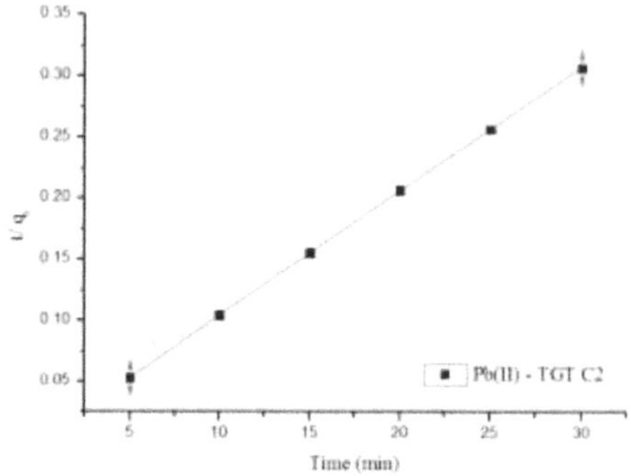

Figure 10: Pseudo second order kinetics.

Conclusion

Stable TGT- C2 - magnetic nanocomposite was synthesized through an efficient auto combustion technique without employing toxic chemicals. Fe_3O_4 encapsulated Treated Goat Teeth matrix possessed excellent nanosized magnetic properties confirmed by cor- responding characteristic analyses. Batch results revealed TGT- C2 as a promising mate- rial to adsorb Pb(II) ions from an aqueous stream. Isothermal studies exhibited the best linear fit for the Langmuir plot suggesting a monolayer sorption mechanism and Pseudo second order kinetic mode was favoured for the chosen system. The more appreciable reusability of the exhausted material was assured by desorption and regeneration experi- ments.

References

Anuradha, J., Andal, N.M. and N.S. Gayathri (2017). Utilization of a material derived from animal waste inthe biosorption of Pb(II) ions: Isothermal/thermodynamic,kinetics and statistical studies. *Current World Environment*,**12(2):** 445-455.

Barakat, M.A. (2011). New trends in removing heavy metals from industrial wastewater.

Arabian Journal of Chemistry,**4(4):** 361-377.

Horst, M.F., Alvarez, M. and V.L. Lassalle (2016). Removal of heavy metals from wastewater using magnetic nanocomposites: Analysis of the experimental conditions, *Separation Science and Technology*, **51(3):** 550-563.

Mohammed A.S., K pri A. and R. Goel (2011). Heavy Metal Pollution: Source, Impact, and Remedies. *In:* Khan, M.,Zaidi A., Goel R. and J. Musarrat (eds) Biomanagement of Metal-Contaminated Soils. *Environmental Pollution*, 20. Springer, Dordrecht.

Ranjithkumar, V., Hazeen, A.N., Thamilselvan, M. and S. Vairam (2014). Magnetic ac- tivated carbon-Fe_3O_4 nanocomposites—Synthesis and application in the removal of acid yel- low dye 17 from water. *Journal of Nanoscienceand Nanotechnology,* **14:** 1-11.

Madhava Rao, M., Chandra Rao, G.P., Seshaiah, K., Choudary, N.V. and M.C. Wang (2008) Activated carbon from Ceiba pentandra hulls, an agricultural waste, as an adsor- bent in the removal of lead and zinc from aqueous solutions. *Waste Management*, **28:** 849-858.

Ristić, Mi, Fujii, T., Hashimoto, H., Mitar, I. and S. Musić (2013). A novel route in the synthesis of magnetite nanoparticles, *Materials Letters,* **100:** 93-96.

Zhang, R., Wilson, V., Hou, A. and G. Meng (2015). Sourceof lead pollution, its influence on public health and the countermeasures. *International Journal of Health, Animal Sci- ence & Food Safety,* **2:** 18-31.

Recovery of Reusable Water and Sludge from Various Wastewater Sources: A Review

Muthulakshmi Andal N

Assistant Professor, Department of Chemistry, PSGR Krishnammal College for Women,Coimbatore-641 004, Tamil Nadu, India

Indhumathy P

Research Scholar, Department of Chemistry, PSGR Krishnammal College for Women, Coimbatore-641004, Tamil Nadu, India

ABSTRACT

Volumes of wastewater and contaminated water tend to increase at a faster rate than the supplies of clean water for several reasons. Few major reasons include: unabated increase in global human population, release of untreated wastewater from industries consuming high volumes of water, contamination of fresh water sources through fouling of catchment areas via industrial activities such as min- ing and /or recreational activities. Several other factors contributing to shortag- es of water for human consumption are: unplanned urbanization and/ or slums, insufficient number of dams to store water during periods of heavy rainfall; al- so, rapid evaporation of stored water in dams is attributed to global warming. The latter is also blamed for poor rainfalls in some regions of the world. It is thus clear that, reclamation of useable water from wastewater is a necessary strategy for replenishment of used- up water. However, the cost of wastewater clean-up is generally high and also results in sludge residues, since disposal of the latter is regarded as waste only and is classed as a menace. On the other hand, if value-added applications can be identified for the sludge, it will indeed be a win-win situation for the recovery of usable water from wastewater and worthwhile uses of the sludge. Since wastewaters from different sources, need varying treatment methods, a review of the studied treatment methods, the cur- rent and potential applications of sludges would be useful in highlighting the areas which would benefit further research.

Keywords: water shortage, treatment methods, sludge residues, potential app plications

1. INTRODUCTION

Water, owing to its indispensable nature, plays a major role in life of living or- ganisms. Fresh water supply which accounts for only 3% of the available water resources is expected to cater the needs of human, animals, industries and agri- culture as the major consumers. Moreover, water utility has been increased as the population rate has doubled in the last century. All day-to-day activities con- sume water in large quantities, from various sources, such as ground water, sur- face water, spring water, bore wells, and harvested rain water. However, ground water is invariably preferred for all domestic and industrial activities. Overex- ploitation of ground water [1] without recharge for the past 60 years has greatly reduced its availability. Every year, 2.7 billion people of the world population, face problems [2,3] due to water scarcity.

Graphical representation of water usage in various sectors of high / low-middle income countries are given below:

Figure 1: Usage of Water in Countries

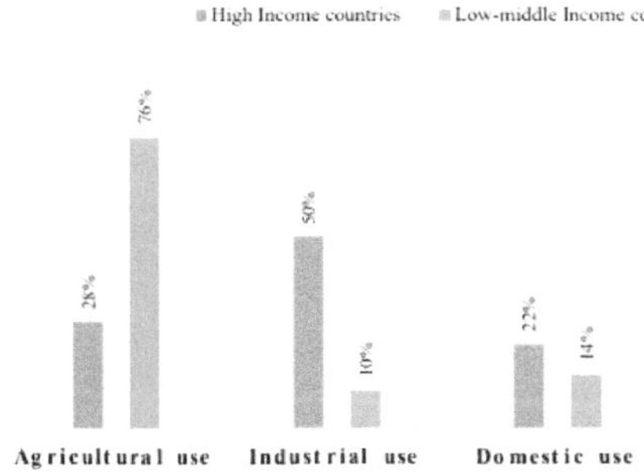

From the bar graph, it is obvious that, comparatively, extensive utilization of wateroccurs as a result of high / rapid industrialization and urbanization by high incomecountries. In low-middle income countries, under development of industrial activi-ties could be due to greater consumption of water in agriculture sector.

Apart from water scarcity, contamination of water resources contributes to the diminishing availability of fresh water [4-6]. Also, pollution of water bodies through point and non-point sources by anthropogenic activities result in inac- cessibility of fresh water. This is because of over population, urbanization, climate change andother natural / man-made factors. Major root causes of water pollution, are:

- Run offs from agricultural lands consisting of pesticides, fertilizer resi-dues that enter fresh water ecosystems, which in turn increase the level of nutrients leading to eutrophication [7] and thereby destroying the habitats of aquatic life
- Industrial sewage, directly dumped into open spaces release contaminants into the ground water table. Industrial effluents often comprise of non- biodegradable / highly persisting chemicals. These toxicants entering wa- ter resources in turn affect the food chain through biomagnification [8] producing serious illeffects.

Water stress developed out of scarcity that would be faced by the future world in 2040 is depicted in figure 2, asper the data suggested by World Resource In- stitute

Figure 2: Water Scarcity – Stress - 2040

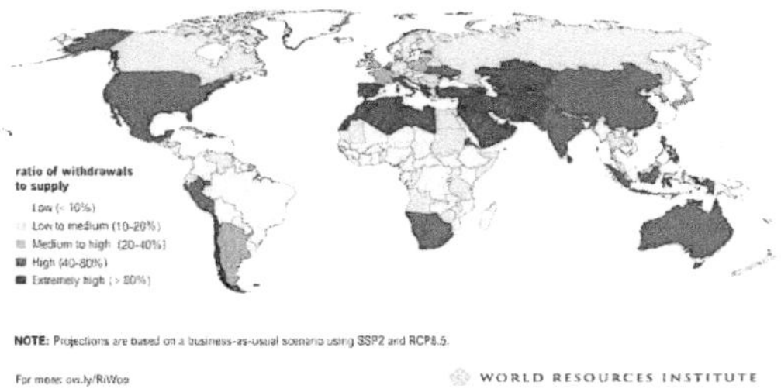

2. WASTEWATER SOURCES

- Municipal Wastewater

Household activities are the prime source of this type of wastewater, enriched with bacteria, viruses and other micro-organisms along with cleaning agents and food waste. These are easily degradable pollutants. Four main types of treat- ments are used for the treatment of municipal wastewater viz., physical [9], physical-chemical [10],biological [11] and sludge use [12].

- Agricultural Wastewater

Agricultural domain consumes enormous amount of fresh water resource but in turn generates polluted water. Recent research discloses that agriculture releases large amount of wastewater when compared to cities and industries which are well known for pollution [13]. Intensified agricultural practices result in soil erosion/increased water salinity[9] and sedimentation loads in water [14]. Ex- cess pesticides and fertilizers after their respective applications are left behind as residues, washed out with water as runoffs [15] from the farm land into the water streams, thereby accumulate as nutrients / agrochemicals, leading to algal blooms. Treatment

[16] of these wastewaters is useful for reuse in agriculture if the recovered water is free of the variety of contaminants mentioned above. However, the cost treatment may be relatively high, compared to other costs involved in

agricul- tural enterprises.

- Abattoir Wastewater

Wastewater from abattoirs consist of roughage that emanates from living organ- isms such as animal faeces, blood, bone and fat, its source being slaughter pro- cessing units. These consist of large amounts of organic matter, pathogens and nutrients. Biological Oxygen Demand, Chemical Oxygen Demand parameters of the polluted water streams are reported to be alarmingly high [17-19]. In the light of the foregoing wastewater from abattoirs, they are not desirable sources of useful water or sludge.

- Mining Wastewater

Large volumes of wastewater are generated due to mining activities in the ex- traction process of several valuablematerials from the earth's surface; in turn re- lease cyanides, sulfates and other heavy metals into water beds. During extrac- tion, heavy metals such as copper, zinc, cobalt, arsenic are leached out. These untreated waters, if disposed into the environment, cause several health implica- tions to all the organisms. These wastewaters need extensive and expensive treatment regimes to regenerate reusable water [20].

- Pharmaceutical Wastewater

Pharmaceutical industries employ a wide range of chemical processes which in- cludes fermentation, extraction, chemical synthesis, formulation and other reac- tions, wherein variety of organic chemicals are employed for the production of various medicinal compounds. If the effluents from these processes used for manufacturing pharmaceuticals are not treated completely, the residual contam- inants may contribute to environmental degradation [21, 22]. This concern is based on the knowledge that these wastewaters contain large amounts of organic chemicals and other salts, declared to be acute toxicants as they are highly reac- tive, non-biodegradable and interfere in the metabolism of plants and animals [23,24]. Due to improper sanitation, human waste and other medical waste such as steroids, hormones and other products serve as agents for potential pollution. Hence, highly effective remedial methodologies are vital to treat these waste products [25]. On account of time- consuming treatment procedures and costs, the prospects for obtaining reusable water from wastewater bearing pharmaceu- ticals is rather bleak.

- Textile Wastewater

A series of processing steps such as pre-treatment, dyeing, printing in textile in- dustries involve an extensive use of chemical reagents, surfactants, fatty acids, natural oils and dyes [26, 27]. These chemical substances are washed out as textile effluents, containing unpleasant odour, colour and excess toxicants which

include reactive dyes, organic compounds and other salts. Dumping of these ef- fluents in open land and discharging into water bodies would have a severe im- pact on the environment. Persistence of these hazardous chemicals in aquatic systems may result in bioaccumulation through food chain; posing a high threat to both human and animal kind [28, 29]. Treatment of textile wastewater through coagulation – flocculation, sedimentation, biological processes and oth- er techniques help in diminishing the lethal effects that result due to the leacha- tes released from textile industries [30, 31]. Complexity of the contaminants and high cost of full treatment does not augur well for recovering useable water from effluents emanating during textile processing.

- Chemical industry Wastewater

Chemical industries are taken to embrace association with petrochemical indus- tries and the paint industries, in addition to those, which manufacture a variety of chemicals for industrial and domestic usage. It is thus clear that,

wastewater from chemical and associated industries will contain a broad spectrum of con- taminants which may be toxic or non-toxic to animal and human health. Water with high chemical contents, after their maximum utilization in these industries may still pose a great threat to life, if they are discharged into the environment, prior to treatment in order to remove the residual chemicals. Methods for treat- ing wastewater from chemical industries, involve inter alia, nano -membranes
[32] and adsorption using resins and ion exchange [33].

3. WASTEWATER TREATMENT METHODS

Diminishing amounts of fresh water resources due to the rise in population and industrialization draws a major concern globally. Treatment of wastewater, which emanates from various anthropogenic activities, helps to address the scarcity of fresh water. However, not all the wastewater produced globally can be treated fully for various reasons, with the availability of the required technology and cost of treatment, being the primary factors militating against recovery of useable water from the array ofwastewater sources.

In general, no single method of wastewater treatment will suffice to recover useable water from wastewater. In the majority of wastewater treatment re- gimes, a step-wise application of a suitable combination of various physical, chemical and biological processes that help to remove most or all the pollutants will be followed. In some cases, pretreatment steps need to be applied, to re- move toxicants which are likely interfere biological processes. These pre- treatment phases involve various physical and chemical procedures.

Due to various research efforts, several physical and chemical water treatment methodologies were developed. Some of the salient methods are listed as fol- lows: [34]

Sedimentation: This is a fundamental and widely used physical technique

[35] for waste water treatment. Heavyparticles suspended in wastewater are removed by gravitational settling. This suspension is a very helpful first step towards crude separation of insoluble matter from wastewater, resulting in clear water that needs to be drained from the 'settling tank' to second tank where further treatment commences. It should be noted that sedimentation efficiency is influ- enced by the temperature of influent wastewater. [36]

Electro Dialysis: This technique employs a membrane applied between the an- ode and cathode. The membranes employed for this technique are of two types, namely, cation-exchange and anion-exchange membranes. Transport of ions across the membranes is aided by means of an electrical current. It is widely used for treatment of effluents for recovery of useful materials from waste wa- ters [37]

Ultrafiltration: Ultrafiltration is a physical water treatment process [38] that us- es membrane technology where colloidal materials are removed by applying pressure onto the membranes. This technique requires low energy for the treat- ment.

Reverse Osmosis: The process of reverse osmosis employs a semi-permeable membrane which allows only the liquid to pass through it and retains the dis- solved contaminants on the other side of it. [39] It is widely

applicable for desalination processes. Effluents from electroplating process can be effectively treated using this technique.

Adsorption: It is an effective, eco-friendly treatment [40] that is used for reme- diation of wastewater. This process uses natural / synthetic adsorbents to adsorb various toxicants. This method is used most widely for the treatment of industri- al effluents due to its easy mode of operation, flexibility and simple design.

Chemical Precipitation: The process of chemical precipitation [41] is compara- tively cheap and effective. It involves the formation of insoluble precipitates of the pollutants by reaction with precipitating agents, which are later removed and disposed. This technique cannot be adopted for wastewater with low metal ion concentrations.

Ion Exchange: This method is employed for treating wastewater containing charged particles. The process involves the substitution of ions using ion exchange resins i.e., cation exchange resins and anion exchange resins. These charged particles are removed by coulombic attraction with the resins thereby reducing the pollutant level in the effluents [42].

Biological Treatment: Biological Treatment methods are carried out to maintain the Biological Oxygen Demand (BOD), Chemical Oxygen Demand (COD), to remove other harmful toxicants from the waste water. Few of the commonly employed biological methods include the following:[43]

- ***Anaerobic Treatment***: This method is mainly used to treat those effluents discharged from agricultural / food processing activities. This technique [44, 45, 46] consumes time, but is highly effective in the removal of biodegradable organic matter.
- ***Bioremediation***: This technique [47, 48] involves the use of microorgan- isms to remove the toxicants. This also aids to maintain the Biological Oxygen Demand in an appropriate level.
- ***Aerobic Treatment:*** This treatment [49, 50] is employed to remove the trace levels of organic pollutants. It also helps in managing the evolution of dissolved gases.
- ***Activated Sludge Method***: This method [51, 52, 53] uses bacterial biomass under aerobic condition. Wastewater with high nutrients and oxygen levels can be condensed as an outcome of their usage for the growth of microorganisms.

All these above-mentioned procedures generate recycled water with appro- priate water quality parametric values making them suitable for salvaging.

4. RECYCLED WATER

Owing to the rapid and enormous population growth / industrialization, recla- mation, of treated wastewaters, contribute a significant amount of water to boost existing water resources. In the light of high environmental concern regarding the uncontrolled disposal of wastewater, the recycled water, from the various water treatment technologies as described above, can be used for wide range of purposes. [54,55]

Few strategies which have been implemented to reduce the need for freshwater areas follows:

- Urban uses of recycled water for sporting activities, automobile washing, irrigation of private / public parks and similar activities can be accom- plished [56,57]
- Reclaimed water is utilized in agricultural practices such as aquaculture, crop fields irrigation, provision for greenhouses, and watering ornamen- tal flowers [58,59]
- Recycled water creation is assisted in the enhancement of wetlands, maintenance of aquatic habitats for flora / fauna, recharging of ground / surface water levels, maintain of streamflow to restore a proper environ- ment for life on earth [60]
- Advancement in treatment methodologies can be implemented to revive the drinking water standard, so as to reduce the need for fresh water [61,62]

- Wastewater with trace levels of nutrients when discharged, pave way for soil nourishment, promoting plantgrowth with avoidance of ferlizers usage
- Thus, water reused from treatment plants help to save and provide a sustaina- ble way for the management ofwater resources.

5. SLUDGE DISPOSAL

During water treatment, huge amounts of waste contaminated with toxi- cants are generated. Hence, more attention needs to be paid to these re- sidual sludges loaded with impurities; and in particular, the techniques for their disposal. Many methods, the principal ones being landfills and in- cineration, have been adopted for dumping of these sludges which pose serious environmental implications [63]. Recent research activities have implemented various strategies to reuse waste materials from landfill sites in order to trim down costs. These waste materials used include sewage sludge, heavy metal residues, laundry sludge and other waste generated during several industrial processes [64]. Most developing countries have started using these waste materials as fertilizers, raw materials for build- ings/ ceramics, chemical substances and so on. But the composition and characteristics of the sludge, decides its mode of reusing. Few approaches that convert waste from various wastewater processing units into useful and compatible materials are summarized below [65].

Table 1: Application of sludges by various researchers - Comparison

S. No.	Name of the Author	Type of Sludge	Application
1.	J. Carneiro etal.[66]	Red mud/ Electroplat-ing sludge	Ceramic pigments
2.	V. Mymrin etal. [67]	Laundry sewage sludge	White ceramics
3.	Shiquan Liu etal. [68]	Electroplating sludge	Glass colourant
4.	S.M. Ashekuzzaman etal.	Dairy Wastewater	Agricultural

	[69]	sludge	plements
5.	Alok Suman etal. [70]	Water treatment sludge	Coagulant
6.	C. M. F. Vieira etal. [72]	Paper effluent sludge	Clay bricks

Recovery and Evaluation of the Worth of Materials from Sludges

Ceramic Pigments

A ceramic pigment is a metal transition complex oxide [73] that shows thermal/ chemical stability and high tinting strength when dispersed and fired with glazes or ceramic matrices [74]. Red mud and electroplating sludge which are rich in Fe and Ni/Cr respectively have been used for pigment production as reported by J. Carneiro et al. [66]. These metals, being transition group elements impart excellent colour to the prepared pigments for use in ceramics [75]. Both the sludge and red mud have been heated at higher temperatures, cooled and mixed in various propor- tions. The colour intensities of the pigments varied according to their mixed proportions [76]. Colour development was checked by adding then to glazes and subjected to UV-Visible spectroscopic studies. Results ex- hibited a better stability and tinting strength proving commercial viability of the waste embedded pigments.

White Ceramics

Sludge from laundry effluents consists mainly of detergents and surfac- tants along with various metals. [77,78,79] V. Mymrin et al., mixed laundry sewage sludge with kaolin clay for the production of white vitre- ous ceramics. Raw materials, namely, kaolin clay and the sewage sludge from the laundry units were examined by X-Ray Fluorescence technique.

These materials were mixed in widely varying

percentages for the manufacturing of the white ceramic material. Thus, ceramic material was prepared and tested for its strength, density and wa- ter absorption property [67] which proved to be in accordancewith Brazil- ian Standard NBR 7170.[80]. The prepared product was also subjected to leaching studies where the leachate concentrations were in agreement with NBR 10004 limits. [81] The incorporation of waste in ceramics is an attractive alternate for the ongoing waste disposal.

Glass Colourant

Coloured glasses in general are produced by adding purified metal salts or pigments and sometimes even by adding powerful oxides/ sulfides which are hazardous in nature. In the progress of developing a better waste disposal technique. [82,83] Shiquan Liu et al [68] used hydrother- mally treated sand (R- sand) and its functionalized form (F- sand) loaded with Cr^{6+}ions from electroplating effluents for the manufacture of col- oured glasses. Metal laden sorbents were collected, filtered and dried. Glass preparation was done by the addiction of clarifying agents with the metal laden sand matrices. [84,85] Glass produced was compared with the glass made from normal sand/ $K_2Cr_2O_7$ colourant and their transmis- sion spectra were analyzed for the colour. It is observed that the sand loaded with Cr^{6+} ions showed a better colour than the commercial one which may be due to the high immobilization of the metal ions in the glass. Thus, the methodology prescribed, showed a promising way for producing coloured glasses in the process of pollution reduction.

Agricultural Supplements

Dairy wastewater treated sludge (DWTS) consists of various elements namely N, K, P, Na, S, Ca, Mg and organic matter which contain major plant growth nutrients such as N, P, K. [86,87] S.M. Ashekuzzaman et al employed DWTS as nutrients for agriculture. DWTS was collected, clas- sified according to its seasonal processing/ nature and compared for its dry matter (DM), organic matter (OM) and nutrients with the commer- cially available fertilizers. The sludge collected from various dairy units were categorized into four types, namely, biochemically treated activated sludge, lime treated air flotation sludge, anaerobically treated sludge, bio- chemical and lime treated sludge. These sludges were freezer dried and pulverized for further studies. Results revealed that the DWTS consisted of a good proportion of DM, OM and other nutrients which are highly vi- tal for plant growth. [88,89,90] Hence, the use of sludge as agricultural supplements is a feasible way for its safe management.[91]

Coagulant

Alok Suman et al [70], have reported the use of water treatment sludge (WTS) as a coagulant for treating dairy wastewater. Coagulation, being a chemical process employs coagulants that usually consist of highly charged cations. [92] WTS consists of metal ions in different forms, mak- ing it compatible to be used as a coagulant. WTS was used as a coagulant along with other conventional coagulants, namely, alum, ferric chloride and ferrous sulphate. Collected waste material was acid-treated with sul- phuric acid and conditioned for use as coagulant with other marketed products. Results displayed a better efficiency of WTS in the removal of turbidity and reduction of other water quality parametric values. There- fore, employment of WTS as coagulant is highly economical and efficient against other conventional coagulants. [93,94]

Building Materials

Water treatment sludges have been used in the production of construction materials. [95,96,97] The clay particles and other raw materials employed during the manufacturing processes held the toxicants firmly making them highly immobilized. A wide range of construction materials have been prepared [98,99,100,101] by Yue Liu et al. and C. M. F. Vieira et al., where drinking water treatment and paper sludges were involved, re- spectively. The prepared materials, namely, concrete paving blocks and clay bricks were tested for their technological properties which were in accordance with the prescribed standard limits. The leachability studies were also assessed confirming that the manufacturing process resulted in eco-friendly materials.

6. CONCLUSION

Treatment of wastewater to regenerate useable water and the processing of residual sludge have attracted attention of researchers from several fields for the problem of waste disposal. Several problems associated with both the recovery of useable water and the fate of the sludge result- ing from wastewater treatment process have been discussed. Recovery of

water, post treatment provides a benefit by reducing the stress on fresh water resources. This helps to provide hope for decreasing the prospects water scarcity for future generations. In this point of view, this brief re- view has focused on the twin issues accessing useable water and finding ways to convert the regenerated sludge into manufacturable marketable products. Sludge materials generated during treatment activities could be deployed for the production of commercial materials, aiding plant growth and treatment methodologies. The progress of research efforts in employ- ing sludges to generate eco-friendly products, provide a basis for the re- duction of pollution caused due to the erstwhile practices. This workable strategy s may reduce the need to utilize raw materials from non- renewable resources, thus, paving way for energy saving processes.

REFERENCES

1. Kohler, Stefan. "Customary Water Rights and Modern Technology: The Overuse of Groundwater Resources in Wādīḥarīb, Yemen." *Etudes Rurales*, no. 155/156, EHESS, **2000**, 167–78, http://www.jstor.org/stable/20122833
2. Falkenmark, Malin. "The Massive Water Scarcity Now Threatening Africa: Why Isn't It Being Addressed?" *Ambio, Springer, Royal Swedish Academy of Sciences* **1989**, 18(2)],. 112–118, http://www.jstor.org/stable/4313541.
3. Rajat K. Chakraborti, Jagjit Kaur, Harpreet Kaur, Water Shortage Challenges and a Way Forward in India, *American Water Works As- sociation*, **2019**, 42-49
4. Amrita Pal, Karina Yew-Hoong Gin, Angela Yu-Chen Lin, Martin Reinhard Impacts of emerging organic contaminants on freshwater resources: Review of recent occurrences, sources, fate and effects, *Science of The Total Environment*, **2010**, 408(24) 6062-6069, https://doi.org/10.1016/j.scitotenv.2010.09.026.
5. Samaneh Ashraf & Amir AghaKouchak & Ali Nazemi & Ali Mirchi & Mojtaba Sadegh & Hamed R. Moftakhari & Elmira Hassanzadeh & Chi-Yuan Miao & Kaveh Madani & Mohammad Mousavi Baygi & Hassan Anjileli & Davood Reza Arab & Hamid Norouzi & Omid Mazdiyasni & Marzi Azarderakhsh &Aneseh

Alborz & Mohammad

J. Tourian & Ali Mehran & Alireza Farahmand &Iman Mallakpour, Compounding effects of human activities and climatic changes on surface water availability in Iran, *Climatic Change*, **2018**, 1-13

6. S. Sharma, A. Bhattacharya, Drinking water contamination and treatment techniques, *Applied Water Science* (2017), **2016**, 7, 1043– 1067
7. Hodgkin, E.P., Hamilton, B.H. Fertilizers and eutrophication in southwestern Australia: Setting the scene. *Fertilizer Research*, **199336,** 95–103 https://doi.org/10.1007/BF00747579.
8. Jiao Ren, Xiaoping Wang, Chuanfei Wang, Ping Gong, Xiruo Wang, Tandong Yao, Biomagnification of persistent organic pollu- tants along a high-altitude aquatic food chain in the Tibetan Plateau: Processes and mechanisms, *Environmental Pollution*, **2017**, 220(A), 636-643, https://doi.org/10.1016/j.envpol.2016.10.019.
9. Jinlu Wu, Y.S. Yang, Jinhua Lin, Advanced tertiary treatment of municipal wastewater using raw and modified diatomite, *Journal of Hazardous Materials*, **2005**, 127(1-3) 196-203, https://doi.org/10.1016/j.jhazmat.2005.07.016
10. Bishop, D. F., O'Farrell, T. P., & Stamberg, J. B., Physical-Chemical Treatment of Municipal Wastewater, **1972**, *Journal (Wa- ter Pollution Control Federation)*, *44*(3), 361–371. http://www.jstor.org/stable/25037323
11. Long term operation of pilot-scale biological nutrient removal pro-cess in treating municipal wastewater, Bioresource Tech- nology, **2009**, 100(13), 3180-3184,

https://doi.org/10.1016/j.biortech.2009.01.062

12. Xue Han, Yu-Ting Zuo, Yu Hu, Jie Zhang, Meng-Xuan Zhou, Mo Chen, Fei Tang, Wen-Qing Lu, Ai-Lin Liu, Investigating the per- formance of three modified activated sludge processes treating mu- nicipal wastewater in organic pollutants removal and toxicity reduc- tion, *Ecotoxicology and Environmental Safety*, **2018** (148), 729-737, https://doi.org/10.1016/j.ecoenv.2017.11.042.

13. Largus T. Angenent, Khursheed Karim, Muthanna H. Al-Dahhan, Brian A. Wrenn, Rosa Domíguez-Espinosa, Production of biochem- icals from industrial and agricultural wastewater, *Trends in Biotech- nology*, **2004**, 22(9), 477-485,https://doi.org/10.1016/j.tibtech.2004.07.001.

14. Mustapha A., Identification of anthropogenic influences on water quality of jakara river, northwestern nigeria, *Journal of Applied Sci- ences in Environmental Sanitation*, 7 (1): 11-20

15. Ruimin Liu, Jiawei Wang, Jianhan Shi, Yaxin Chen, Chengchun Sun, Peipei Zhang, Zhenyao Shen, Runoff characteristics and nutri- ent loss mechanism from plain farmland under simulated rainfall conditions, *Science of The Total Environment*, **2014**, (468-469), 1069-1077, https://doi.org/10.1016/j.scitotenv.2013.09.035.

16. Alexandra EV Evans, Javier Mateo-Sagasta, Manzoor Qadir, Eline Boelee and Alessio Ippolito, Agricultural water pollution: key knowledge gaps and research needs, *Current Opinion in Environ- mental Sustainability,* **2018**, 36, 20–27.

17. MT Adam and M Dahiru, A Review on Abattoir Wastewater Treat- ment for Environmental Health Improvement, *Journal of Environ- mental Bioremediation and Toxicology*, **2020**, 3(2), 26-

31.

18. Olugbenga Oluseun Elemile, Davids Olorungbon Raphael, David Olugbenga Omole, Elizabeth Omoladun Oloruntoba, Elijah Oluwasegun Ajayi1 and Nyore Adams Ohwavborua, Assessment of the impact of abattoir effluent on the quality of groundwater in a resi-dential area of Omu- Aran, Nigeria, *Environmental Sciences Europe*,

2019, 31(16), 1-10

19. Terrumun Kenneth Kwadzah, Oliver Terna Iorhemen, Assessment of the Impact of Abattoir Effluent on the Water Quality of River Ka- duna, Nigeria, *World Journal of Environmental Engineering*, **2015**, 3(3), 87-94.
20. Lorenzo Reyes-Bozo, Alex Godoy-Faúndez, Ronaldo Herrera-Urbina, Pablo Higueras, José Luis Salazar, Héctor Valdés-González, Eduardo Vyhmeister, Blanca Antizar-Ladislao Greening Chilean copper mining operations through industrial ecology strate- gies, *Journal of Cleaner Production*, **2014**, 84, 671- 679,https://doi.org/10.1016/j.jclepro.2014.03.088.
21. Heloise Beatriz Quesada, Aline Takaoka Alves Baptista, Luís Fernando Cusioli, Daiana Seibert, Charleston de Oliveira Bezerra, Rosangela Bergamasco, Surface water pollution by pharmaceuticals and an alternative of removal by low-cost adsorbents: A review, *Chemosphere* 2019, 222 766-780,https://doi.org/10.1016/j.chemosphere.2019.02.009
22. Milina de Oliveira, Breno Emanuel Farias Frihling, Jannaina Velasques, Fernando Jorge Corrêa Magalhães Filho, Priscila Sabioni Cavalheri, Ludovico Migliolo, Pharmaceutical's residues and xeno- biotics contaminants: Occurrence, analytical techniques and

sustain- able alternatives for wastewater treatment, *Science of the Total Envi-ronment*, **2020**, 705, 1 -72

23. Xueqing Shi, Kwok Yii Leong, How Yong Ng, Anaerobic Treat-ment of Pharmaceutical Wastewater: A Critical Review, *Biore-source Technology*, **2017**, 1-24
24. Rasna Gupta, Bindu Sati, and Ankit Gupta, Treatment and Recycling of Wastewater from Pharmaceutical Industry, *Advances in Biologi- cal Treatment of Industrial Waste Water and their Recycling for a Sustainable Future*, Applied Environmental Science and Engineer-

    ing for a Sustainable Future, **2019**, 267 – 302
25. Lawrence Mzukisi, Madikizela Nikita Tawanda, Tavengwa Luke Chimuka, Status ofpharmaceuticals in African water bodies: Occur-rence, removal and analytical methods, *Journal of Environmental Management* **2017**, 193 211-220, https://doi.org/10.1016/j.jenvman.2017.02.022
26. Rummi Devi Saini, Textile Organic Dyes: Polluting effects and Elimination Methods from Textile Waste Water, *International Jour- nal of Chemical Engineering Research.*, **2017**, 9(1), 121-136

27. Sujata Mani, Pankaj Chowdhary, and Ram Naresh Bhara-gava,Textile Wastewater Dyes: Toxicity Profile and Treatment Ap- proaches, Emerging and Eco-Friendly Approaches for Waste Man- agement, **2018,** 219 – 244 https://doi.org/10.1007/978-981-10-8669- 4_11
28. Dina A. Yaseen & Miklas Scholz, Treatment of synthetic textile wastewater containing dye mixtures with microcosms

Environmen- tal Science and Pollution Research, **2018**, 25, 1980 – 1997 https://doi.org/10.1007/s11356-017-0633-7

29. Khadija Siddique Muhammad Rizwan Munazzam Jawad Shahid Shafaqat Rehan Ahmad Hina Rizvi, Textile Wastewater Treatment Options: A Critical Review, *Enhancing Cleanup of Environmental Pollutants*, **2017**, 183-207
30. Rajat Pratap Singh, Pradeep Kumar Singh, Rasna Gupta and Ram Lakhan Singh, Treatment and Recycling of Wastewater from Textile Industry, *Advances in Biological Treatment of Industrial Waste Wa- ter and their Recycling for a Sustainable Future*, Applied Environ- mental Science and Engineering for a Sustainable Future, **2019**, 225 – 266
31. Textile Wastewater Treatment for Water Reuse: A Case Study, by Hua Yin, Peiwen Qiu ,Yuange Qian ,Zhuwen Kong ,Xiaolong Zheng ,Zhihua Tang and Huafang Guo, Processes **2019**, 7(1), 34; https://doi.org/10.3390/pr7010034
32. Roxana Cristina Popescu, Mariana Oana Mihaela Fufă, Alexandru Mihai Grumezescu, Alina Maria Holban, Nanostructured mem- branes for the microbiological purification of drinking water, *Water Purification,Academic Press*, **2017,** 421-446
33. Dietmar Rolf Kammerer, Judith Kammerer, Reinhold Carle, Resin Adsorption and Ion Exchange to Recover and Fractionate Polyphe- nols, *Polyphenols in Plants, Academic Press*, **2014**, 219-230
34. Manar Elsayed Abdel-Raouf, Nermine E Maysour, Reem Kamal Farag and Abdul-Raheim Mahmoud Abdul- Raheim, Wastewater

Treatment Methodologies, Review Article, *International Journal of Environment & Agricultural Science*, **2019,** 3(1), 018, 1-25

35. Yunkun Qian, Yue Hu, Yanan Chen, Dong An, Paul Westerhoff, David Hanigan, Wenhai Chu. Haloacetonitriles and haloacetamides precursors in filter backwash and sedimentation sludge water during drinking water treatment, *Water Research*, **2020**, 186
36. Athanasia M. Goula, Margaritis Kostoglou, Thodoris D. Karapantsios, Anastasios I. Zouboulis. The effect of influent temperature variations in a sedimentation tank for potable water treatment—A computational fluid dynamics study, *Water Research*, 2008, 42(13), 3405-3414
37. Efficient techniques for the removal of toxic heavy metals from aquatic environment: A review,
    C. Femina Carolina P. SenthilKumar A. Saravanana G. Janet Joshi- baa Mu. Naushad, *Journal of Environmental Chemical Engineering*, 5(3), **2017**, 2782 – 2799.
38. Mika Manttari, Arto Pihlajamaki, Marianne Nystrom Comparison of nanofiltration and tight ultrafiltration membranes in the filtration of paper mill process water, *Desalination*, **2002**, 149(1-3)131-136
39. Valentina Colla, Teresa Annunziata Branca, Felice Rosito, Carmelo Lucca, Beatriz Padilla Vivas, Vanesa Menéndez Delmiro, Sustaina- ble Reverse Osmosis application for wastewater treatment in the steel industry, *Journal of Cleaner Production*,**2016**, 130, 103-115
40. Alexander E. Burakov, Evgeny V. Galunin, Irina V. Burakova, Ana- stassia E. Kucherova, Shilpi Agarwal, Alexey G. Tkachev,

Vinod K. Gupta, Adsorption of heavy metals on conventional and nanostruc- tured materials for wastewater treatment purposes: A review, *Eco- toxicology and Environmental Safety*, **2018**, 148, 702-712

41. W. De los Santos Ramos, T. Poznyak, I. Chairez, I. Córdova R., Remediation of lignin and its derivatives from pulp and paper indus- try wastewater by the combination of chemical precipitation and ozonation, *Journal of Hazardous Materials*, **2009**, 169(1-3)
42. Andriy Malovanyy, Halyna Sakalova, Yosyp Yatchyshyn, Elzbieta Plaza, Myroslav Malovanyy, Concentration of ammonium from mu- nicipal wastewater using ion exchange process, *Desalination*, 2013, 329, 93-102
43. Mohamed Samer, Wastewater Treatment Engineering - Biological and Chemical Wastewater Treatment Processes, **2015**
44. Tran, Q.K., Jassby, D., Schwabe, K.A., The implications of drought and water conservation on the reuse of municipal wastewater: Rec- ognizing impacts and identifying mitigation possibilities, *Water Re- search*, **2017**, 1 – 45
45. Amin Goli, Ahmad Shamiri, Susan Khosroyar, Amirreza Ta- laiekhozani, Reza Sanaye, Kourosh Azizi, A Review on Different Aerobic and Anaerobic Treatment Methods in Dairy Industry Wastewater, Journal *of Environmental Treatment Techniques* **2019**, 6(1), 113-141
46. Korbinian Kaetzl, Manfred Lübken, Tito Gehring, Marc Wichern, Efficient Low-Cost Anaerobic Treatment of Wastewater Using Bio- char and Woodchip Filters, *Water* **2018**, *10*(7), 818;

47. Djamel Ghernaout, Increasing Trends Towards Drinking Water Rec- lamation from Treated Wastewater, *World Journal of Applied Chem-istry*, **2018**; 3(1): 1-9
48. Jemal Fito1, Nurelegne Tefera and Stijn W. H. Van Hulle, Sugar-cane biorefneries wastewater: bioremediation technologies for envi- ronmental sustainability, *Chemical and Biological Technologies in Agriculture* **2019** 6:6, 1-13
49. Mashallah Rezakazemi, Mohsen Maghami & Toraj Mohammadi, Wastewater's treatment containing phenol and ammonium using aerobic submerged membrane bioreactor, *Chemistry Central Jour- nal*, **2018**, 12, 1-10
50. R.Campo, G.Di Bella, Petrochemical slop wastewater treatment by means of aerobic granular sludge: effect of granulation process on bio-adsorption and hydrocarbons removal, , *Chemical Engineering Journal*, 2019, 378
51. Britt-Marie Wilén1 & Raquel Liébana1 & Frank Persson1 & Oskar Modin1 & Malte Hermansson, The mechanisms of granulation of activated sludge in wastewater treatment, its optimization, and im- pact on effluent quality, *Applied Microbiology and Biotechnology* **2018**, 102, 5005–5020
52. Marianna Garfí, Laura Flores, Ivet Ferrer, Life Cycle Assessment of wastewater treatment systems for small communities: Activated sludge, constructed wetlands and high-rate algal ponds, *Journal of Cleaner Production*, **2017**,161 211-219
53. Fuqiang Fan, Ronghua Xu, Depeng Wang, Fangang Meng, Applica- tion of activated sludge for odor control in wastewater treatment plants: Approaches, advances and outlooks, *Water Research*, **2020**, 181, 1 -16

54. Djamel Ghernaout, Water Reuse (WR): The Ultimate and Vital So- lution for Water Supply Issues, *International Journal of Sustainable*

Development Research, **2017**; 3(4):

55. Miquel Salgot and Montserrat Folch, Wastewater treatment and wa- ter reuse, *Current Opinion in Environmental Science & Health*, **2018**, 2, 64–74

56. Jonathan L. Bradshaw and Richard G. Luthy, Modeling and Optimi- zation of Recycled Water Systems to Augment Urban Groundwater Recharge through Underutilized Stormwater Spreading Basins, *En- viron. Sci. Technol.* **2017**, 51(20), 11809–11819

57. Brian D. Richter, Kendall Benoit, Jesse Dugan, Gabriella Getacho, Natalie LaRoe, Bailey Moro, Tyler Rynne, Maria Tahamtani and Al- len Townsend, Decoupling Urban Water Use and Growth in Re- sponse to Water Scarcity, *Water* **2020**, 12, 2868, 1-14

58. Sarah M. Markland, David Ingram, Kalmia E. Kniel and Manan Sharma,Water for Agriculture: the Convergence of Sustainability and Safety, *Microbiol Spectrum*, **2016**5(3), 1-11

59. Bruno J. L. Pitton, Charles R. Hall, Darren L. Haver, Sarah A. White & Lorence R. Oki, A cost analysis for using recycled irrigation run- off water in container nursery production: a Southern California nursery case study, *Irrigation Science,* 2018, 36, 217-226

60. Declan Page,Elise Bekele,Joanne Vanderzalm and Jatinder Sidhu, Managed Aquifer Recharge (MAR) in Sustainable Urban Water Management, *Water*, **2018,** 10(3), 1-16

61. Casey Furlong Jega Jegatheesan, Matthew Currella, Usha Iyer, Ran- igaa Tehmina Khana Andrew S.Balla, Is the global public willing to drink recycled water? A review for researchers and practitioners, *Utilities Policy*, **2019**, 56, 53-61

62. SujanaAdapa, Factors influencing consumption and anti-consumption of recycled water: Evidence from Australia, Author links open overlay panel, *Journal of Cleaner Production*, **2018**, 201, 624-635

63. Yousif algamal, N. M. Khalil, Qayid M. Saleem, Usage of the Sludge from Water Treatment Plant in Brick-Making Industry, *Journal of Chemical Technology and Metallurgy,* **2016**, 53, 3, 504-510

64. Wojciech Franus, Grzegorz Jozefaciuk, Lidia Bandura and Malgorzata Franus, Use of Spent Zeolite Sorbents for the Preparation of Lightweight Aggregates Differing in Microstructure, *Minerals*, **2017**,7, 25, 1-14

65. Joan A. Cusido, lazaro V. Cremades, Environmental Effects of Us- ing Clay Bricks Produced with Sewage Sludge: Leachability and Toxicity Studies, *Waste Management*, **2012**, 32, 1202– 1208

66. J. Carneiro, D.M. Tobaldi, M.N. Capela, R.M. Novais, M.P. Seabra, J.A. Labrincha, Synthesis of Ceramic Pigments from Industrial Wastes: Red Mud and Electroplating Sludge, *Waste Management,* **2018**, 80, 371–378

67. V. Mymrin, Camila F.G. Santos, Kirill Alekseev, Monica A. Avanci, Marcio A. Kreusch, Tiago Borga, Olaf Graupmann, Fabio L. Cava- lin, Renata A. Monteiro, Valter A. Ruy, Influence of Kaolin Clay on Mechanical Properties and on the Structure, Formation Processes of White Ceramics with Inclusion of Hazardous Laundry Sewage Sludge, *Applied Clay Science*

2018,155, 95–102

68. Shiquan Liu, Shivani Bhardwaj Mishra, Yun Zhang and Ling Qi, Uptake of Hexavalent Chromium in Electroplating Wastewater by Hydrothermally Treated and Functionalized Sand and its Sustainable Reutilization for Glass Production, ACS *Sustainable Chem. Eng.*, **2017**, 5, 1509–1516
69. S.M. Ashekuzzaman, Patrick Forrestal, Karl Richards, Dairy Indus- try Derived Wastewater Treatment Sludge: Generation, Type and Characterization of Nutrients and Metals for Agricultural Reuse, *Owen Fenton Journal of Cleaner Production* **2019**, 230, 1266-1275
70. Alok Suman, Tarique Ahmad, Kafeel Ahmad, Dairy Wastewater Treatment Using Water Treatment Sludge as Coagulant: A Novel

    Treatment Approach, *Environ Dev Sustain*, **2018**, 20, 1615–1625
71. Yue Liu, Yan Zhuge, Christopher W.K. Chow, Alexandra Keegan, Danda Li, Phuong Ngoc Pham, Jianyin Huang, Rafat Siddique, Uti- lization of Drinking Water Treatment Sludge in Concrete Paving Blocks: Microstructural Analysis, Durability and Leaching Proper- ties, *Journal of Environmental Management* **2020**, 262 110352, 1-9
72. C. M. F. Vieira, Regina M. Pinheiroa, Ruben J. Sanchez Rodriguez, Veronica S. Candido, Sergio N. Monteiro, Clay Bricks added with Effluent Sludge from Paper Industry: Technical, Economical and Environmental Benefits, *Applied Clay Science* xxx (2016) xxx–xxx, **2016**, 1-7
73. Youssef El Jabbar, Hind Lakhlifi, Rachida El Ouatib, Lahcen Er-Rakho, Sophie Guillemet- Fritsch, Bernard Durand, Preparation

and characterisation of green nano-sized ceramic pigments with the spi- nel structure AB_2O_4 (A = Co, Ni and B = Cr, Al) *Solid State Com- munications*, **2021**,334-335

74. Zhongtao Chen Yi Du Zhongfu Li Dandan Sun Chaofeng Zhu, Syn- thesis of black pigments containing chromium from leather sludge, *Ceramics International*, **2015**, 41(8), 9455 - 9460
75. Ljiljana Matovi, Radojka Vujasin, Ksenija Kumri, Sanja Krsti, Yi- nan Wu, Daniel Manaye Kabtamu, Aleksandar Devečerski, Design- ing of technological scheme for conversion of Cr-rich electroplating sludge into the black ceramic pigments of consistent composition, following the principles of circular economy, *Journal of Environ- mental Chemical Engineering* **2021**, 9
76. Minxing Du, Yi Du, Zhongtao Chen, Zhongfu Li, Kai Yang, Xingjie Lv, Yibing Feng, Synthesis and characterization of black ceramic pigments by recycling of two hazardous wastes, *Appl. Phys. A* **2017** 123:575, 1-7
77. Luciana C.S. Herek, Carla Eponina Hori, Miria Hespanho, Miranda Reis, Nora Diaz Mora, Celia Regina Granhem Tavares, Rosangela

    Bergamasco, Characterization of ceramic bricks incorporated with textile laundry sludge, *Ceramics International*, **2012**, 38, 951–959
78. Fernanda Carla Camilo, Thiago Peixoto de Araújo, Heloise Beatriz Quesada, Alexandre Amado Moura, Murilo Pereira Moises, Rosan- gela Bergamasco, Sergio Henrique Faria, Maria Angelica Simoes Dornellas de Barros, Synthesis of hydrochars derived from industrial laundry sludge and its application in the removal of cationic dye, *Journal of Water Process Engineering*

2021, 40, 1– 8

79. Silva, T.L., Ronix, A., Pezoti, O., Souza, L.S., Leandro, P.K.T., Be- din, K.C., Beltrame, K.K., Cazetta, A.L., Almeida, V.C., Mesopo- rous activated carbon from industrial laundry sewage sludge: adsorp- tion studies of reactive dye Remazol Brilliant Blue *R. J. Chem. Eng.*, **2016**303, 467–476. http://dx.doi.org/10.1016/j.cej.2016.06.009.
80. NBR 7170/1983. Ceramic Solid Brick for Masonry. (Rio de Janeiro, Bra- zil).
81. NBR 10004/2004. Solid waste. Classification. (Rio de Janeiro, Brazil).
82. Ashokkumar, M.; Narayanan, N. T.; Gupta, B. K.; Reddy, A. L. M.; Singh, A. P.; Dhawan, S. K.; Chandrasekaran, B.; Rawat, D.; Tala- patra, S.; Ajayan, P. M.; Thanikaivelan, P. Conversion of industrial bio-waste into useful nanomaterials. *ACS Sustainable Chem. Eng.* **2013**, 1, 619–626.
83. Mella, B.; Glanert, A. C.; Gutterres, M. Removal of chromium from tanning wastewater and its reuse. *Process Saf. Environ. Prot.* **2015**, 95, 195–201.
84. Shelby, J. E. Glass melting in Introduction to Glass Science and Technology, 2nd ed.; *RSC press: London*, **2006**,26–50.
85. Dyer, T. D. Glass Recycling. *In Handbook of Recycling; Elsevier*, **2014**,191–209.
86. Gil-Pulido, B., Tarpey, E., Almeida, E.L., Finnegan, W., Zhan, X.,Dobson, A.D.W., O'Leary, N., Evaluation of dairy processing wastewater biotreatment in an IASBR system: aeration rate impacts on performance and microbial ecology., *Biotechnol. Rep.*,

2018,19, https://doi.org/10.1016/j.btre.2018.e00263.

87. MP Ryan, G Walsh, The Characterisation of dairy waste and the po- tential of whey for industrial fermentation - Limerick (Ireland): Uni- versity of Limerick. *EPA Research Report*, 2016
88. Hukari, S., Hermann, L., Nattorpa, A., From wastewater to fertilisers d technical overview and critical review of European legislation governing phosphorus recycling. *Sci. Total Environ.* **2016**, 542, 1127-1135.
89. Benchmarking resource efficiency in Ireland's dairy processing sector. Ireland, Enterprise, **2011**.
90. M.ELópez-Mosquera. C Moirón, E Carral, Use of dairy-industry sludge as fertiliser for grasslands in northwest Spain: heavy metal levels in the soil and plants, *Resources, Conservation and Recycling* **2000,** 30(2), 95-109
91. Lucid, J.D., Fenton, O., Healy, M.G., Estimation of maximum bi- osolids and meat and bone meal application to a low P index soil and a method to test for nutrient and metal losses. *Water, Air, Soil Pollut.*, **2013**, 224, 1464 - 1475.
92. Duan, J., & Gregory, J., Coagulation by hydrolyzing metal salts, *Ad- vances in Colloid and Interface Science*, **2003**, 100–102, 475–502.
93. Tchamango, S., Nanseu-Njiki, C. P., Ngameni, E., Hadjiev, D., & Darchen, A., Treatment of dairy effluents by electrocoagulation us- ing aluminium electrodes, *Science of the Total Environment*, **2010**, 408, 947–952.
94. Vourch, M., Balannec, B., Chaufer, B., & Dorange, G.,Treatment of dairy industry wastewater by reverse osmosis for water reuse, Desal-ination, **2008**, , 219, 190–202.
95. A.Benlalla, M.Elmoussaouiti, M.Dahhou, M.Assafi, Utilization of

water treatment plant sludge in structural ceramics bricks, , *Applied*
Clay Science, **2015**, 118, 171 – 177

96. Samuel De Carvalho Gomes, John L.Zhou, Wengui Li, Guangcheng Long, Progress in manufacture and properties of construction mate- rials incorporating water treatment sludge: A review, *Resources, Conservation and Recycling*, **2019**,145, 148 – 159
97. Monteiro, S.N., Vieira, C.M.F., 2014. On the production of fired clay bricks from waste materials: a critical update. *Constr. Build. Mater.* **2014**,68, 599–610.
98. Raut, S.P., Ralegaonkar, R., Mandavgane, S.A., Utilization of recy- cled paper mill residue and rice husk ash in production of light weight bricks. *Archiv. Civil Mech Eng.* **2013**, 13, 269–275.
99. Wong, H.S., Barakat, R., Alhilali, A., Saleh, M., Cheeseman, C.R., Hydrophobic concrete using waste paper sludge ash. *Cem. Concr. Res.* **2015**, 70, 9–20.
100. Chung-HoHuang, Shun-YuanWang, Application of water treatment sludge in the manufacturingof lightweight aggregate, *Con-struction and Building Materials*, **2013**, 43, 174 – 183
101. Buruberri, L.H., Seabra, M.O., Labrincha, J.A., 2015. Prepa- ration of clinker from pulp industry wastes. *J. Hazard. Mater.* **2015**, 286,252–2

Sequestration of Cu(II) ions From Aqueous Media using Natant Magnetite – Chitosan Composite

N. Muthulakshmi Andal[1]. Dharani[2*], G. Preethi[2*],**

[1]Department of Chemistry, Assistant Professor, PSGR Krishnammal College forWomen, Coimbatore – 4, India

[2] Department of Chemistry, Research Scholar , PSGR Krishnammal College forWomen, Coimbatore – 4, India

**Corresponding author, Email address: sdharani1005@gmail.com,preethiyasathy@gmail.com*

***Corresponding author, Email address: muthulakshmiandal@psgrkcw.ac.in*

Abstract

The present study is focused on the synthesis of Hydroxy Apatite (HAP) compo- site and magnetite composite using Native Chitosan (NC) as precursor by co- precipita- tion method,abbreviated as HAPNCC (HAP – Natant Chitosan Composite) and MHAPNCC (Magnetite –HAP - Natant Chitosan Composite) to chelate the copper ions from aqueous matrices. Quantitative efficiencies of NC, HAPNCC and MHAPNCC for sequestrating copper metal are verified through Batch Equilibration mode. Pre and post run composites are characterized using sophisticated techniques viz., FTIR, SEM and EDAX toanalyze the involvement of functional groups, changes in surface mor- phology and elemental composition. Impact of operating parameters viz., sorbentdosage, agitation time,Cu(II) initial concentration, pH, temperature and co-ions upon the systems Cu – NC, Cu – HAPNCC, Cu – MHAPNCC are optimized, since they play a vital role as controlling factors. A judicious comparison is made for the verified systems. It is con- cluded that MHAPNCC possesses excellent sorption nature with a maximum Cu(II) removal of 90%.

Keywords

Precursor, Chitosan, Magnetite, Batch equilibration, Operating parameters

1. Introduction

In the world economy, water plays a significant role, owing to the increase in its re- quirement day by day. The main sources of water pollution include pathogens, chemi- cals and also elevated temperatures[1]. Water pollution is caused due to the backflow of wastewaters into the environment, in an untreated manner. Industries discharge wastewaters loaded with huge amount of toxic chemicals, along with heavy metals like nickel, copper, lead, chromium, arsenic in the form of sulphates, phosphates and sili- cates into the water streams [2]. In general, metals are released from their endemic spheres to different environmental sources, mainly due to the natural weathering pro- cesses. The present work aims at arriving upon the remedial measures to control the pollution caused due to copper ions. The sources of this divalent metal entry into the en- vironment include mining industries, construction, plating and electrical equipment manufacturing units apart from natural sources [1]. Water quality deteriorates, when it contains copper concentration greater than 1.3 mg/L. Utilization of such polluted sources leads to common complaints of nausea and diarrhoea. This is because; only 50μg/L concentration of copper is permittedto be prevalent in industrial wastes as per EPA norms. Acute presence damages liver cirrhosis in children), kidney (inhibits urine production), heart (cardiovascular defects), brain (Wilson's disease [3], Alzheimer's [4] and other neurodegenerative diseases) and even fatality [5]. Most of the copper compounds bound to water sediment. Such solu- bleparticles result in greatest threatto humans [6]. Coagulation [7], membrane operation, precipitation, adsorption, etc are frequently adopted methods for the removal of toxicants [8]. Adsorption is preferred for the present study amongstall the suggested methods, due to its selectivity / effectiveness to sorb even minimal concentration of organic and inorganic pollutants [9].

Objectives

Pilot studies were planned using chitosan, a linear polysaccharide i.e., deacetylatedform of chitin, wherein C-2 hydroxyl groups are replaced by acetamido groups [10].

Chitosan has high reactivity, stability towards heat or radiation, sorption capacity, bio- compatibility with living tissues, non-toxic and biodegradability [11]. It also exhibits bio-inert and bacteriostatic properties, good adhesion,coagulation ability and immune- stimulating activity [12]. HAP and Magnetite composites weresynthesized from native chitosan, followed by planning of Batch experiments to verify their potential to se- questrate copper ions from aqueous media.

2. Methodology

Chitosan [$(C_6H_{11}O_4N)_n$], was purchased from Pelican Biotech & Chemical Labs Pvt. Ltd.,Kuthiathodu, Kerala (**Figure 1**). Raw chitosan was pulverized and used as a pre- cursor in the synthesis of HAP (**Figure 2**) and Magnetite composites (**Figure 3**).

2.1 Preparation of Composites (Co-precipitation Technique)

$Ca(NO_3)_2.4H_2O$ and [$(NH_4)_2HPO_4$] were dissolved as 5:1 ration in 10 mL ethanol and 3mL doubly distilled water separately. These solutions were thoroughly mixed, result- ing in the formation of white residue, labeled as Hydroxy Apatite (HAP). Ferrous and ferric chloride salts were taken up in 1:2 ratio, and thoroughly dissolved in 5 mL dis- tilled water followed by the addition of 4g NaOH pellets and ensured complete disso- lution through continuous stirring. The black magnetite precipitate obtained was dried. Prepared HAP and Magnetite were mixed at 1:1 ratio to form magnetite - HAP

(MHAP), later, dried in an oven at 250°C for 24 hours. 4g each of pulverized raw chi- tosan was transferred into two beakers labeled as 1& 2. 10 mL of 0.1N acetic acid and 5g of dried HAP and MHAP were added correspondingly to the beakers.

The contents were stirred well to obtain homogeneous mixtures, referred to as HAP - Natant Chitosan(HAPNCC) (**Figure 4**) and Magnetite - HAP- Natant Chitosan Composites (MHAPNCC). The filtered, oven dried (300°C) sorbent materials stored in air tight containers (**Figure 5**).

Figure 1. **Native Chitosan (NC)** Figure 2. **HAP** Figure 3. **Magnetite**

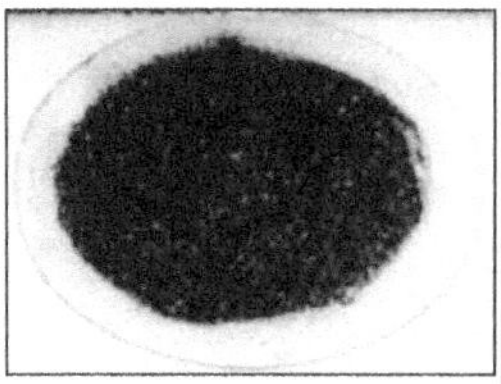

Figure 4. HAPNCC Figure 5. MHAPNCC

2.2 ***Preparation of Adsorbate Solutions***

A stock solution of 1000 mg/L was prepared by dissolving appropriate amount of $CuCl_2 . 2H_2O$ in double distilled water. Standard (100 mg/L) and aliquots (1 to 10 mg/L) solutionswere prepared usingstock solution, respectively.

2.3 ***Characterization Studies***

FTIR, SEM & EDAX analyses were carried out for NC and MHAPNCC to assess the peak variationspertaining to functional groups, surface texture andelemental composi- tion correspondingly.

2.4 Batch Equilibrium Studies

Batch mode experiments were carried out to resolute the sorption capacities of the de- rived sorbent materials. Parameters with fixed ranges verified for Cu(II) - NC, Cu(II) - HAPNCC and Cu(II) - MHAPNCC systems include: Initial metal concentration (2– 10 mg/L: 2mg/L); material dosage (2 – 10 mg: 2 mg); contact time (2-10 minutes: 2 minutes). 50 mL of aqueous Cu(II) solutions of specified concentrations were trans- ferred into 250 mL Erlenmeyer flasks loaded with appropriate doses of synthesized materials and agitated at 120 rpm in a mechanical shaker (KEMI model). The flasks solutions were filtered at preset time intervals, followed by analysis of Cu(II) residual concentrations along with initial concentration values using Atomic Absorption Spec- trophotometer (AA 6200) at a specific wavelength (324.8 nm) and slit width (0.5 nm). Influence of variable factors like pH, temperature and co-ions were registered for MHAPNCC system under optimized data of initial concentration, dosage and con- tact time. pH values of the initial and final sample solutions were measured using CO LI-120) pH meter.

3. Results and Discussion

Recorded values for Cu(II) – HAPNCC and Cu(II) – MHAPNCC systems reveal marked sorption nature of the latter in preference to the former. Hence, the characteri- zation studies were restricted to magnetite-based sorbent and discussed in detail. How- ever, the performance of theaforesaid material is compared with that of native precur- sor.

3.1 FTIR Analysis

Broad bands at 3410 cm^{-1} and 2818 cm^{-1} correspond to the presence of -NH and -OH groups in MHAPNCC (**Figure 6**). Shifts in these bands are envisaged at 1673 cm^{-1} and 858 cm^{-1} indicative of amidecarbonyl stretching and involvement of HPO4 $^{2-}$ (HAP). Peak stretching at 1051 cm^{-1} shows the participation of –

C=C in MHAPNCC towards adsorption. The presence of aromatic ring in native precursor is evidenced by C-H bending at 661 cm^{-1}.

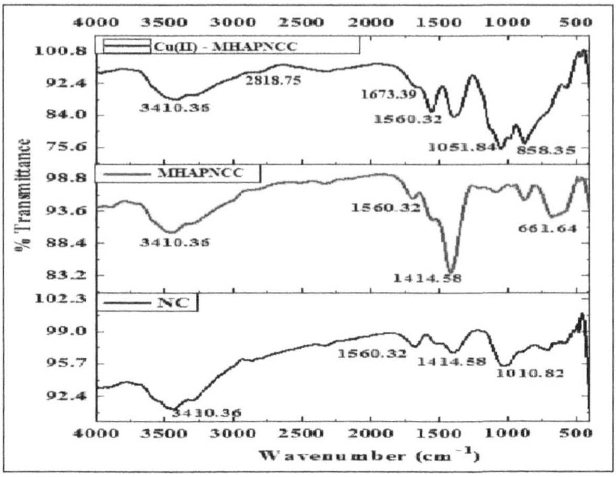

Figure 6. FT-IR Spectra - NC, MHAPNCC and Cu(II) – MHAPNCC

3.2 Scanning Electron Microscopy

SEM images (**Figures 7a, 7b**) exhibit rod shape / coarse surface texture with multiple orifices of different patterns for NC and unloaded MHAPNCC. However, a close look on Cu(II)- MHAPNCC micrograph (**Figure 7c**) reveal a smooth cubic pattern with pores being uniformly occupied by Cu(II) ions. EDAX spectra of Cu(II) - MHAPNCC (**Figure 8c**) registers a new peak (8 KeV) confirming theoccurrence of Cu(II) sorption. Also, a peak at 6 KeV in **Figure 8b** shows the presence of iron in the prepared magnet- ite against NC (**Figure 8a**).

3.3 Effect of Adsorbent Dosage, Preset time and Initial Concentration

Dose of MHAPNCC, predetermined time and Initial Cu(II) concentrations frames play a vital role in the fixation of best sorption conditions. In this regard, the experimental run has revealed that maximum Cu(II) removal had occurred at

10 mg/L initial Cu(II) concentration; 8 mg dose with a contact time of 6 minutes for all the studied three sys- tems (**Figures 9-13**).

Figure 7a. SEM- Native Chitosan (NC)

Figure 7b. SEM-Unloaded MHAPNCC

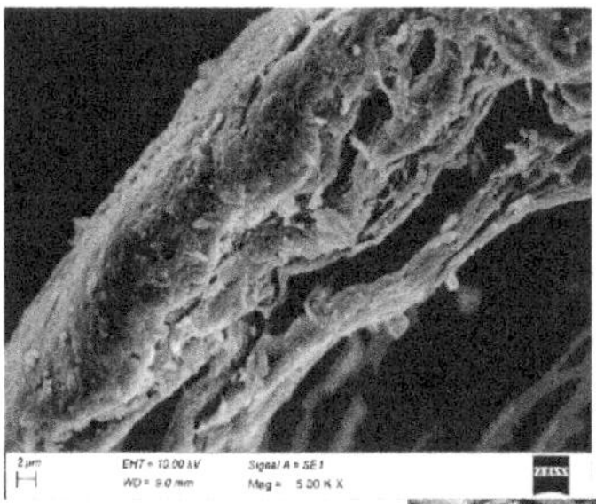

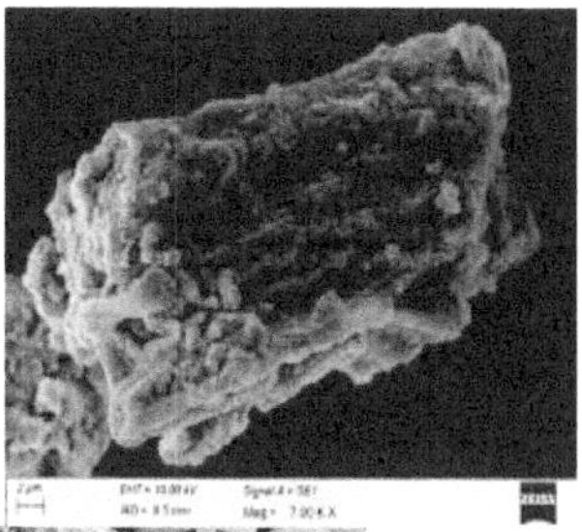

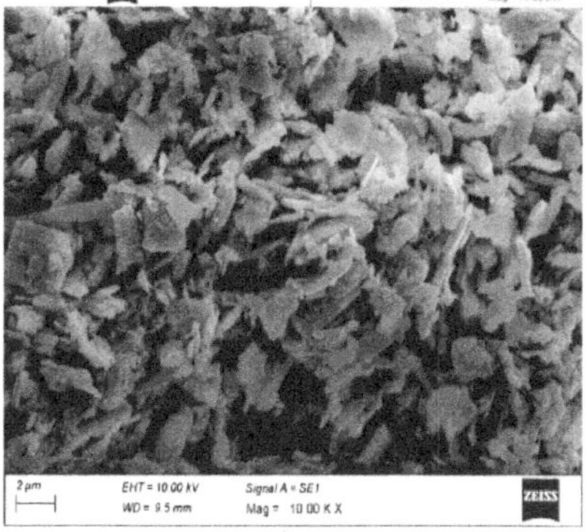

Figure 7c. SEM- Cu(II)- MHAPNCC

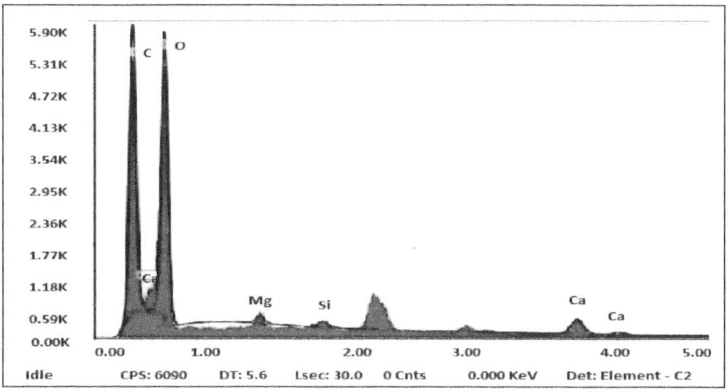

Figure 8a. EDAX – Native Chitosan (NC)

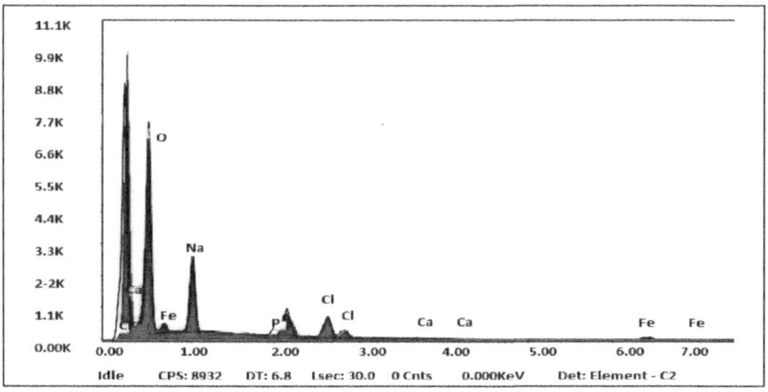

Figure 8b. EDAX-Unloaded MHAPNCC

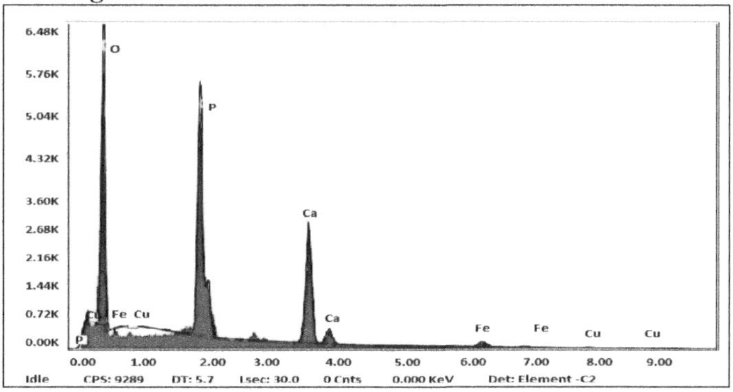

Figure 8c. EDAX-Cu(II)-MHAPNCC

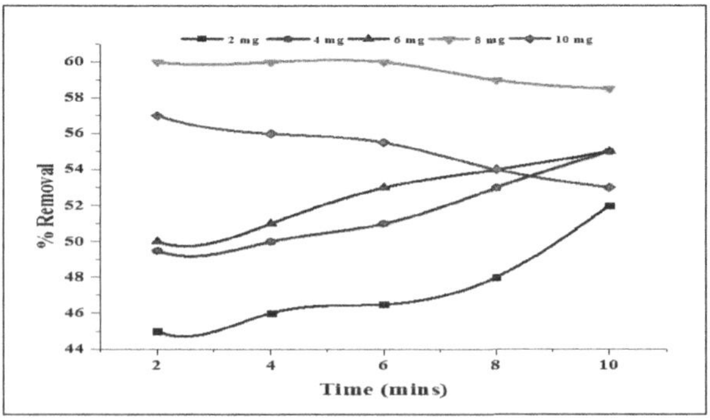

Figure 9. Effect of NC Dosage and Contact Time

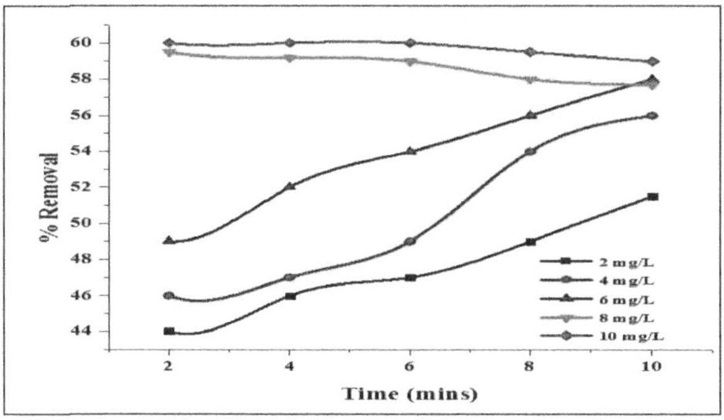

Figure 10. Effect of Initial Cu (II) Concentration onto NC

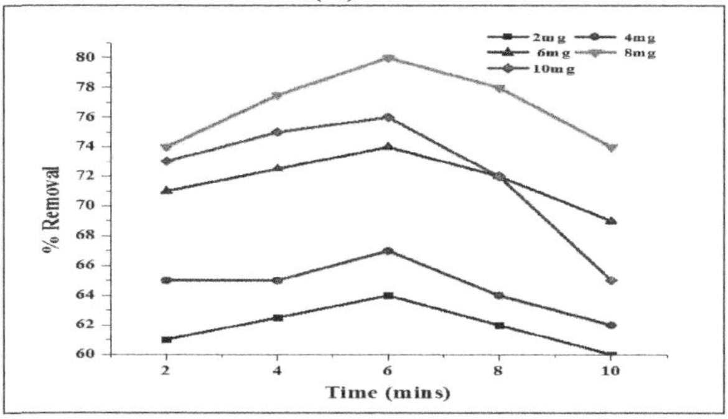

Figure 11. Effect of HAPNCC Dosage and Contact Time

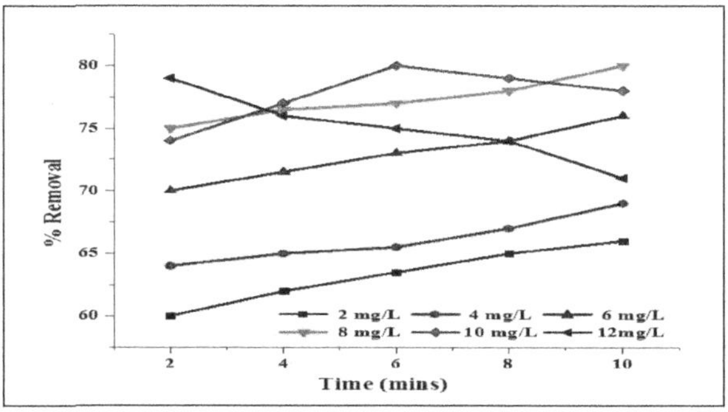

Figure 12. Effect of Initial Cu (II) Concentration onto HAPNCC

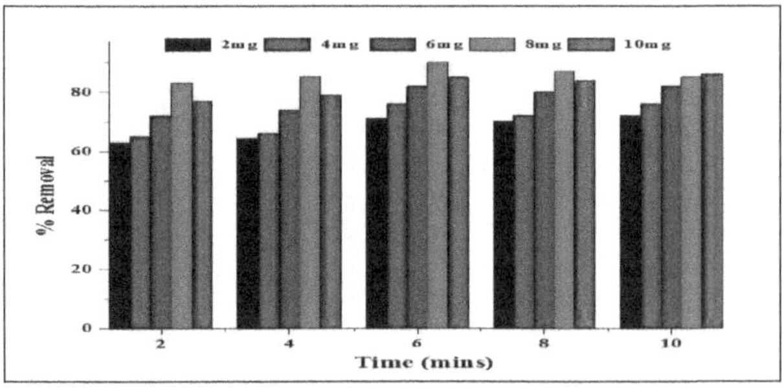

Figure 13. Effect of MHAPNCC Dosage and Contact Time

3.4 Comparison

A judicious comparison made among the studied systems viz., NC, HAPNCC, MHAPNCC reveal that MHAPNCC system exhibits better sorption characteristics against HAPNCC & NC (**Figure 14**). Thereason may be the contribution of unpaired electrons in Fe atom to exhibit magnetic moment, which attract the sorbate cations [13]. Thence, MHAPNCC possess better surface ability to aggregate Cu(II) ions than HAPNCC:

Order of preferential: **MHAPNCC** > HAPNCC > NC

Percentage removal: **90%** > 80% > 60%

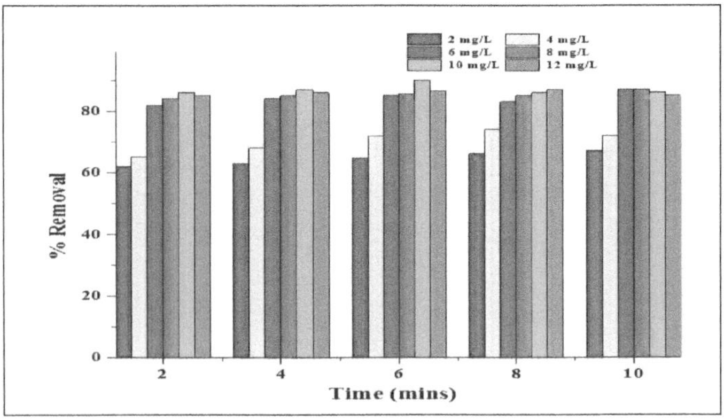

Figure 14. Effect of Initial Cu (II) Concentration onto MHAPNCC

*3.5 **Effect of pH***

Influence of pH environment on Cu(II) - MHAPNCC revealed that maximum trapping ofsorbate species had occurred at pH 7. This is evident from the inverted parabola as depicted in **Figure 15**. Competency of H^+ ions against Cu^{2+} ions is greater due to the smaller ionic radii of the former in acidic pH. Also, the sorbate species tend to form complexes with OH^- ions preferentially, thus minimizing their availability to get sorbed on MHAPNCC surface.

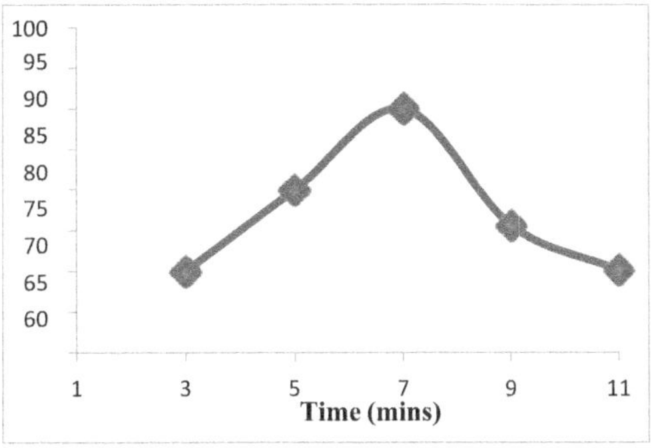

Figure 15. Effect of pH in MHAPNCC

*3.6 **Effect of Temperature***

The influence of temperature holds to be an important role in the sorption process. The bar graph (**Figure 16**) implies an inverse relation between the experimented tempera- tures and removal of the divalent ion. The reason for the above statement shall be due to the rapid mobility of the sorbate ions at higher temperatures leading to diminishing removal.

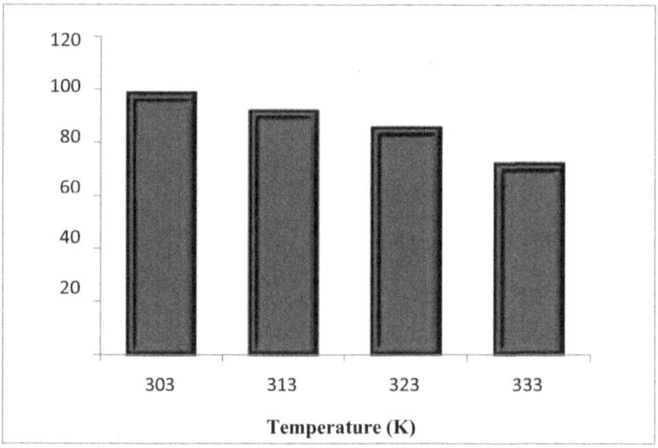

Figure 16. Effect of Temperature in MHAPNCC

3.7 Effect of Co-ions

The prime idea of studying the impact of cations / anions / co-ions is due to their pres- ence inthe effluentdischarges, thereby assessing the sorption efficiency of MHAPNCC under their influence. **Figure 17** shows a notable reduction in Cu(II) sorption due to inhibition caused by Mg , Na , K , NO_3 , Cl ions on the studied system. Mg^{2+} & Cl^-

ions registered pronounced hindrance upon Cu(II) sorption byMHAPNCC, due to their lower ionic radii which facilitating preferential sorption against the other verified cati- ons and anions, whose ionic radii values are marginally higher. As far as, co-ionic study isconcerned, the metal binding property of MHAPNCC is little affected, as evi- dent from the minimal inhibition exhibited by Ni (II)ions in the bar chart.

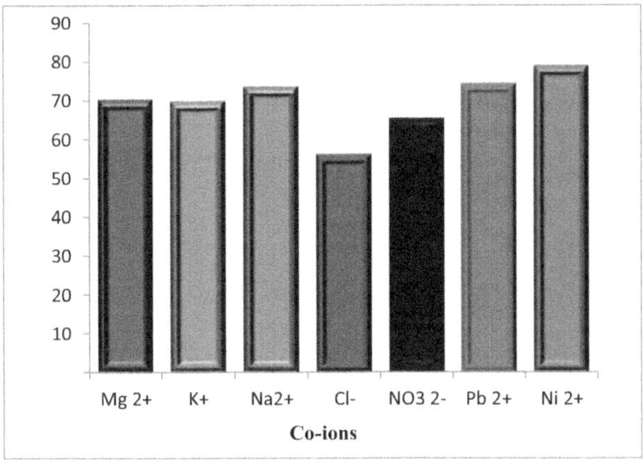

Figure 17. Effect of Co-ions

Conclusion

Batch experiments were conducted for three systems involving Cu(II) chelation by NC, HAPNCC andMHAPNCC to determine the enhanced sorption efficiencies of the func- tionalized materials against their native precursor. The experimental results reveal that MHAPNCC exhibited maximum Cu(II) removal (90%) from aqueous solutions. Under optimized factors viz., 10 mg/L initial metal concentration, 8 mg dosage, 6 mins agita- tion time; pH 7 and temperature 303 K, MHAPNCC is found to be a potential eco- friendly sorbent against other prepared samples in sequestering toxic copper ions,since magnetite has large specific surface area,reversible adsorption behavior and high reac- tivity, chitosan is used as a precursor which also manifest good sequestering property against Cu(II) ions.

Conflict of Interest: The authors declare that there are no conflicts of interest.

References

[1] Jebin Ahmed, Abhijeet Thakur and Arun Goyal, "Industrial Wastewater and Its Toxic Effects",

Royal Society of Chemistry, (2021), 1-14; 10.1039/9781839165399-00001

[2] Michel Phair; "Effects of Water Pollution and Control Measures, *Journal of Pol- lution Effects and Control,* 10 (2022) 326

[3] Beata Kasztelan-Szczerbinska and Halina Cichoz-Lach, "Wilson's Disease: An Update on the Diagnostic Workup and Management", *Journal of Clinical Medi- cine,* 10(21) (2021) 5097; https://doi.org/ 10.3390/jcm10215097

[4] Fábio C. Coelho, Rosanna Squitti, Mariacarla Ventriglia, Giselle Cerchiaro, Jo- ão P. Daher,

Jaídson G. Rocha, Mauro C. A. Rongioletti and Anna-Camilla Moonen, "Agricul- tural Use of Copper and Its Link to Alzheimer's Disease", *Biomole- cules* 10(6) (2020)897; https://doi.org/10.3390/biom10060897

[5] A. Alicia, Taylor, S. Joyce Tsuji, R. Michael. Garry, E. Margaret. McArdle, Wil- liam L. Goodfellow Jr., William J. Adams and Charles A. Menzie , ''Critical re- view of exposure and effects: implications for setting regulatory health criteria for ingested copper''; *Environmental Management* 65 (2020) 131–159

[6] Jessica Briffaa, Emmanuel Sinagrab, Renald Blundell, "Heavy metal pollution in the environment and their toxicological effects on humans", *Heliyon,* 6(9) (2020)
E04691; https://doi.org/10.1016/j.heliyon.2020.e04691

[7] Maryarn Gholami, Mohammad Taghi Ghaneian, Seyedeh Salehe Fatemi, Parvaneh Talebi, ArashDalvand, "Investigating the Efficiency of Coagulation and Floccula- tion Process in Wastewater Treatment of Paper and Cardboard Recycling Indus- try", *International Journal of EnvironmentalAnalytical Chemistry*, (2020)

[8] Shaikh A. Razzak, Mohammed O.Faruque, Zeyad Alsheikh, Laila Alsheikhmo- hamad, Deem Alkuroud, Adah Alfayez, S. M. Zakir Hossain, Mohammad M. Hossain, "A comprehensive review on conventional and

biological-driven heavy metals removal from industrial wastewater", *Environmental Advances*, (2022)100168;

[9] Liping Liang, Fenfen Xi, Weishou Tan, Xu Meng, Baowei Hu & Xiangke Wang, "Review of organic and inorganic pollutants removal by biochar and biochar- based composites", *Biochar*, 3 (2021) 255-281; https://doi.org/10.1007/s42773- 021-00101-6

[10] Anwar Jardine, "Amino-functionalized polysaccharide derivatives: Synthesis, Properties and Application", *Current Research in Green and Sustainable Chemis- try,* 5 (2022) 100309, https://doi.org/10.1016/j.crgsc.2022.100309

[11] Rayssa de Sousa Victor, Adillys Marcelo da Cunha Santos, Bianca Viana de Sou- sa, Gelmires de Araújo Neves, Lisiane Navarro de Lima Santana, and Romualdo Rodrigues Menezes, "A Review on Chitosan's Uses as Biomaterial: Tissue Engi- neering, Drug Delivery Systems and Cancer Treatment", *Materials*, 13(21) (2020) 4995; 10.3390/ma13214995

[12] Patricia Garcia Ferreira, Vitor Francisco Ferreira, Fernando de Carvalho da Silva, Cyntia Silva Freitas, Patricia Ribeiro Pereira and Vania Margaret Flosi Paschoal- in, "Chitosans and Nanochitosans: Recent Advances in Skin Protection, Regen- eration and Repair", *Pharmaceutics,* 14 (2022) 1307. https://doi.org/10.3390/ pharmaceutics14061307

[13] A. Ali, H. Zafar, M. Zia, I. ul Haq, A. R. Phull, J. Sarfraz Ali and A. Hussain, "Synthesis, characterization, applications, and challenges of iron oxide nanoparti- cles", *Nanotechnology, Science and Applications*, 19(9) (2016) 49-67, https://doi.org/10.2147/nsa.s99986

Organophosphorus Pesticide Chelation by Modified Pod Shells – Batch Equilibration and Isothermal Studies

N. Muthulakshmi Andal [1*] ***and Preethi. G*** [2]

[1*]*Assistant Professor, Department of Chemistry, PSGR Krishnammal College for Women, Coimbatore- 4*

[2] *Research Scholar, Department of Chemistry, PSGR Krishnammal College for Women, Coimbatore- 4*

*Email: muthulakshmiandal@psgrkcw.ac.in

ABSTRACT

Extensive loads of pesticides are utilized to increase the crop productivity in modern agricultural practices. Slow degradation of pesticides into the environment and exten- sive or inappropriate use by farmers can lead to contamination of water, soil, air, sev- eral types of crops and in turn human beings. Malathion pesticide has been widely used as fumigant, in agriculture and in residential landscapes to replace more toxic organochlorine pesticides, even otherwise result in a serious setback to human health causing skin and nervous defects. Adsorption technique is one of the most intensively availed and reliable method adopted due to its ease of operation, non- energy intensive and high removal competence of organic pollutants like pesticides. *Delonix regia* Pod Shells (DRPS), an eco-based material is employed to chelate malathion from aqueous media through batch mode under varying operational factors viz., pH (3 – 11), initial malathion concentration (50 – 300 mg/L), DRPS dosage (50 – 250 mg), contact time (10 – 60 mins) and temperature (293 – 333 K). Optical Microscope, FTIR, SEM are utilized to characterize the selected sorbent's nature. Isothermal data fitted well with the Langmuir plot in preference to Freundlich plot. From the observations made, DRPS is found to be a promising ecofriendly, economically viable material for the sequestration of Malathion.

Keywords: pesticides, adsorption, malathion, isotherms

1. INTRODUCTION

Pesticides form a group of organics like insecticides, herbicides, nematicides, ro-
denticides, bactericides etc which are extensively utilized for crop protection, as it is an important challenge for the maintenance of country's welfare. [1] Synthetic/ Non- syn- thetic insecticides like organophosphorus pesticides are mostly non – biodegradable, highly toxic, carcinogenic and their residues have accumulated in soil, rivers, ponds, underground water streams due to agricultural run- off, leaching and improper applica- tions and disposal. Pesticide pollution is framed as one of the reasons to adopt food safe- ty issues wherein veggies and fruits yield to higher dosages of pesticides. [2]

Malathion (O,O, - dimethyl thiophosphate diethyl mercaptosuccinate) is a non- systematic, wide spectrum of an organophosphorous insecticide; applied to control pests on vegetables/ fruits in agro sectors/ to kill mosquitoes, ectoparsites, flies, lices in do- mestic sectors, [6] also used as disinfectant in industrial sectors [3] Malathion is procura- ble in liquid formulations, wettable/ dustable powder from agrochemical stores. Mala- thion is an Acetylcholinesterase - AChE (esteratic activity) inhibitor, in which the phos- phoester group (Figure 1.1), combine strongly to the cholinesterase and irrevocably neu- tralizes the enzyme initiates the building up of acetylcholine at the synapse.

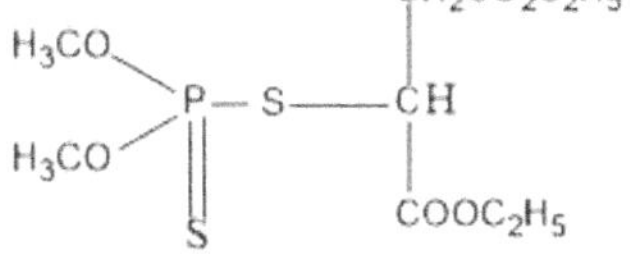

Figure 1.1 Structure of Malathion

Recent findings report that the prevalence of toxic malathion in surface waters, via ingestion/ inhalathion/ dermal exposure for non target organisms, leads to overstimu- lation of glandular cells, Central Nervous System (depression/ anxiety/ reduced respira- tions/ seizures/ coma) and eye irritation. [4]

The toxicity of malathion and its degraded product Malaxon, is a threat to food chain through water/ soil contamination.[5] Therefore, minimization of malathion con- centration from water streams and soil systems is an inevitable environmental concern nowadays.

Several independent and combined methods [6] have been adopted for the removal of malathion from aqueous/ wastewater/ spiked veggies like ozonation , photo – fenton treatment, advanced oxidation, aerobic degradation, coagulation and adsorption . Pres- ently, there is an increasing trend to evaluate some indegenious cheaper, locally availa- ble materials like industrial/ agrowastes to remove malathion.[7] Materials reported as adsorbents for malathion removal includes: fly ash from industrial waste, chitosan from marine industry waste, spent tea/ coffee grounds, egg shells, rice husk, drumstick peel powder, utherani/ nela usiri herbal leaves powder, waste jute fiber etc., The primacy of using agrowastes as biosorbents leads to conversion of wastes to potential sorbent mate- rials, thereby disposal costs of these solid litters are minimized, further curbing of unre- alized environmental issues.[8]

The present work deals with the removal of malathion from aqueous solutions us- ing plant waste ie., *Delonix regia* pod shells (DRPS), a litter material least research work has been reported elsewhere in exploring its sorption nature against malathion.

MATERIALS AND METHODS

2.1 Collection/ Preparation/ Modification of the adsorbent

The identified material, *Delonix regia* pod shells were collected from the localities of Sathyamangalam. The seeds were removed from the hulls and latter were hammered into smaller pieces (Figure 2.1 a), washed off the impurities with DDW, completely sun dried and arid in hot air oven at 100°C for 3 hours to ensure moisture elimination. The dried pieces were crushed, milled in an electrical mixer (Figure 2.1 b) and sieved into varying mesh sizes (0.18mm –

0.71mm) using Scientific Test Molecular Sieves pro- cured from Jayant Scientific Instruments Co., Mumbai. The milled DRPS of varying particle sizes soaked in 0.1N HCl for 3 hrs, was boiled and the clear solution was de- canted. Similar process was repeated with 0.1N NaOH, further several standing washings with DDW was done to attain a neutral pH range and to ensure enhanced surface bulkiness of the prepared material. This chemically modified sorbent DRPS is expected to possess sorption sites for protonation labeled as TDRPS (Figures 2.1 c), stored in air

tight containers and utilized for further experiments.

Figure 2.1 (a): Hammered DRPS

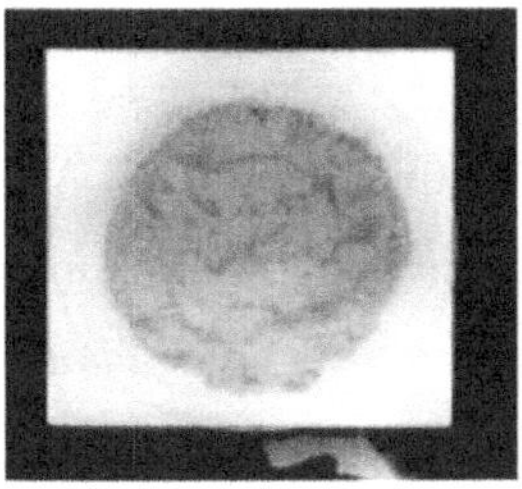

Figure 2.1 (b): Powdered DRPS

Figure 2.1 (c): Treated DRPS

2.2 Adsorbate

Technical grade 95% pure malathion, the chosen adsorbate species was purchased from Kongu Agro Chemicals Pvt Ltd., Sathyamangalam, Tamilnadu, India. The chemical structure of malathion is depicted in figure 1.1. A stock solution of 1000 mg/L was pre- pared by dissolving an appropriate quantity of malathion pesticide in glaciel acetic acid and further diluted with doubly distilled water in a 1000mL standard flask. Correspond- ing aliquots were prepared by diluting the stock solution as per the experimental re- quirements. N-bromosuccinimide (NBS) and Rhodamime – B, the complexing agents required for the planned experiments of AR grade were procured from Loba Chemie Pvt Ltd, Mumbai, India. Standard solutions of HCl and NaOH with 0.1M strengths were used for pH determinations.

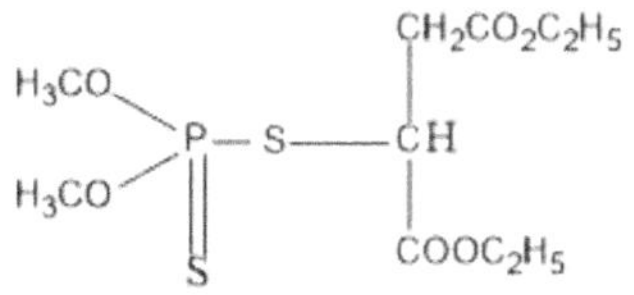

Figure 1.1 Structure of Malathion

2.3 Equipments Employed

The instrument utilized for various studies in the present work are listed below:

- Binocular microscope equipped with Ocular micrometer (OLYMPUS make, Model CX21I) for particle size determination of TDRPS
- Automated JOEL JEM – 6390 Scanning Electron Microscope under a vacuum of 1.33* 10^{-6}m Bar for examination of surface morphological changes on TDRPS
- Shimadzu Infrared Spectrophotometer for identification of functional groups in sorb- ate/ sorbent species
- Elico LI120 pH meter for pH measurements of raw, treated and pesticide loaded sorbent
- Uv-Vis Spectrophotometer (Lab India UV 3000^{+}) for recordance of initial and final concentrations of malathion species (prior and post experimental solutions) by complex-ation method.

2.4 Adsorption Experiments

Adsorption studies (Malathion- TDRPS), were conducted by batch equilibration meth- od. 50 mL of the aliquot solutions (pH: 7) with initial concentration ranging from 50- 300mg/L :L 50mg/L along with 25-150mg: 25 mg each of

TDRPS were added to 250mLErlenmeyer flasks and stoppered. The contents of the flasks were agitated in mechanical shaker (KEMI model) at a speed of 140 rpm, setting up a time frame of 50- 30 mins at the rate of 5 mins interval, to decant and analyze the experimented solutions. 1.5mL NBS, 1 mL glaciel acetic acid, 0.5mL HCl and 0.7mL Rhodime B solution were used in complexing the residual malathion solution species and the corresponding data

were registered in UV- Vis spectrophotometer at 550 nm against the reagent blank. The

% removal of malathion and the amount adsorbed (mg/g) were calculated using the fol-lowing equations.

% Removal = $(C_i - C_f /$

$C_i) * 100Q_e = (C_i - C_f/$

$W) * V$

where,

C_i and C_f are initial and final concentrations of malathion in aqueous solutions (mg/L),W = weight of the adsorbent (mg) and V = volume of sorbate species (mL)

3. RESULTS AND DISCUSSION

3.1 Optical Microscopic Studies

The lengths and breadths of by and large 15 granular particles with varying di- mensions for the mesh size (85 BSS) were resoluted under the microscope and a multi- plication factor was applied to approximate the average values, thence the particle size calculated was 0.18mm against the mesh size 85 BSS for native and TDRPS are depict- ed as microscopic images in figures 3.1 a and b. The distinct and enlarged open porous nature of TDRPS is in favor of extended sorption capacity of the modified material.

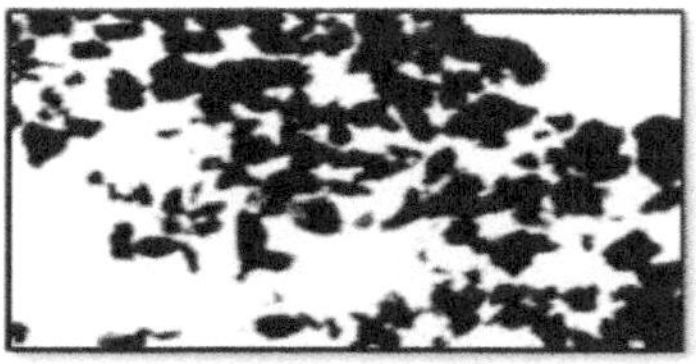

Figure 3.1 (a): Microscopic image- DRPS

Figure 3.1 (b): Microscopic image-TDRPS

3.2 FTIR Spectra

The recorded FTIR spectra for bare TDRPS and pesticide laden TDRPS is shown in Figure 3.2. Occurrence of OH- and C=O- groups at 3438.16 cm^{-1} and 1146.92 cm^{-1} re- spectively in the spectra indicates the presence of these groups in the pod shells. How- ever, a shift from a peak at 1739.86 cm^{-1} to 1146.92 cm^{-1} representing C-O-H stretching, corresponds to the ethyl acetate groups'

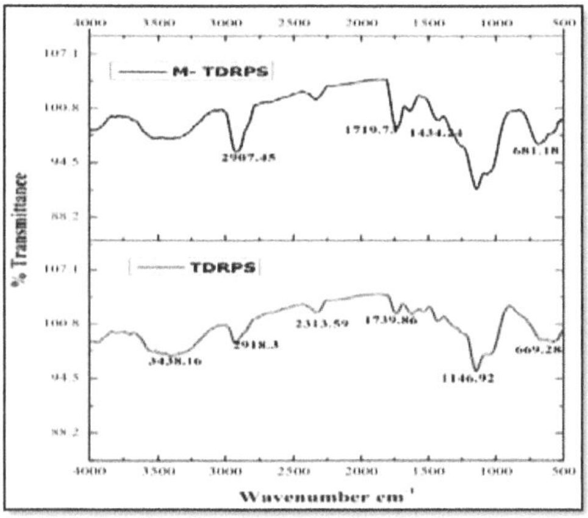

Figure 3.2: FTIR spectra of unloaded / loaded TDRPS

involvement in the sorption process.

3.3 SEM Analysis

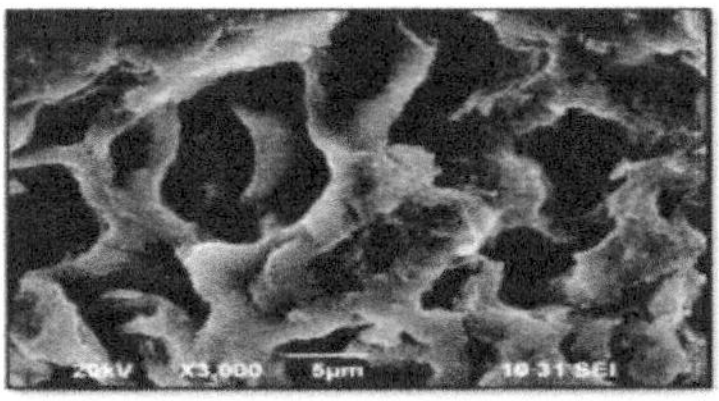

Figure 3.3 (a): SEM image: TDRPS

Figure 3.3 (b): SEM image : M- TDRPS

Surface morphological images of TDRPS as shown in Figure 3.3 a, reveal well defined highly porous honey comb structure. Smooth appearance of Figure 3.3 b, support the coverage of the pores as indicated by the binding of Malathion onto the active sites of TDRPS.

3.4 Adsorption Studies

3.4.1 Effect of Particle Size:

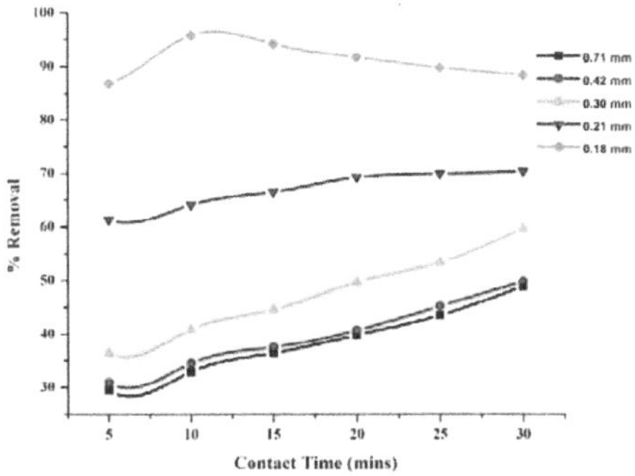

Figure 3.4 Effect of Particle Size

The pictorial results (Figure 3.4) of particle size impact for malathion- TDRPS system imply a maximum malathion removal (95.8%) from 100mg/L concentration for 0.18 mm particle size. This indicates larger number of active binding sites. However, a di- minishing trend of sorption is evident for increased particle size (071 mm- 0.21 mm) reveal greater surface area for 0.18 mm size. Thence, 0.18mm particle size being fixed for further experiments.

3.4.2 Effect of Initial Concentration and Agitation Time:

The adsorption pattern at varying initial malathion concentration against their respective agitating time intervals as in figure 3.5, registers a maximum of 95.8% adsorbate species of 100mg/L initial concentration at a contact time of 10 mins. Also, a steep decline is obvious for other concentrations which shall be due to partial desorption. In view of this, 100 mg/L initial malathion concentration and 10 mins contact time were optimized for

the forthcoming studies.

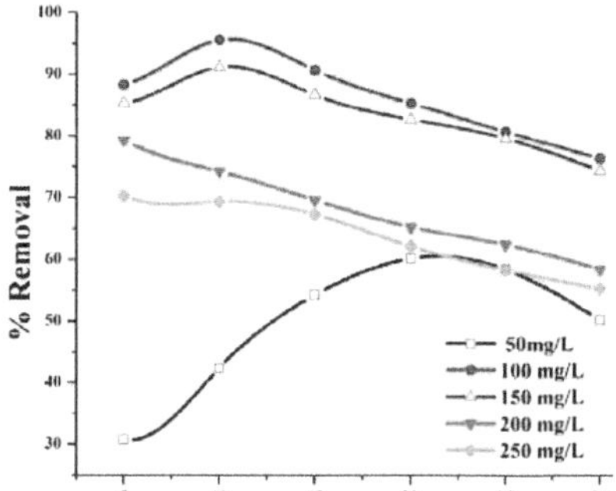

Figure 3.5 Effect of Initial Concentration

3.4.3 Effect of Dosage:

The influence of varying TDRPS dosages (figure 3.6) depict a steady rise in the curve upto 100mg at the optimized time frame of 10 mins, wherein 95.8% sorbate removal is recorded for 100mg. Further decline in the sorption at higher doses may be attributed to the reduction in the sorbent surface area per unit mass, in favor of overlapping / agglom-erating nature of TDRPS.

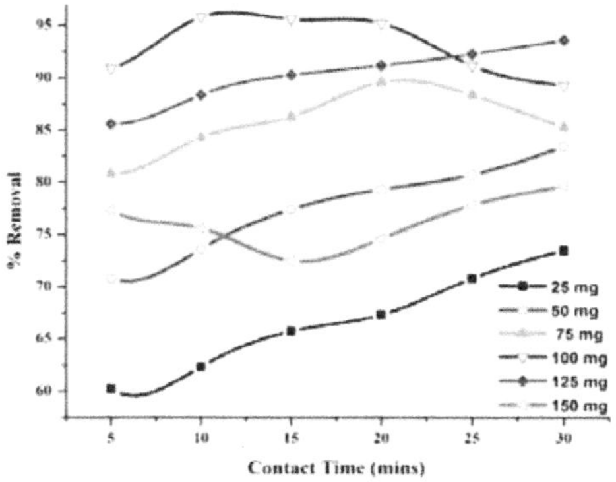

Figure 3.6 Effect of Dosage

3.4.4 Effect of pH :

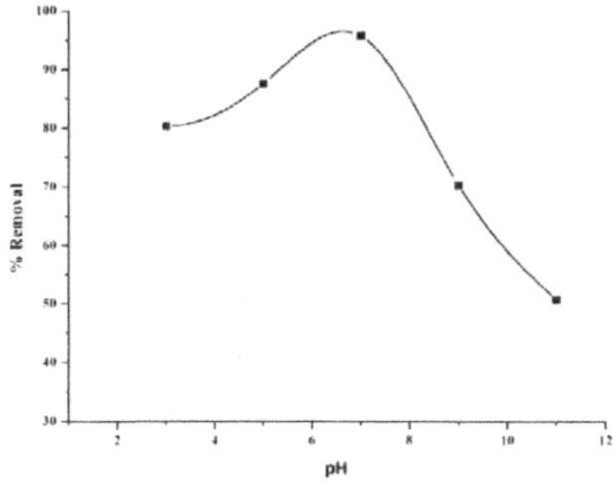

Figure 3.7 Effect of pH : TDRPS - Malathion

The most significant variable, dominating the sorption of pesticides is the solution pH. The maximum removal of malathion was registered at pH 7 (figure 3.7), whereas at higher/ lower pH ranges a drastic decline is observed. This can be because of a competi- tive dissociation between malathion functional groups (methoxide and phosphoester groups) and hydronium/ hydroxyl ions

3.4.5 Effect of Temperature:

It is evident from the figure 3.8 that a maximum sorption had occurred at 303K, indica- tive of the inverted parabola. Further rise in temperature registered a declined sorption, the reason shall be due to the extended collision between the sorbate species making them less available to adhere onto the sorbent surface.

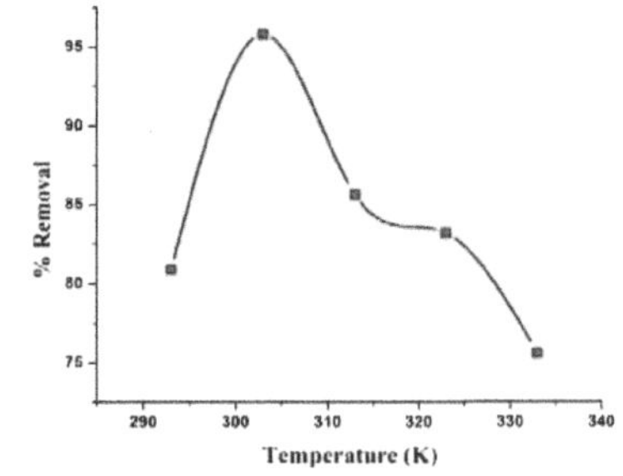

Figure 3.8 Effect of Temperature

3.5 Biosorption Isothermal studies

The parameters pertaining to Langmuir and Freundlich isotherms for the Malathion - TDRPS system are listed in table 1. Langmuir (qm, B) and Freundlich (k, 1/n) constant values calculated from the slopes and intercepts of the straight lines (Figures 3.9 a and b). Correlation co-efficient value (R^2) of Langmuir model exhibits a better linearity of points against Freundlich model,

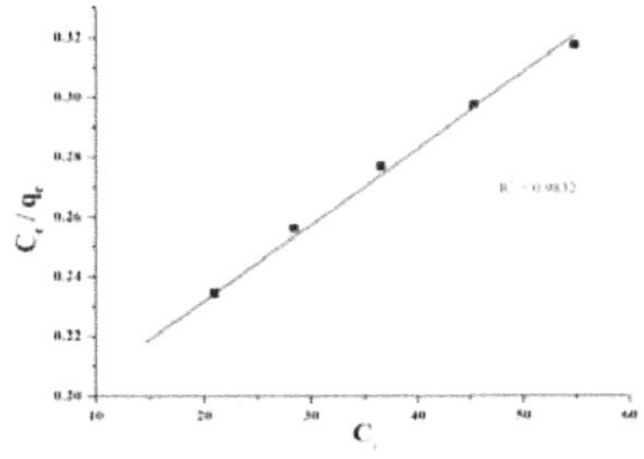

favoring monolayer adsorption.

Figure 3.8 (a): Langmuir Plot

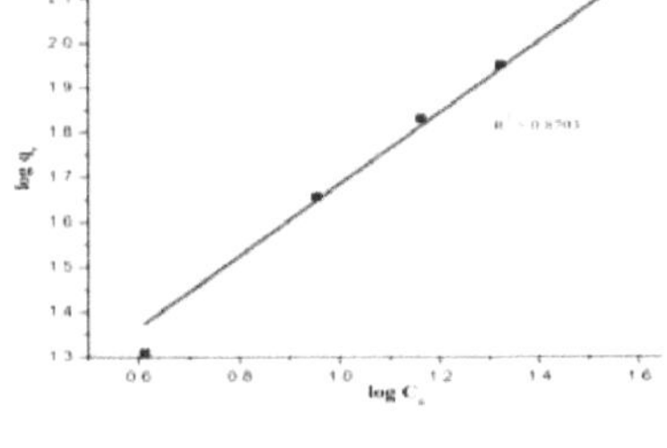

Figure 3.8 (b) : Freundlich Plot

Table 1 - Langmuir and Freundlich Constants

Langmuir	**q_m (mg/g)**	**B (L/g)**	**R^2**
	20.12	0.15	0.9832
Freundlich	**K (mg/g)**	**1/n**	**R^2**
	15.92	0.68	0.8703

CONCLUSION

Treated *Delonix regia* pod shells (TDRPS), a novel plant litter material was employed as an effective adsorbent for the confiscation of organophosphorous pesticide (Malathion) from aqueous environs under variable concentrations. The sorption conditions were op- timized through batch equilibration method for a maximum malathion removal (95.8 %)

: 0.18mm particle size, 100 mg dosage of TDRPS, 100 mg/L initial malathion concen- tration, , 10 mins agitation time, pH 7, temperature 303K. The characterization of the modified pod shells was studied using different techniques namely, Microscopic, SEM, FT-IR to assess the particle size, surface morphology and presence of functional groups. Isothermal models were validated where, the linearity of the Langmuir plot is indicative of monolayer adsorption of the studied system. From the made observations, it is con- cluded that *Delonix regia* pod shells, serve as an excellent alternative to chelate the cho- sen toxic organophosphorous pesticide from the aqueous matrices. Also, the nature of the identified material makes it promising for extensive studies.

REFERENCES

[1] S.Senthilkumaar et al., 2010. Kinetic Approach for the Adsorption of Organophos- phorous Pesticides from Aqueous Solutions using waste jute Fiber Carbon. *E J. of Chem.*, 7(S1): S511- S519.

*[2]*Subhankar Chatterjee et al., 2010. Interaction of malathion, an organophosphorous pesticide with *Rhizopus oryzae* biomass. *J. of Hazardous*

Materials., 174: 47-53

[3] Ahmad Jusoh et al., 2014. Removal of pesticide in agricultural runoff using granu- lar- activated carbon: a stimulation study using a fixed bed column approach. *Desal-ination and Water Treatment.,* 52: 861-866

[4] Vinay K Singh et al., 2010. Removal of malathion from aqueous solutions and wastewater using fly ash. *J. Water Resource and Protection.,* 2: 322-330

[5] Battala Gangadhar et al., 2016. Tea (*Camellia sinensis*) waste for the reduction of malathion and phorate in aqueous solutions. *Int. J. of Env. And Agric. Res.,* 2(6): 38-44

[6] S. E. Aggary et al., 2015. Bioadsorption of 2,6 dichlorophenol from aqueous solu- tion onto plantain pineapple peels mixture used as adsorbent: optimization studies based on taguchi method, batch equilibrium and kinetic modeling. *Chem. And Mate-rials Res.,* 7(3): 1-14

[7] Tharakeswar Yadamari et al., 2011. Bioadsorption of malathion from aqueous solu- tions using herbal leaves powder. *American J. of Analytical Chem.,*2: 37-45.

[8] G. Zuhra Memon et al., 2008. Adsorption of methyl parathion pesticide from water using watermelon peels as a low- cost adsorbent. Chem. Eng. J., 138: 616-621.

[9] Aparajita Shankar et al., 2020. Removal of pentachlorophenol pesticide from aque- ous solutions using modified chitosan. *Arabian J. of Chem.,* 13(1) : 1821-1830

[10] Z. Abdeen et al., 2014. Study of the Adsorp- tive efficiency of an eco-friendly carbohydrate polymer for contaminated aqueous solution by organophosphorus pesticide. *Open J. of Organic Polymer Materials,*4 : 16-28.

Pulp and Paper Industrial Discharges on Soil Environs in Crop Nutrients' Management

N. Muthulakshmi Andal[1] ,S. Karthika[1*]

[1]Research & Development Centre, Bharathiar University, Coimbatore – 641046, Tamil Nadu, India.

[1*]Assistant Professor, Department of Chemistry, PSGR Krishnammal Collegefor Women, Coimbatore – 641004,Tamil Nadu, India.

Abstract

Pulp and paper industry is considered as one of the major industries contributing to the economic growth of our country. However, nowadays they are also iden- tified as the industry discharging heavy volumes of pollutants into water bod- ies and soil environment. These industries generate varieties of contaminants depending upon the manufacturing process. Especially, disposal of these indus- trial wastes to water streams and land sources remainsa critical concern pertain- ing to the localities in their vicinity, wherein the aforesaid resources are vital for the healthy farming practices. The present work emphasizes on the study of af- fected agricultural soil structures in Sathyamangalam area, contaminated by the release of excess heavy metals from pulp and paper industrial effluents. Four soil samples were collected at specific seasons throughout the year (February to March, June to July, September to October and November to December) and es- timated for the nutrients' levels as per the standard quality procedures, in order to assess the growth impact of seasonal crops cultivated. Dissimilarities in the uptake of primary and secondary nutrients were observed in the soil samples, there as on being low and high nutrient availabilities (against prescribed Indian Agricultural Standards for crop development) in Rabi and Kharif seasons, which shall be attributed to the rainfall, in turn promoting toxic leachates into the soil. The soil performances due to the percolation of industrial discharges reflected in

the complexities of crop yields.

Keywords: Pulp and paper industries, Wastewaters, Soil, Crops, Seasons

1. Introduction

Environment is a primary concern and it plays a vital role in human health, society, plants, animals, andaquatic species. If there is imbalance created in any of these environmental factors, it changes the entire bio system, causing tremendous setback. Generally, environment gets deteriorated by the consump- tion of resources in an enormous volume by discharging huge unwanted toxic wastes. This consumption and wastage of resources in an unlimited manner pol- lutes air, water, land and soil. If any one sort of the pollution occurs then it con- tinues as an association to form other kinds of pollution. Major reason for pollu- tion is due to the establishment of industries in rural as well as in urban areas. Industries utilize water for their production. At the same time, large amount of wastewater gets emitted from these industries[1].Treatment of these wastewaters is mandatory before discharge, however certain industries directly discharge the effluents into nearby water streams or land thereby contaminating the resources. Pulp and paper industry consume huge volumes of water and emanate wastewaters, mostly without proper discharge treatment, leading to an adverse impact on the environment, one of the major sources being aquatic pollution[2] which are fortified as toxic to the environment. The present work is focused on the impact created by the pulp and paper industry effluents in soil structures col- lected from the cultivable lands. Methodical investigations pertaining to the soil structure were analysed during four seasons of the year for the occurrence of varied nutrient levels and toxic metal ions.

1. **Experimental**

Pulp and paper industries consume natural resources and energy in an alarming amount for its paper manufacturing processes. The conventional flow chart of paper making in pulp and paper industry is indicated in Fig.1. The prime requirement includes a source of cellulosic fibre containing materials, additives, chemical agents and water[3].

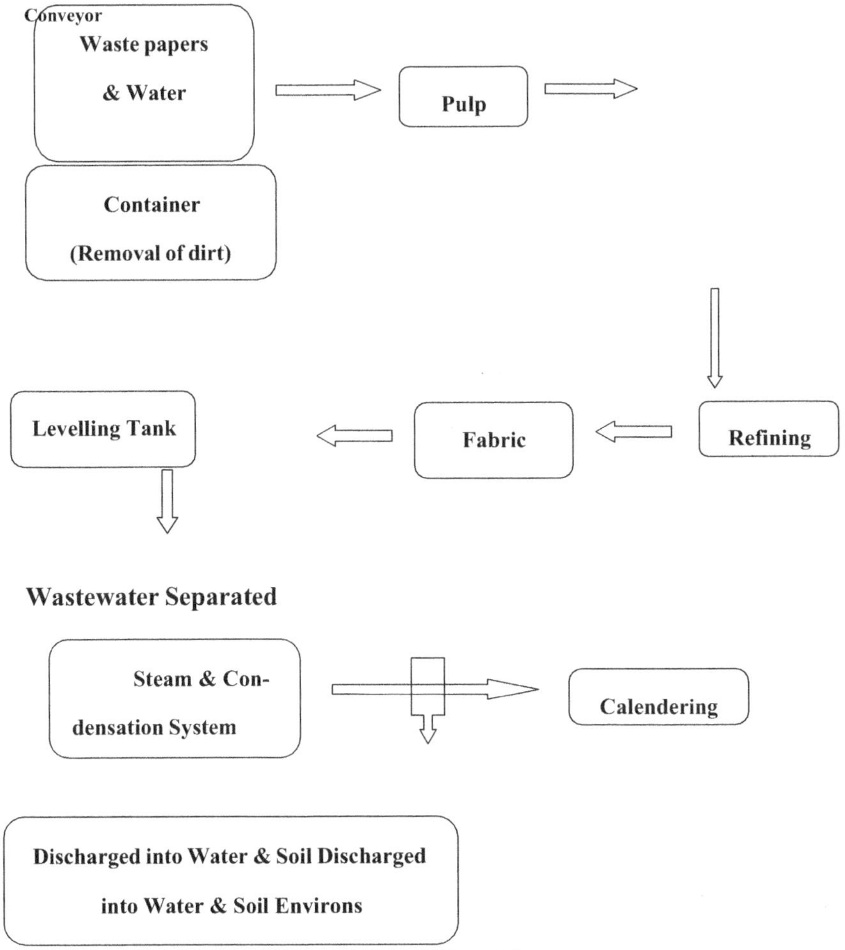

Fig 1 : Conventional Flow Chart

Cellulose is the fibre component of wood, a natural existence[4]. The wood materials are chipped well by means of a chipper, which are digested with the chemicals and separated into fibre / lignin, the materials responsible for cellulose binding. The cellulosic sources are blended with water and impuri- ties present in the mixture are removed by passing through screeners, then through the conveyor to form pulp like stuff. This mixture is blown out into a tank for cleaning and turning it into the form of pulp. The obtained pulp is subjected to refining process followed by addition of fabric like substances (agaltite / talc) and bleaching agents (calcium hypochlorite, sodium hypo- chlorite). The resulting mixture is allowed to be in levelling tank for few hours, passed to steam condensate system for moisture removal from the re- sultant. Finally, it is subjected to calendaring process, where paper products are obtained leaving behind the wastewaters, mostly discharged without proper treatment into nearby water and soil environments.

2.1 Emerging pollutants routes from Loams to Human Intake

Soil, an upper cover of earth's surface is much important for human health in a number of ways. These are major source of nutrients, which are natu- ral filters to remove contaminants from water. They possess well- developed structure, adequate organic content, and other properties that are helpful to sus- tain plant growth promoting improved yield[5]. Soil degradation, loss of soil structure and nutrient content, directly affects crop production and pose a threat to food safety[6]. This degradation in soil may be due to the presence of domestic wastes, agro waste, accrual of heavy metals, chemicals from industries, waste discharges reflecting a negative impact on human health and other living organ- isms. Farm practices with contaminated water sources have a viable chance to get exposed to heavy metals. Accumulation of heavy metals even in trace amounts while exceeding their tolerant limits, lead to many health disorders, such as cancers (arsenic), neurological damage, skin damage (lead, arsenic), kidney disease (lead, mercury, cadmium), skeletal and bone diseases (lead, cadmium)[7]. This accumulation of toxic metal ions by soils and its transport to plants is dependent on many factors such as metalforms, soilstructureand the

crop variety. Uptake of heavy metals by plants via soils tends to increase with metal concentrations. When the concentration becomes intolerable, the plant roots get injured because of low absorption capability, leading to pollution as a result of soil quality deterioration.

2.2 Choice of Study Region

Sathyamangalam region had been selected for assessing the agricultural soil contamination by the pulp and paper industry discharges and is represented in the map (Fig. 2). This region had been identified due to the presence of nu- merous small and large scale industries that are emitting wastewaters into the land sources.

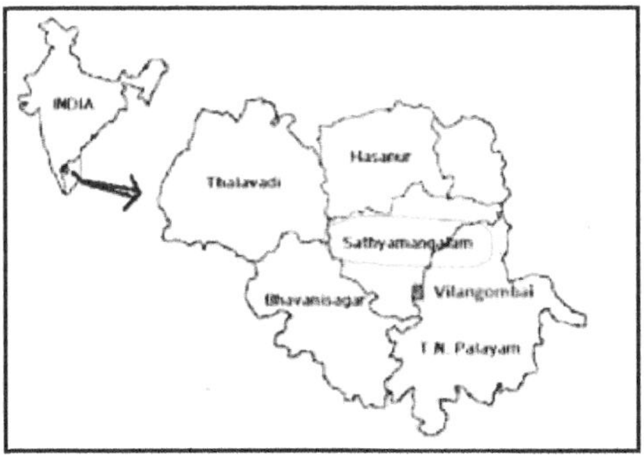

Fig 2 - Study region (Sathyamangalam)

3.0 Material and Methods

Soil samples (I1, I2, I3, I4, I5, I6, I7, I8) from the environs of two pulp and paper industries, agricultural field samples (Fig 3) declared as contaminated by the aforesaid industrial effluents (A1, A2, A3, A4, A5, A6, A7, A8) were collected in the vicinity of Sathyamangalam region. Sampling was done in steri- lized containers at a depth of 10 cm from the ground level during February - March, June - July, September - October and November - December. Evaluation

of the contaminated samples was carried out to assess the occurrence of varied nutrient levels and heavy metal concentrations viz., pH, nitrogen (kjeldahl method), phosphorus(Olsen's method); potassium, sodium (Ammonium ace- tate); calcium, magnesium (EDTA method), sulphur (Sodium peroxide); zinc, chromium, lead, copper ions (Atomic Absorption Spectrometry) by analysing through digestion methods and absorption technique.

4.0 Results and Discussion

4.1 pH

pH plays a predominant quality control factor in soil samples. The rec- orded pH values for I1 and I2 samples collected during different seasons is in- dicative of alkaline nature, which shall be due to the excess bleaching additives in the second industry (I2, I4, I6, I8). However, Table 2 shows comparatively a noticeable change in the soil pH, lying within the prescribed standards. Seasonal changes in soil pH values are perceptible during pre-summer and spring seasons as per Tables 1 and2.

4.2 Nutrients

Almost all the macro / micro nutrients and the heavy metal ions are found to be varied from the prescribed standard values for all the 16 samples analyzed. However, the variations in the registered values (Table 1, 2) amidst the sam- ples, I1 – I8 and A1 – A8 during the seasons are detailed asfollows:

N, P, K present in the soil samples act as "Primary Nutrients" for the plant growth. However, presence

of Ca, Mg and S are considered as "Secondary Nutrients". Primary nutrients in the soil are much vital for the replenishment of soil nutrients. Nitrogen plays an important role in soil fertility. Phosphorus content of soil serves as a constituent of cell nucleus, essential for cell division and development of meristematic tis- sues at the growth points. Potassium, an indispensable constituent for the overall plant development, termed as "Quality nutrient" is an essential factor for several

biological and chemical processes occurring inplants.

Table 1 - Parametric Analyses - Soil Samples - Pulp and Paper Industrial Area

S.No	Months	Sample	pH	Parameters										
				N	P	K	Na	Ca	Mg	S	Zn	Cr	Pb	Cu
IAS (Indian Agriculture Standards)			5.8-	2.3 – 2.8 %	0.1 – 1 %	1.3 – 2 %	0.1 %	<1 %	< 0.5	0.5 %	300 – 600 mg/kg	400 mg/kg	250 – 500 mg/kg	135 – 270 mg/kg
1.	FEB – MAR	I1	8.36	1.18	0.11	0.38	0.20	9.88	4.56	0.02	52.0	65.4	12.1	61.5
		I2	9.15	1.77	0.10	0.54	0.71	8.90	3.29	0.02	52.4	64.0	20.4	67.0

2.	JUN – JUL	I3	6.85	1.40	0.54	0.42	0.34	8.13	3.96	0.04	51.8	67.1	13.3	64.0
		I4	7.82	1.16	0.36	0.66	0.32	8.08	4.10	0.02	52.3	65.7	16.0	64.8
3.	SEP – OCT	I5	7.33	1.62	0.66	0.48	0.31	8.24	5.14	0.04	52.6	66.2	17.1	67.0
		I6	8.52	1.22	0.41	0.71	0.34	8.21	5.24	0.03	52.9	67.4	18.0	68.0
4.	NOV– DEC	I7	8.35	1.30	0.42	0.55	0.36	9.12	5.44	0.04	54.4	69.8	22.0	69.2
		I8	8.60	1.24	0.43	0.47	0.36	8.45	4.31	0.03	53.4	61.8	21.0	69.4

Table 2 - Parametric Analyses - Soil Samples - Agricultural Area

S.No	Months	Sample	pH	Parameters										
				N	P	K	Na	Ca	Mg	S	Zn	Cr	Pb	Cu
IAS (Indian Agriculture Standards)			5.8-	2.3 – 2.8 %	0.1 – 1 %	1.3 – 2 %	0.1 %	<1	< 0.5 %	0.5 %	300 – 600 mg/kg	400 mg/kg	250 – 500	135 – 270 mg/kg
1.	FEB – MAR	A1	6.54	1.32	0.50	0.35	0.31	8.08	0.39	0.04	42.4	363.3	11.2	165.3
		A2	7.36	1.71	0.38	0.49	0.32	7.94	0.38	0.03	41.8	365.6	13.0	168.8
2.	JUN – JUL	A3	8.21	1.69	0.56	0.37	0.09	9.61	4.48	0.04	42.4	460.4	14.2	167.4
		A4	6.95	1.22	0.35	0.51	0.04	8.70	4.31	0.06	42,2	370.7	11.0	169.5

3.	SEP – OCT	A5	7.59	1.36	0.52	0.35	0.32	8.46	4.41	0.04	43.4	369.6	18.4	169.0
		A6	7.48	1.22	0.42	0.52	0.36	7.98	4.40	0.03	52.6	468.9	8.8	169.2
4.	NOV– DEC	A7	8.23	1.42	0.38	0.31	0.24	8.54	0.41	0.04	42.8	364.0	8.9	151.1
		A8	7.55	1.30	0.47	0.49	0.25	8.38	0.40	0.03	43.4	265.4	11.0	159.8

Relative record of primary and secondary nutrients in industrial and agri- cultural soil samples reveals that, The nitrogen and phosphorus contents are ob- served to be comparatively higher in the agricultural soil samples(A1-A8) than their industrial counter parts(I1-I8)for the first two seasons and the last one. Similar results have been revealed during the irrigation of sugarcane crops using combined pulp and paper mill / ginning mill effluents[8,9]. As far as potassium content is concerned, it is lesser in the agricultural samples than those men- tioned in Table 1. Anomaly of nitrogen, phosphorus and potassium in the soil samples leave behind the leaves eventually turning brown, and retarded plant growth leading to chlorosis, as depicted in the image

(Fig.3).

Fig 3 - Image of affected Crops by Effluent discharge

Crop nutritional requirements depend on calcium, magnesium and sul- phur concentrations. The existence of magnesium and sulphate ions is essential for plant metabolism[10,11].

Magnesium was observed to be higher in A3, A4, A5, A6 samples and sulphate ions was found to be lower from the prescribed standard. This may be due to leaching of the ions into the soil structures during rainy seasons, as it is highly soluble, being predominantly absorbed by plants. This process is reflect- ed in the nutritional imbalance of the crops. Proper functioning of plant cell

walls, membranes are ensured by calcium ions. Minimal calcium content is ob- vious in all seasons except the second one. This marked variation could be at- tributed to the fact of high levels of exchangeable sodiumion. Crop damage oc- curs when the sodium concentration is higher, resulting in poor yield. In the pre- sent study, sodium impact is greater during first and third seasons for A1 – A8 samples, in preference to I1- I8. This shows a marked inhibition in the photo- syntheticprocesses[12,13]

.4.3 Heavy Metals

Trace levels of heavy metal ions have been considered as crucial for the plant growth as it is evident from the lower limits of prescribed standard, below which the specific ions shall lead to deficiencies of themselves, affecting soil quality. Analyses of heavy metals viz., zinc, chromium, lead, copper imply that the concentrations of zinc and lead fails to fall within the permissible limits for I1 – I8 samples. However, their concentrations in A1 – A8 samples are found to be lower than standard values, in support of the zinc / lead deficiencies in those samples. Copper and chromium concentrations are found to be in excess for A1

– A8 samples than those of I1 – I8 samples, in lieu with Indian Agricultural Standards

The above implications for the total samples holds good invariably for all seasons. This disparity signifies to create their hazardous nature as biomagnifi- cation through food chain.

(Polluted water for irrigation of plants→ Bioaccumulation in ediblecrops, Acutereflection of concentration in living organisms)

5. Conclusions

Soil quality parameters were analysed during the four seasons, according to standard operating procedures for the samples collected from Industrial and Agricultural region in Sathyamangalam, Tamil Nadu, India. Marked dissimilari- ties were noticed in the parameter values of the agricultural soil samples against their industrial precursors as per Indian Agricultural Standards. Several dispari- ties reflected in the growth of the crop, evident from the leaves, as an indicative of the affected soil fertility. It is therefore concluded that the careless disposal of industrial wastes without proper pre-treatment should be curbed as it paves way to utmost contamination of the soils which in turn imply the insight of the risks associated with the environmental characteristics of soils.

References

1. Point sources of pollution: Local effects and its control, Shi Hanchang, Encyclopedia of Life SupportSystems (EOLSS), 1: 1-6(2002).

2. Richa Sharma, Subhash Chandra, Amrita Singh and Kriti Singh. Degra- dation of Pulp and Paper Mill Effluents. *Inst. of Integrative Omics and Appl. Biotechnol.*, 5 (3): 6 -12(2014).

3. Lwako Kamoga Omar M, Byaruhanga Joseph K, Baptist Kirabira John. A Review on Pulp Manufacture from Non Wood Plant Materials. *Int. J of Chem. Eng and Appl.*, 4 (3): 144-148(2013).

4. P. K. Henneberger, J.R. Ferris and R.R. Monson. Mortality among pulp and paper workers in Berlin. *BritishJ. of Ind. Medicine.*, 46(9): 658 – 664(1989).

5. S. S. Rana and R. S. Rana, Advances in crop growth and productivity. De- partment of agronomy, Himachal Journal of Agricultural Research, CSK Hima- chal Pradesh Krishi Vishvavidyalaya, Palampur.(2014).

6. R.Bhattacharyya,B.N.Ghosh,P.K.Mishra,B.Mandal,C.S.Rao,D.Sarkar, K. Das, K.S. Anil, M.Lalitha, K. M.Hati. Land Degradation in India: Challenges and Potential Solutions.*Sustainability.*, 7: 3528 – 3570 (2015).
7. J. Godt, F. Scheidig, C. Grosse – Siestrup, V. Esche, P. Brandenburg,A. Reich, A. Papandreou, C.J. Stournaras and D. Panias. Copper and cadmium adsorption on pellets made from fired coal fly ash.*J. of Hazard. Mat.*, 148: 538 – 547 (2007).
8. M. Nagaraju, G. Narasimha, V. Rangaswamy. Impact of effluents of sugar- cane industry on Physicochemical and biological properties. *J. of Ind. Poll. Control.*, 23: 73 – 76(2007).
9. G. Narasimha, G. V. A. K. Babu, B. Rajashekarreddy. Physico-chemical and biological properties of soil samples collected from soil contaminated with ef- fluents of cotton ginning industry. *J. of Envtl Biol.*, 20: 235 – 239 (1999).
10. A. Gransee, H. Fuhrs. Magnesium mobility in soils as a challenge for soil and plant analysis, magnesium fertilization and root uptake under adverse growth conditions. *Plant and Soil.*, 368 (1-2): 5 -21 (2013).
11. J. Gerendás, H. Fuhrs. The significance of magnesium for crop quality. Plant and Soil. *Plant and Soil.*, 368(1-2): 101 – 128(2013).
12. M. Ashraf, P. J. C. Harris. Photosynthesis under stressful environments: An overview. *Photosynthetica.*, 51: 163 – 190(2013).
13. Shrivastava Pooja and Kumar. A serious environmental issue and plant growth promoting bacteria as one of the tools for its alleviation. *Saudi J. of Bio. Sci.*, 22(2): 123 – 131(2015).

SEQUESTRATION OF ZINC ION USING SUGAR INDUSTRY RES-IDUES: AN AVENUE TO ENVIROSAFE

N. MUTHULAKSHMI ANDAL[1*] and S. KARTHIKA[1] *[1]Research and Development Centre, Bharathiar University, Coimbatore641 046, Tamil Nadu, India.*

[1]PSGR Krishnammal College for Women, Peelamedu, Coimbatore 641 004, Tamil Nadu, India*

(Received 6 December, 2020; Accepted 18 January, 2021)

ABSTRACT

The present study focuses on the removal of zinc from aqueous solution utilis- ing a sugar industry residues (untreated) as an adsorbent. Adsorption exper- iments were performed by following batch techniques, sorption capacity of the adsorbent determined as the function of pH, initial concentration of the metal solution, contact time and particle size. Various kinetic parameters were evaluated using the observed experimental data from different tech- niques. The morphological investigations were carried out using Scanning Electron Microscope (SEM), Energy Dispersive X- ray Spectrometer (EDAX). The results indicate that the sugar industry residues, a new cost ef- fective industrial based sorbent material holds great potential for the extrac- tion of Zn(II) from the aqueous solution with maximum removal of 95.55% at pH value of 5.

KEY WORDS : Zinc removal, Sugar industry waste, Solution, Sorbent

INTRODUCTION

Industrialization is a big challenging issue nowadays and is mainly related to environmental pollution. The waste water discharged from theindustries carries a lot of toxic heavy metals which are harmful to the environment (Barakat,

2011). Heavy metals which are fatal to the environment include lead, mercury, cadmi- um, arsenic, copper, zinc, and chromium at high concentrations. The main sources of heavy metals to the environment are mining operations, secondary metal pro- duction, coal combustion, rubber tyre wear and phosphate fertilizers (Amalina Amirah Bt Abu Bakar *et al*., 2016). In recent studies, it was pointed that the heavy metals were found to remain high in drinkingwater, soil by means of direct usage and indirect transfer. Direct usage includes consumption by the human so- ciety in every area of modern consumerism like construction materials, cosmet- ics, medicines, processed foods. Indirect transfer of heavy metals includes the discharge from the industries to the water sources. Some metals like copper, iron play an important role for the functioning of critical enzyme systems. Most other metals have no proper role in living organisms and remain toxic even at trace levels of exposure. The metals which are considered to be toxic turn to be harmful in nature by their increased dosage (Hazrat Ali, 2019). They degrade the environment by the contamination with the natural resources, particularly wa- ter. The presence of heavy metals in the waste waters has a tendency to bio accu- mulate through the food chain and remain persistent in the environment. Their presence not only affects the human beings but also the aquatic species too. Ow- ing to these environmental, biological, communal health issues in view, it is considered necessary to attempt and provide an easy, feasible, economical, relia- ble method for the removal of heavy metals. Several methods have been devised for the treatment of heavy metals which include chemical precipitation, lime coagulation, ion exchange, reverse osmosis, ozonation, solvent extraction, elec- troplating and electrodialysis (Tonni Agustiono Kurniawan, 2006). In certain cas- es, biosorption is used as natural method for the removal of heavy metals by utilizing the microorganisms, microphytes, macrophytes as adsorbents (Sri Lakshmi Ramya Krishna Kanamarlapudi *et al.,* 2018). A cost effective, en- vironmentally friendly technique i.e., adsorption is used as a best method for the removal of heavy metals (Subhas Sarkar, 2018) where natural adsorbents, in- dustrial by-products can be used as asorbent material either with or without modification

The objective of the present research is to figure out the sorption capacity of the raw sugar industry residues (SIR) as an adsorbent for the removal of heavy metal, Zinc [Zn(II)] from aqueous solutions so as to facilitate its adsorption ability in com- parison with other adsorbents. The optimal condition for increasing the efficiency ofthis removal process wasalso investigated.

Experimental Methods

Sugar industry waste (semi-solid material after processing) was collected from the nearby sugar industry located in Palani area, Coimbatore district, Tamilnadu, India and sun-dried for 15 days. The dried material were well grounded, sieved to dif- ferent particle sizes and sealed in a tight lid separate container for further adsorption studies. Deionised distilled (DD) water was employed throughout the study for the batch adsorption studies.

Adsorbate solution was prepared by dissolving 4.5502 g of $Zn(NO_3)_2$ in 1000 mL distilled water. The stock solution was then diluted to the required concentrations. All experiments were carried out at room temperature and clear solutions obtained during the study were measured using Atomic Absorption Spectrophotometer (Shimadzu AA 6200). The pH of the solutions was monitored and adjusted to the desired value using 0.1M NaOH/ HCl as required. The surface morphology of the adsorbents was visualized on JEOL 6360 model scanning electron microscope (SEM).

RESULTS AND DISCUSSION

Characterisation of the sorbent material

The investigation on adsorbent is presented in Table

1. Surface morphology of the sorbent material was examined by SEM and is represented in Fig. 1, 1(a). It was observed that the sorbent material morphology was found to vary before and after adsorption process i.e., dense surface texture with little grooves on the surface gets shifted to smooth surface. EDAX analysis of SIR indicates that carbon, oxygen are the major constituents in the sorbent and the results suggest that they are capable of bindingmetal ions.

Table 1. Characterisation of the adsorbent

Properties	Adsorbent
pH	6.1
Surface area (m^2/g)	3.282
Carbon content (%)	41.03
Moisture content (%)	3.42
Ash content (%)	10.66
Bulk Density (g/cc)	0.56

Batch Studies

The particle size of the sorbent material has greaterinfluence on the adsorption studies. The adsorptionof Zn(II) studied for various particle sizes viz., 0.18

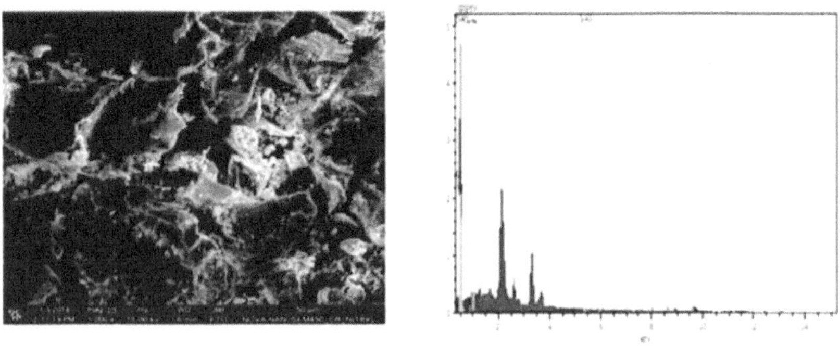

Fig. 1. SEM & EDAX images of SIR.

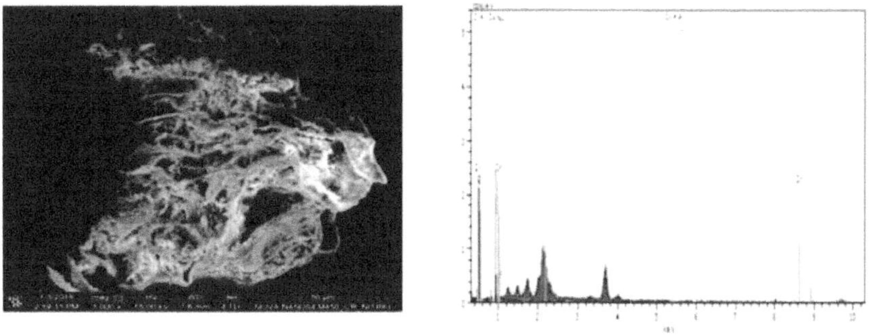

Fig. 1 (a). SEM & EDAX image of Zn(II) loaded SIR.

mm, 0.24 mm, 0.30 m, 0.42 mm, 0.71 mm reveals that the adsorption of metal ion in- crease with decrease in particle size of the adsorbent as smaller size of the material takes shorter time to equilibrate and is associated with larger surface area which removes more metal ions (Al-Anber, 2011). Consequently, adsorbent particle sizes of 0.18 mm have been selected for the batch experiments.

pH of the metal solution is a very critical parameter in the abstraction of Zn(II) from aqueoussolution. The effect of pH on the percentage removalof Zn(II) by SIR is shown in Figure 2. When pH is raised from 3 to 11, removal of Zn(II) was increased slowly from 3 to 5, and thereafter drops slowly. This may be attributed to the compe- tition between the hydrogen and metal ions on the sorption sites, at low pH values (Mamatha *et al.*, 2013). Formation of soluble hydroxy complexes makes sorption studies impossible at higher pH values. Consequently, the working pH value for Zn(II) removal was chosen as 5.0 and the adsorption experiments were performed at this pH value.

The effect of agitation time at different initial Zn(II) concentration, specific uptake by the adsorbent is shown in Figure 3. It was observed that the maximum adsorption occurs at 100 mg/L and further increase in the concentration of the metalions results in saturation of the adsorbent sites. It was evident (Figure 2) that the maximum sorptionhad occurred at 30 min for Zn(II) and smooth curvesobtained further endorsing no more metal sorptionfrom the solutions (Kannan, 2001; Madu *et al.*, 2011). The effect of adsorbent dosage was studied at various dosages of the sorbent material like 50 – 200 mg in the removal of Zn(II). It was observed that with the increase in dose of the adsorbent, adsorption efficiency was found to rise due to largenumber of binding sites (Hawari *et al.*, 2009). Theresults indicate that high adsorption efficiency occurs with the increase in dose of the adsorbent upto a certain limit, then it remains almost dropped

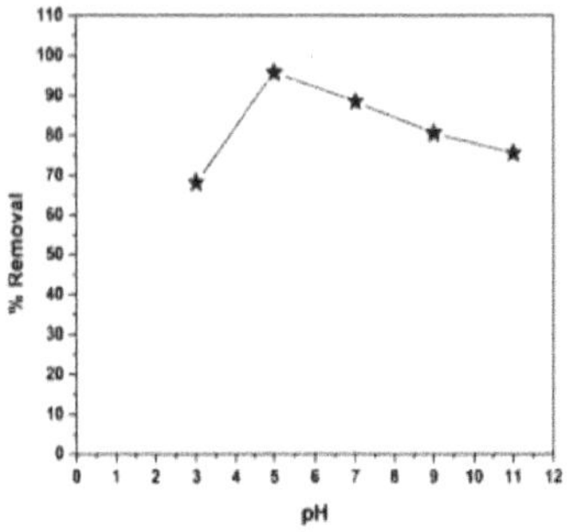

Fig. 2. Effect of pH

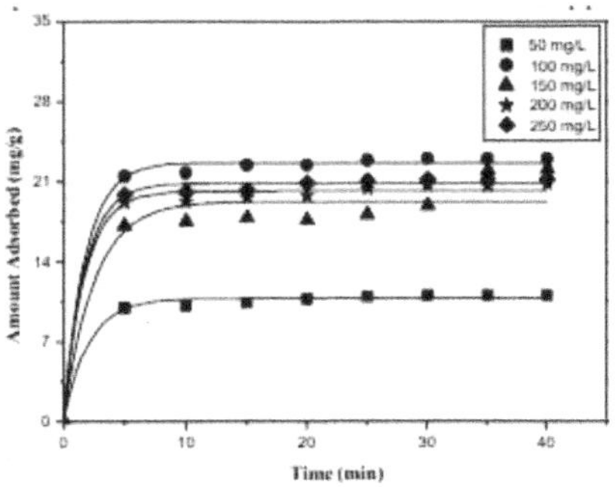

Fig. 2. Effect of pH **Fig. 3.** Effect of agitation time and initial concentration

due to agglomeration and 200 mg of adsorbent dosage is fixed as optimum for stud- ies.

Adsorption Isotherms

Adsorption isotherm explains the interaction between the adsorbate and ad- sorbent. An equilibrium is established when the amount of solute being adsorbed onto the adsorbent is equal to the amount being desorbed. The mathematical models developed by Langmuir, Freundlich and Tempkin isotherms were applied to the equilibriumdata in order to understand the sorption mechanismand surface character- istics. Langmuir isotherm is based on the adsorption happening at specific adsorp- tion sites. The Langmuir isotherm equation (Silkeschiewer, 2009; Gupta, 2008) is expressed as follows:

$$C_e/q_e = C_e/q_m + 1/bq_m$$

where, q_m is maximum adsorption capacity (mg/ g); b is the Langmuir sorption con- stant (mg/L); C_e is equilibrium metal ion concentration in solution (mg/l); q_e is equi- librium metal ion concentration on the sorbent (mg/g). A linear plot is obtained from Ce/qe against Ce over the entire concentration range of metal ions where maximum sorption capacity (q_m) calculated as 49.26 represents monolayer coverage of the sorbent with sorbate. The investigated correlation coefficient (R^2) values indicate the suitability of Langmuir model to the selected system.

Freundlich isotherm is applicable to the adsorption processes that occur on heterogeneous surface. The linear form of the Freundlich equation (Kalavathy, 2005) is as follows,

$$\log q_e = \log K_f + \frac{1}{n}\log C_e$$

whereqe is the amount of metal ion adsorbed perunit weight of adsorbent (mg/g); $C_e$is equilibrium concentration (mg/l); b is heat of adsorption (J/ mol). Tempkin isothermal data (Table 2) reveals that the heat of adsorption (B_T) remains higher whereas tempkin isotherm constant (A_T) remains lower. This nature indicates a weaker interaction between the sorbate and sorbent in the process (Yuh-Shan Ho, 2000).

$$q_e = \frac{RT}{b}\ln AT + \frac{RT}{b}\ln C_e$$

Table 2. Isothermal Constants

	Parameters	Zn(II) - SIR
Langmuir Isotherm	q_m (mg/g)	49.26
	b (L/g)	0.07
	R^2	0.9904
Freundlich Isotherm	K_f(mg/g)	1.83
	1/n	0.49
	R^2	0.8064
Tempkin Isotherm	A_T(L/g)	1.14
	B_T (J/mol)	5.12
	R^2	0.9287

Adsorption Kinetics

The kinetic study of an adsorption plays an important role because it affords an important insight into the reaction pathways and mechanismof the reaction. Pseudo first order, Pseudo second order kinetic models, Elovich models are applied to the experimental data.

The pseudo first-order kinetic model has been widely used to predict the metal ad- sorption kinetics. The metal adsorption kinetics following the pseudo first-order model (Ho, 1999a) is given by,

$$\frac{dq}{dt} = k_1 (q_e - q_t)$$

Where q_e is amount of metal ion adsorbed per Where k is rate constant of the pseudo first-order,equilibrium metal ion concentration (mg/l); K_f is adsorption capacity; 1/ n is adsorption intensity. The calculated K_F and 1/n values from the freundlich adsorption isothermal plots are listed in Table 2. Smaller K_F values and $1/n$ values (lying between 0 and 1) suggests that the adsorption of metal ionsonto the surface of sorbent supports physisorptionprocess.

Tempkin isotherm model considers the effect of adsorbate / adsorbent interac- tions. The linear form of Tempkin isotherm model (Demirbas, 2008) is given by the following equation, (mg/g); q_e is the amount of adsorption at equilibrium (mg/g).

The adsorption kinetic data can be further analyzed using pseudo second-order kinetics (Mckay, 1999 b,c). This is represented by,

$$\frac{dq}{dt} = k_2(q_e - q_t)^2$$

The linear plots corresponding to pseudo first order and pseudo second order ki- netic models are shown in Figure 4. The equilibrium rate constants along with SSE are listed in Table 3. The data (Table 3) reveals that the pseudo second order model fitted the experimental record well with a correlationcoefficient (R^2) close to unity at low concentration and SSE values are found to be minimum.

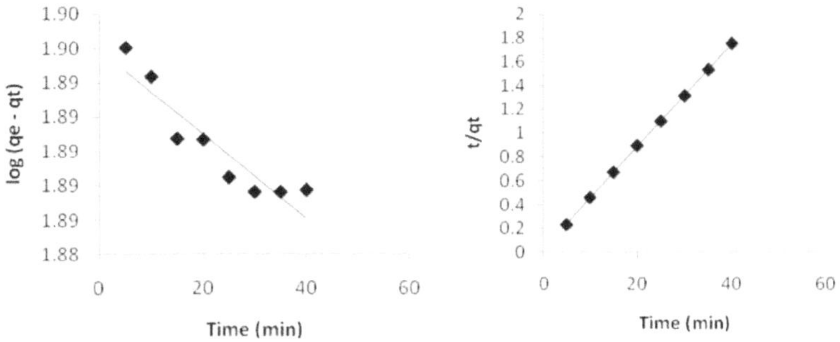

Fig. 4. Pseudo First and Second Order Plot – Zn(II)

The Elovich equation has been applied if the adsorbing surface remains heteroge neous. The Elovich model equation is expressed (Suresh Gupta,2009) as

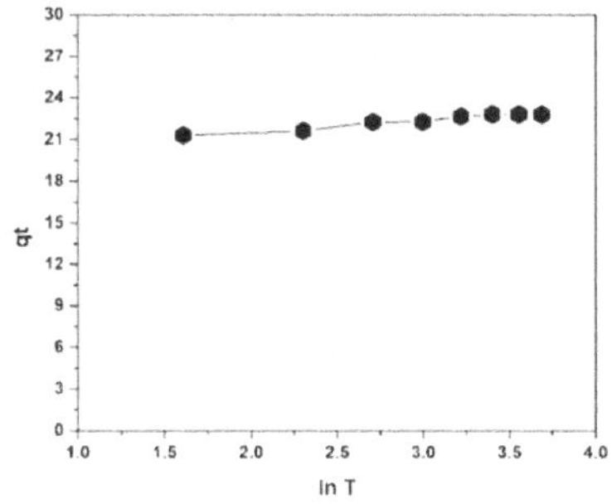

Fig. 5. Elovich Plot

$$q_t = \frac{1}{\beta}\ln(\alpha\beta) + \frac{1}{\beta}n(t)$$

where alpha is initial adsorption rate (mg/g min); beta is adsorption constant (g/mg).The adsorption rate alpha for Zn(II) shows notable change in their fixed concentration followed by thedecline at higher concentration range.

CONCLUSION

The ability of sugar industrial residues, an industrial based sorbent material for re- moving Zn(II) metal ions from aqueous solution was studied with various pa- rameters such as pH, initial concentration, adsorbent dosage, particle size. The maximum removal percentage was obtained at pH 5 with a dose of 200 mg. The ex- perimental data was found to follow the monolayer coverage of the metal ion on the sorbent surface and best fit the sorption process. Adsorption isotherm and intensity value showed that there is greater affinity for zinc extraction from aqueous solu- tion that initiates a challenge in the water treatment process. Further, this study may extend to detoxify the pollutants

Table 3. Kinetic Constants

Conc. of Metal Ions (mg/L)	$q_e$exp (mg/g)	Pseudo First Order Kinetics				Pseudo Second Order Kinetics			
		q_{e1} (mg/g)	$k_1 \times 10^{-2}$ (min^{-1})	R^2	SSE	q_e^2 (mg/g)	$k_2 \times 10^{-2}$ (min^{-1})	R^2	SSE
Zn(II) - SIR									
50	11.03	3.04	0.92	0.8967	1.63	10.07	0.11	0.9997	0.39
100	22.80	3.72	0.46	0.8588	2.52	18.79	0.11	0.9999	0.81
150	21.80	4.78	0.23	0.7055	2.38	18.05	0.24	0.9999	0.79
200	20.53	4.78	0.12	0.9456	2.29	4.21	1.18	0.9994	1.64
250	21.03	5.14	0.03	0.8733	2.30	17.63	0.10	0.9998	0.75

values are found to be minimum.

from the industrial effluents.

REFERENCES

Al-Anber, M.A. 2011. Thermodynamics approach in the adsorption of heavy met-als, in Thermodynamics - Interaction Studies - Solids, Liquids and Gases. J.

C. Moreno - Pirajan, Ed. Intech Open, London, UK.

Amalina Amirah Bt Abu Bakar., Khairul Ammar B Muhd Ali., Nurul Athirah BtAhmad Tarmizi., Ahmad Zia Ul-Saufie B. Mohamad Japeri. and Nor Janna Bt

Tammy. 2016. Potential of using Bladderwort as a Biosorbent to remove Zinc in Wastewater. AIP Conference Proceedings. 1774 : 030023.

Barakat, M.A. 2011. New trends in removing heavy metals from industrial wastewater.Arabian Journal of Chemistry. 4(4) : 361-377.

Demirbas, E., Kobya, M. and Konukman, A.E.S. 2008. Error Analysis of Equilibri- um studies for the Almond shell Activated Carbon Adsorption of Cr(VI) from aqueous solutions. Journal of Hazardous Materials. 154 : 787–794.

Gupta, V.K. and Rastogi, A. 2008. Biosorption of lead from aqueous solutions by green algae Spirogyra species: Kinetics and Equilibrium studies. Journal of Hazardous Materials. 152 : 407 – 414.

Hawari, A., Rawajfih, Z. and Nsour, N. 2009. Equilibrium and thermodynamic anal- ysis of zinc ions adsorption by olive oil mill solid residues. J. Hazard. Mater. 168: 1284–1289.

Hazrat Ali and Ezzat Khan, 2019. Environmental Chemistry and Ecotoxicolo- gy of Hazardous Heavy Metals : Environmental Persistence, Toxicity and Bio- accumulation. Hindawi Journal of Chemistry. 1-14.

Ho, Y.S. and Mckay, G. 1999a. A kinetic study of dye sorption by biosorbent wasteproduct pith.Resour. Conserv. Recycl. 25 : 171–193.

Kalavathy, M.H., Karthikeyan, T., Rajgopal, S. and Miranda, L.R. 2005. Kinetic and isotherm studies of

Cu(II) adsorption onto H_3PO_4 - activated rubber wood sawdust. Journal of Colloid and Interface Science. 292 (2) : 354–362.Kannan, N. and Xavier, A. 2001. New composite mixed adsorbents for the removal of acetic acid by adsorption from aqueous solutions a comparative study. Toxicological and En- vironmental Chemistry. 79 (1-2) : 95–107.

Madu, P.C., Akpaiyo, G.D. and Ikoku, P.J. 2011. Biosorption of Cr^{3+}, Pb^{2+}, and Cd^{2+} ions from aqueous solution using modified and unmodified millet chaff. J. Chem. Pharm. Res. 3 (1) : 467-477.

Mamatha, M., Aravinda, H.B., Puttaiah, E.T. and Manjappa, S. 2013. Adsorbent from Pongamia Pinnata Tree Bark for Zinc Adsorption. International Jour-nal of Scientific Engineering and Technology. 2(1) : 22-25.

Mckay, G. and Ho, Y.S. 1999b,c. Pseudo second-order model for sorption process-es. Process Biochem. 34: 451–460.

Silkeschiewer and Ankit Balaria, 2009. Biosorpiton of Pb^{2+} by original and proto- nated citrus peels: Equilibrium, Kinetics and Mechanisms. Chemical Engineer- ing Journal. 146 : 211– 219.

Sri Lakshmi Ramya Krishna Kanamarlapudi, Vinay Kumar Chintalpudi and Sud- hamani Muddada, 2018. Application of Biosorption for Removal of Heavy Metalsfrom Wastewater. Intech Open, http://dx.doi.org/10.5772/intechopen.77315

Subhas Sarkar and Subhendu Adhikari, 2018. Adsorption Technique for Removal of Heavy Metals from Waterand Possible Application in Wastewater-Fed Aq- uaculture. In : Wastewater Management ThroughAquaculture. 235-251.

Suresh Gupta and Babu, B.V. 2009. Utilization of waste product (Tamarind seeds)for the removal of Cr(VI) from aqueous solutions: Equilibrium, Kinetics and Regeneration Studies. Journal of Environmental Management. 90 : 3013-3022. Tonni Agustiono Kurniawan, Gilbert, Y.S. Chan, Wai- Hung Lo. and Sandhya Ba-bel, 2006. Physico– chemical treatment techniques for wastewater laden with heavy metals.Chemical Engineering Journal. 118 (1–2) : 83-98.

Yuh-Shan Hoand Gordon Mckay, 2000. The Kinetics of Sorption of Divalent Met- al Ions Onto Sphagnum Moss Peat. Water Research. 34 (3) : 735-742.

Investigation on Physico-Chemical Characteristics of Pulp and PaperIn-dustrial Discharges into the Environs

,N. Muthulakshmi Andal[1]*and S. Karthika[1]

[1]Research and Development Centre, Bharathiar University, Coimbatore - 641046, Tamil Nadu, India.

[1*]Assistant Professor, Department of Chemistry, PSGR Krishnammal College for Women, Peelamedu, Coimbatore - 641004, Tamil Nadu, India.

*Corresponding Author E-mail: yenmmaa@gmail.com

ABSTRACT:

Pulp and paper industries by and large contribute to the economic growth of our country. However, it causes threat to ecological balance through its discharging of large volumes of organics/inorganics into the water bodies and soil environ- ment. These toxic effluents from pulp and paper industries frame one of the ma- jor sources of pollutant bodies. The present study is focussed on the analyzing Physico –chemical parameters, toxic metal concentrations [Pb(II), Cu(II), Cr(VI), Zn(II)] of the paper mill effluents discharged from industries located at Pollachi, Udumalpet and Sathyamangalam of Tamil Nadu, India. The parametric values registered notable deviations from permissible limits [CPCB standards] and showed variations in their metal concentrations during the four seasons, re- flecting the adverse impact of the toxicants loaded in the mill outlets.

KEYWORDS: Paper Industry, Effluent, Physicochemical Parameters, Seasons.

1. INTRODUCTION:

Water pollution is a huge problem since it is relatively concerned with rapid indus- trialization and release of excess amount of toxic pollutants into the environment. Demand and need for expansion of industries are the outcome of population growth to cater the essentialities of humans. But on the other side, they are declared as menace to the environment, depleting the natural resources[1]. Pulp and paper indus- tries being one of the vital manufacturers of paper products, generate huge volumes

of toxicants[2], which are led into the water and soil environs, predominantly untreat- ed[3,4]. Statistical data refers to that 55% of industries in India do not have proper dis- posal treatment plants and 20% of industries seldom record partial treatment. These effluents are rich in organic halides, bleaching agents and other deadly toxins. The volume and the nature of the effluent produced depend on the

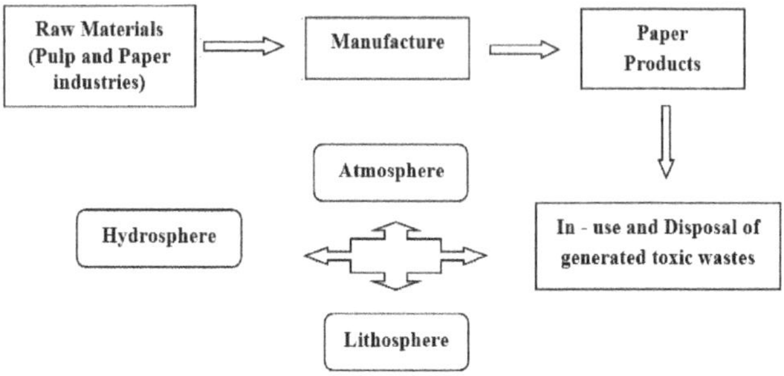

types of raw materi-
als employed, chemical compositions and manufacturing processes.

A flowchart depicting the environmental impact caused by pulp and paper industries is shown below, Enormous quantity of water viz., 270 - 455 m^3 is utilized during the processing of a single ton of paper in paper industries (Tamil Nadu), where about 300 m^3 of wastewaters are released every year[5,6]. Effluents originated from these indus- tries are found to possess high contents of biochemical oxygen demand (BOD), chemical oxygen demand (COD), organic compounds, inorganic salts and total dis- solved solids. Use of lignin or polysaccharides as prime material during the paper manufacture contributes to the dark colouration of the effluent which creates an im- pact on the rheological nature of the wastewaters[7,8,9]. Occurrence of heavy metal concentrations in effluents are due to the impart of colour shades to the paper and pollution (heavy metal) remain as a serious environmental issue[10]. The excess of toxic metal salts, also contribute to their prevalence in the wastewaters, which has a serious threat to the irrigation systems[11]. Thepresent work deals with the collection of five effluent samples from paper industries during the four seasons (February to march, June to July, September to October, November

to December) located at Pollachi, Udumalpet and Sathyamangalam of Tamil Nadu, India. Physico – chemical characteristics and occurrence of trace metal concentrations in the collected samples were determined.

2. MATERIALS AND METHODS:

2.1 Study Sector and Sample Collection:

Pulp and Paper industries of Pollachi, Udumalpet and Sathyamangalam regions had been selected to assess the impact and analysis of the characteristics of their liquid discharges. The choice of these regions lie in the fact that acute pollution is caused by several small and large scale industries leading to adverse environmental implications. Three large scale and two small scale paper industries were identified for the study. Newsprint, writing/tissue/printing papers, posters, crafts and duplex boards are the products generated from large scale industries, where chemical pulp- ing method through utilisation of wastepapers, wood pulp, bamboo is followed for the manufacturing of aforesaid products. The small scale agro – based industries in- clude writing papers, newsprint, cardboards as their produce, wherein wastepapers, water along with sawdust and coal (fuel) are involved in chemical pulping. Liquid waste contaminants discharged from both small / large scale industries carry heavy organic and inorganic loads. Wastewater samples from these five industries situat- ed in Pollachi (I1), Udumalpet (I2) and Sathyamangalam area (I3, I4, I5) were col- lected in the period between February - March, June - July, September - October and November-December months following specific sampling procedures.

Table 2 - Physico – chemical Studies.Study on Physico-chemical characteristics:

The operating parameters viz., pH, Total alkalinity, Total Hardness, Biochemical Oxygen Demand (BOD), Chemical Oxygen Demand (COD), Total Solids, Total

Suspended Solids, Total Dissolved Solids were analyzed to assess the physico – chemical nature of the five samples following standard methods of water/wastewater analysis as per APHA[12] (American Public Health Association). ELICO pH meter with standard buffers, Titrimetric, EDTA, Dichromate reflux methods were followed for the analyses of pH, Total Alkalinity, Total Hardness and COD factors. BOD and Total Solids loads were determined using standard incubation/Gravimetric methods. Occurrence of trace concentrations of heavy metals (Pb, Cu, Zn, Cr) were examined by spectrophotometric method (AA 6200 Spectrophotometer).

3. RESULTS AND DISCUSSION:

Results of the analysed parametric values are listed in Table 2, heavy metal ions in Table 3 and compared with Central Pollution Control Board (CPCB) standard val- ues(Table 1).

Table 1 – Permissible limits – CPCB, Tamil Nadu

S. No	Parameter	*Permissible Limit (mg/L)
1.	pH	5.5 – 9.0
2.	Total Alkalinity	200
3.	Total Hardness	500
4.	COD	350
5.	BOD	100
6.	Total Solids	4000
7.	Total Suspended Solids	100 – 200
8.	Total Dissolved Solids	2000
9.	Pb	0.1
10.	Cu	3.0
11.	Cr	0.1
12.	Zn	5.0

Table 2 - Physico – chemical Studies.

Parameters	February – March					June – July				
	I1	I2	I3	I4	I5	I1	I2	I3	I4	I5
pH	7.86	7.80	8.35	6.70	7.50	8.20	8.15	8.60	7.35	7.71
TA	84	124	95	68	179	67	164	63	87	165
TH	780	690	594	638	728	668	687	359	730	576
COD	892	792	687	1615	1025	660	488	520	940	854
BOD	230	187	142	285	319	237	240	163	230	392
TS	2044	2254	1499	2434	2126	1449	2122	1241	2254	2029
TSS	153	201	108	276	213	181	197	124	219	188
TDS	1891	2053	1391	2158	1913	1268	1925	1117	2035	1841

Continue

September – October					November - December				
I1	I2	I3	I4	I5	I1	I2	I3	I4	I5
7.58	7.95	8.10	7.85	8.25	7.98	8.95	7.90	8.65	8.95
72	98	71	65	137	62	84	78	60	113
558	610	425	687	540	510	630	652	468	715
610	505	475	890	815	730	708	675	710	880
240	319	106	184	172	221	119	287	110	223
1563	1559	1384	1607	1985	1401	2024	1695	2451	2162
176	162	119	197	171	130	213	179	260	209
1387	1397	1265	1410	1814	1271	1811	1516	2191	2162

Table 3 – Heavy Metal Analysis.

Parameters	February – March					June – July				
	I1	I2	I3	I4	I5	I1	I2	I3	I4	I5
pb	-	1.71	-	0.54	-	-	0.40	-	0.95	-
Cu	3.50	-	3.90	2.80	-	8.10	-	4.10	3.65	-
Cr	1.65	-	1.90	-	-	0.15	-	0.05	-	-
Zn	9.80	3.40	-	3.55	8.20	4.50	3.65	-	2.25	5.50

Continue

September – October					November - December				
I1	I2	I3	I4	I5	I1	I2	I3	I4	I5
-	1.01	-	0.09	-	0.05	-	0.15	-	-
4.20	4.85	2.73	3.18	-	5.16	4.34	-	4.75	5.24
0.65	-	1.10	-	-	-	-	0.05	-	-
5.65	4.90	-	-	6.15	3.14	5.80	-	3.88	5.39

Fig 1 and 2, depict the bar diagrams of pH variations and alkalinity fluctuations amongst the studied samples during different seasons. pH and Total alkalinity val- ues for the analysed samples of five industries during different seasons registered well between 6.70 - 8.95, 60 -179 mg/L respectively falling within the permissible limits. The bar chart (fig 2), indicative of total alkalinity values imply greater alka- line nature of I5 sample whichis due to the excess addition of bleaching agents dur ing the manufacturing process.

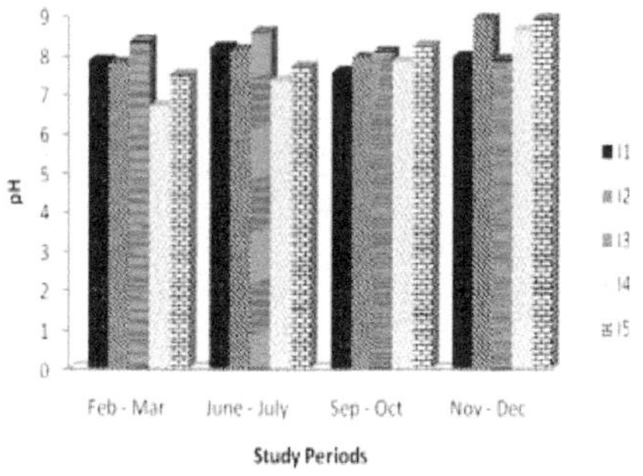

Fig 1 – pH

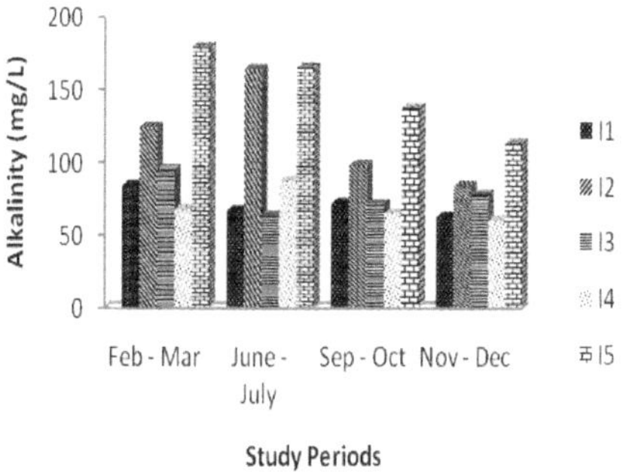

Fig 2 – Total Alkalinity

The calculated total hardness values are shown in figure 3. Almost all the samples recorded total hardness values greater than the permissible limit, invariably during all seasons. This is evidenced from the usage of lightening agents, sizing substances and chemicals at various processing steps[13]. Of which, bar diagram correspond- ing to February – march periods show inclined hardness for the five sources, the reason being prevalence of early summer leading to increase in Ca^{2+} and Mg^{2+} con- centrations.

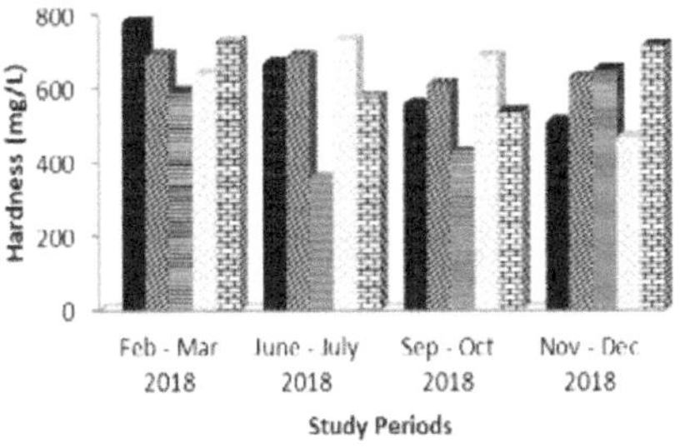

Fig 3 – Total Hardness

Higher COD and BOD values (> permissible limits) were observed in all the sam- ples during the specified seasons, as obvious from the figures (fig 4 and 5). Exist- ence of oxidisable organic compounds[14] and excess lignin composition[15] could at- tribute to the enhanced oxygen demand values which pose difficulty in degrada- tion.

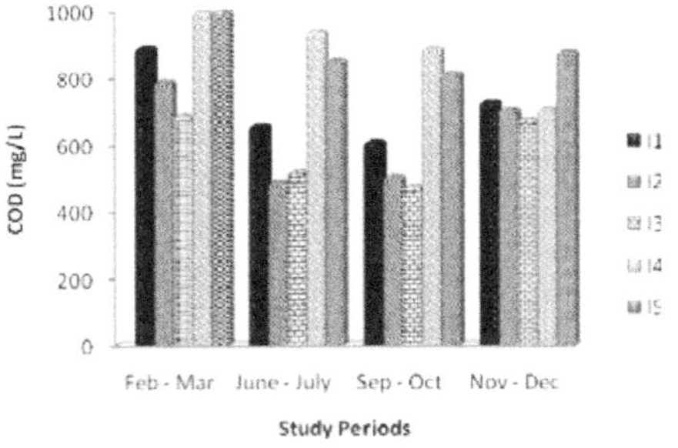

Fig 4 – COD

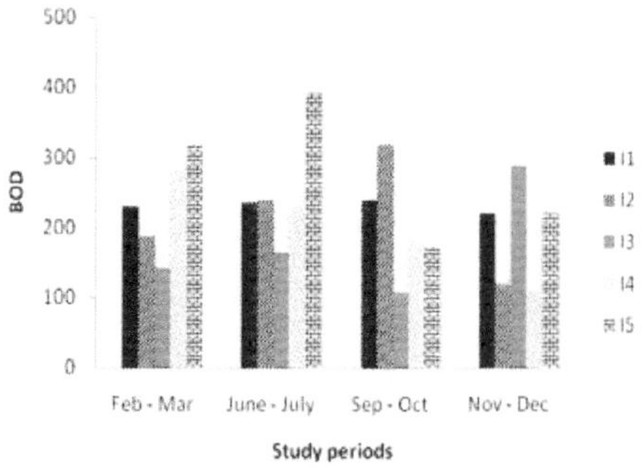

Fig 5 – BOD

The total solids at all the seasonal variations range from 1241mg/L to 2451mg/L, within the acceptable standards. Total dissolved solids varied between 1117mg/L and 2191mg/L (Table 2), exhibiting desirable limit, with the exceptions of I2, I4 (February – March); I4 (June–July); I4 (November –December) samples. This shall be attributed to the fact of chloro lignin compounds solubilisation[16]. The readings (Table 2) imply that TSS value to fall between 108 to 276mg/L (Fig. 6) against the permissible limit of 100–200mg/L. However, the deviations recorded for I4, almost in three seasons, reveal the incomplete treatment procedures adopted for the wastewater outlet for the sathyamangalam industry.

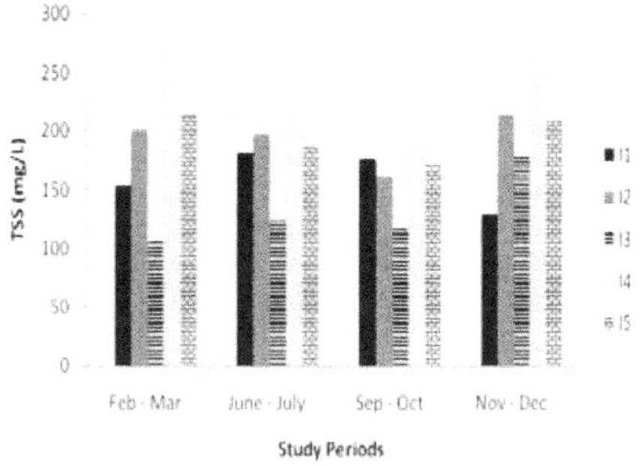

Fig 6 – Total Suspended Solids

Table 3 shows the divalent heavy metal concentrations with by and large variations throughout the seasons for all the analysed samples. Inspite of greater deviations, the chromium concentrations are marginally lesser than other toxic metal ions. However, the occurrence of higher concentration of these ions exceeding the toler- ance levels implicit the heavy metal contamination of the environment through thesedischarges.

CONCLUSION:

Pulp and paper industrial wastewater samples were collected from five sources in the regions of Pollachi, Udumalpet and Sathyamangalam subjected to systematic physicochemical characteristic analysis and heavy metal concentrations following standard operating procedures. It was observed from the results, that the parame- ters viz., Total hardness, Chemical Oxygen Demand, Biological Oxygen Demand, Total Solids, Total Suspended Solids and Toxic metal ions registered greater con- centration than the prescribed permissible limits as per CPCB standards of Tamil Nadu, India. Variations in the quality of wastewater discharged from the identified sources is assessed not only upon its dependence on seasonal influences, but also the type of manufacturing processes followed at the industries, in view of the pro- duce. The concluding remarks include the insistence / follow – up of appropriate mandatory measures with suitable treatment methods to the pulp and paper indus- tries, so that the discharges can comply with set standards.

3. REFERENCES:

1. Wastewater: The untapped Resource. The United Nations world water Devel- opment Report, UNESCO Publisher. 2017.
2. Energy from Toxic Organic Waste for Heat and Power Generation. WoodheadPublishing. 2019.
3. Kuzhali SS, Manikandan N and Kumuthakalavalli R. Physicochemical and bio- logical parameters of paper Industry effluent. Journal of Natural Product and

Plant Resources. 2 (3); 2012:445–448.

4. Kumar V, Chopra AK, Kumar S, Singh J and Thakur RK. Effects of pulp and paper mill effluent disposal on soil characteristics in the vicinity of Uttaranchal Pulp and Paper Mill, Haridwar, India. International Journal of Agricultural Sci- ence and Research. 4 (6); 2015:117–125.

5. Subrahmanyam PVR. Waste management in pulp and paper industry. Journal of Indian Association for Environmental Management. 17; 1990: 79–94.

6. Devi NL, Yadav IC and Shihua QI. Physicochemical characteristics of paper in- dustry effluents - a case study of South India Paper Mill (SIPM). Environmental Monitoring and Assessment. 177; 2011: 23–33.

7. Cardoso M, De Oliveira ED and Passos ML. Chemical composition and physi- cal properties of black liquors and their effects on liquor recovery operation in Brazilian pulp mills. Fuel. 88(4); 2009: 756–763.

8. Ingle ST. Pollution potential of pulp and paper mill effluent. Bulletin of Envi- ronmental Science. 18; 2000: 21–24.

9. Yedla S, Mitra A and Bandyopadhyay M. Purification of pulp and paper mill ef- fluent using eichornia crassipes. Environmental Technology. 23; 2002: 453–465.

10. Rajesh. A Samant, Vijay L Gurav. A Biosorption of Heavy Metal Ions from ef- fluent using Waste Fish Scale. Asian Journal of Research in Chemistry. 11(5); 2018: 775 – 777.

11. Reyes F, Chamorro S, Yeber MC and Vidal G. Characterizations of E1 Kraft mill effluent by toxicity identification evaluation methodology. Water, Air and Soil Pollution. 199; 2009: 183–190.

12. American Public Health Association / American Water Works Associa- tion/Water Environment Federation. Standard Methods for the Examination of Water and Wastewater, Washington, DC, USA.1998.

13. Hubbe MA and Gill RA. Fillers for papermaking: A review of their properties, usage practices, and their mechanistic role. Biological Research. 11(1); 2016: 2886–2963.

A study on the bioaccumulation and screening of toxicmetal ions in effluents, fod-der and milk samples collected from Erode, Tirupur, Coimbatore districts Muthulakshmi Andal N.* and Anuradha J., Gayathri N.S

Department of Chemistry, PSGR Krishnammal College for Women, Coimbatore-641004, Tamil Nadu, INDIA
*muthulakshmiandal@psgrkcw.ac.in

Abstract

Water pollution is a serious problem of concern due to the potential health im- pacts of bioaccumulation in the food chain. Based on this fact, a field study is conducted in the areas of three major polluting industries viz. dyeing, tannery and plating located at Coimbatore, Tirupur and Erode pertaining to ground wa- ter table contamination due to heavy metals from industrial discharges. The out- come of the field study formed a baseline to investigate the accumulation of toxic metal ions: Cd, Cu, Zn, Pb, Fe, Co, Cr, Mn and Ni in the effluent, fodder and milk samples. A marked extremity of pollution is recorded against lead and chromium concentrations in comparison to WHO permissible limits for effluents, diminish- ing reflecting in the inherent increase of 7 and 5 fold times for fodder/ milk sam- ples. Also, reclamation studies are performed using naturally occurring eco- friendly materials to minimize the extent of pollution in all the chosen samples. The analysis and assessment of the results conclude that thus the materials pos- sess excellent metal chelating ability and serve as better choice in theheavy metal bioaccumulation by treating the effluents even before being discharged into the soil environment.

Keywords: Bioaccumulation, heavy metals, fodder, milk, Coimbatore.

Introduction

Extensive industrial activities pave way for intense heavy metal ion concentra- tions in water/ soil systems[10]. Unlike water system, leached out metal ions are not readily available for treatment from soils/ sediments. But under changing en- vironmental conditions of the overlying water, the adsorbed toxic metals from soils are released back to the aqueous phase[8]. Plants acquire the capabilities of bio- concentrating the toxic metals in their tissues during their growth on these metals laden polluted soils/ water posing great threat to the living beings consum-ing them[3].

This may include serious threats like renal failure, liver damage, cardiovascular diseases and in some cases even death.[1] In the last few decades, the studies on heavy metal concentrations in water streams/ sediments/flora/ fauna havebeen ex- tensively dealt in different parts of the world. Milk is promoted as "the perfect food" for humans due to its nutritious factor[4]. There are no global recommenda- tions ofmilk for daily consumption. The incumbent of toxic metals in milk through their biomagnifications in food chain can counteract their beneficial ef- fects leading to several adverseeffects on human health.[6] The present study deals with the study on bioaccumulation of toxic metals via the route: Effluents en- riched with heavy metals discharges as landfills:Growth of fodder on these pollut- ed soils: Fodder consumption by mulching animals and Bio- concentration oftoxic metal distribution to humans via milk consumption.

Material and Methods

Field mapping and Sampling: Coimbatore, Tirupur and Erode districts belong- ing to the western parts of Tamil Nadu,known for their industrialization, have been identified as the sampling areas. Field mappings of these cities viz. Coimbatore, Erode and Tirupur lying between 10° 10′ and 11° 30′ of the northern latitude and 76° 40′ and 77° 30′ of eastern longitude at an elevation of 432 metres from sea lev- elcorresponding to their state mapping are represented in figures 1 and 2 respec- tively.

Fig. 1: Tamil Nadu Mapping

Sample Collection: Sampling collection schemes adopted include: (i) Effluent samples from 6 dyeing units/ 2 tanneries/ 3 electroplating units located in Coim- batore/ Tirupur /Erode districts directly at the effluent discharged points of re- spective units and stored in air tight containers for further analysis (Table 1) (ii) fodder samples identified within a region in the vicinity of 50 meters of the chosen industries and (iii) Raw milk samples from the domesticated cows grazing over the sampling fodder sites.

Fig. 2: Sampling Area Mapping

Fodder/milk samples were collected in polyethylene bags/ bottles respectively and were preserved in refrigerator with proper labels prior to estimation. Analyses of heavy metal content present in the collected samples were carried out within 24 hours from the time of collection.

Metals' content in samples are compared against the recommended WHO's maxi- mum permissible limit (MPL) to assess the quality of milk for human consumption. Also, the determined heavy metals in effluent samples were sequestrated by treated tree waste and its prepared activatedcarbon as adsorbents.

Samples Preparation

Dilution of Effluent Samples: The effluents collected werefiltered to remove insol- uble impurities followed by dilution (1:50) using doubly distilled water in order of equivalence with respect to AAS standards.

Digestion of Fodder samples: 5 grams of the fodder samples were washed well with deionized water to remove sediments/ soil and other impurities. Later, they were dried in a hot air oven (Make- Genuine) at a temperature of 65^0C to remove the moisture content, cooled, shredded into small pieces and again dried up to 80^0C to ensure complete drying.samples were cooled, filtered and diluted to 50 ml with doubly distilled water and stored in dry containers.

Digestion of milk samples: 5ml of acetic acid was added to the 50 ml of lukewarm milk samples, stirred, cooled and thenfiltered. The filtrate was extracted with ethanol followed by ether where it solidifies and the solid milk samples were dried and ground well using ceramic mortar/ pestle. 5gramsof the prepared milk powders were taken in a porcelain crucible and ashed in the muffle furnace (Yorco Make) for 30 minutes at 200°C. The heating was continued till the ashed samples turned colorless, later left overnight for cooling.

The ashed samples were diluted with 25ml deionized water and 5ml con. HNO3, gently stirred, filtered, diluted to 50ml and labeled, the precipitate being discarded[9]. The images ofthe different types of collected samples are depicted in fig. 3.

Sample Analysis: The processed fodder/milk samples were subjected to pre- dilutions several times to match the operating optimized standards as in the case of effluent samples. pH of the diluted samples of five types was measured using Deluxe pH meter (LABTRONICS) and the heavy metal concentrations of lead, co- balt, chromium, zinc, manganese, iron, nickel, copper and cadmium were analysed using Atomic Absorption Spectrophotometer (Shimadzu- AA6200) (fig. 4).

Adsorption Studies/ Choice of Adsorbent: Samples containing heavy metals were screened for their removal via adsorption method using bio-waste material as an ad- sorbent. *Prosopis juliflora* is a fast-growing small tree in India. This tree is widely grown in dry land, arid regions. It is one of themost tolerant species for saline, alka- line soil and also capable of growing in water-logged areas. The barks of the tree carry no significance importance except as manipulated wastes, then these barks (PJB) were employed for adsorptionstudies.

Preparation and treatment of Bio -waste: PJB was washedwell with distilled water, sun dried, pulverized using electrical mixer, sieved into different mesh sizes using Scientific Test Molecular Sieves and subjected to modification using 0.1 N HCl, thereafter being referred to asTreated *Prosopis juliflora* Bark (TPJB).[2]

Preparation of Activated Carbon: PJB was also activated by soaking in 35% H PO for 24 hours, dried, carbonized atThe dried fodder samples were placed in a desic- cator and allowed to cool for 24 hrs. 100 mg of dried samples were weighed in an an- alytical balance and dissolved in 8ml of 65% HNO3, the dissolved samples were sub- jected to digestion on a hot plate (70^0C) for 8 hrs and again for another 4 hrs with in between cooling[6].

20ml of 30% H2O2 were added to the digested samples which refer to completion of digestion. The prepared500 C for 1 hour and 800 C for 10mins in the Muffle fur- nace. The prepared activated carbon was washed with doubly distilled water, dried and further referred as *Prosopis juliflora* Bark Activated Carbon (PJBAC). Several washingswith DD water were done for the pH values of both TPJB and PJBAC to at- tain neutrality.[5] Fig. 5 refers to the materialin treated and activated forms respective- ly.

Sample 1 D1	Sample 2 D2	Sample 3 D3	Sample 4 D4	Sample 5 D5	Sample 6 D6

Sample 1T1	Sample 2T2	ample 1P1	Sample 2P2	ample 3P3

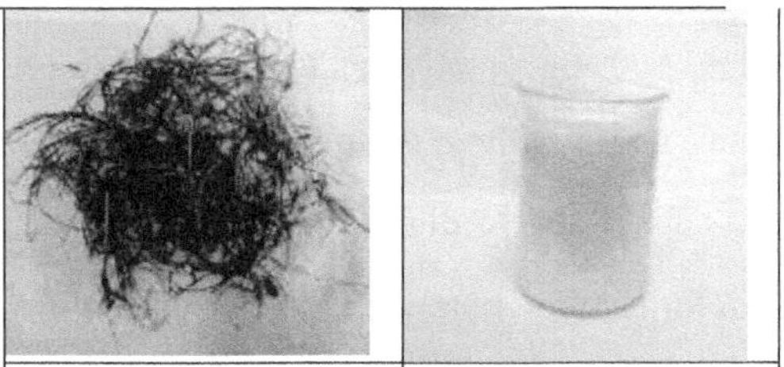

Fodder	Milk

Fig. 3: Different types of samples

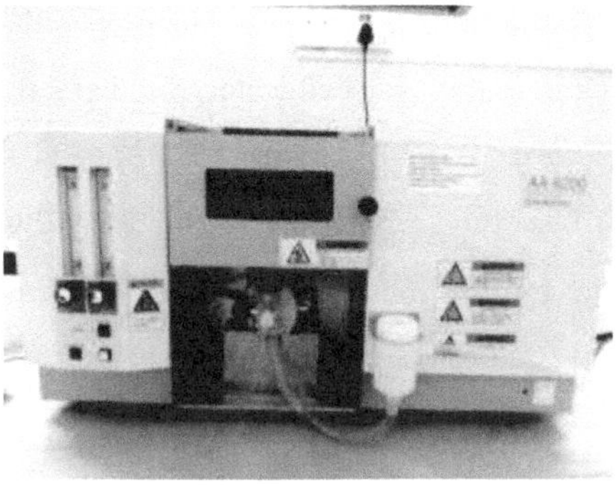

Fig. 4: Atomic Absorption Spectrophotometer - Shimadzu (AA 6200)

Characterization Studies

SEM analysis: SEM images of the modified sorbents, TPJB and PJBAC are illustrated in figures 6(a) and 6(b) respectively (JEOL JFM-6390 Scanning Electron Microscope). The appearance of highly porous morphologyand coarse surface texture with pores of different shapes wasobserved.

Brunauer-Emmet-Teller (BET) analysis: Surface areas and pore characteristics of the TPJB / PJBAC were determined (Micromeritics, BEL, Japan, Inc) and the cor- responding values are listed in table 2.

Batch Equilibration Method: Batch studies were carried out at room temperature in order to investigate the nature ofmetal – adsorbent interaction. Erlenmeyer flasks with testing samples (50ml of prepared samples with optimized particle size (0.18mm), dosage (1000mg) and pH (3.0-5.0) after pilot estimations) were agitated at preset time intervals (140 rpm) using a Kemi Orbital Mechanical Shaker. The agitated samples were filtered and the residual

Table 1
Effluent Samples

Type of Effluent	Areas of Collection	Sample Names
Dye	Tirupur	D1-D5
	Erode	D6
Tannery	Erode	T1, T2
Electroplating	Coimbatore	P1 – P3

Table 2
BET plot parameters

	TPJB	PJBAC
Surface area(m^2/g)	3.281	185.261
Mean Pore diameter (nm)	6.56	2.8594

concentrations of the corresponding metal ions were recorded.

The amounts of heavy metals adsorbed onto the TPJB/PJBAC at the time of equilib-rium (qe) were calculatedusing the following expression:

$$qe = (Ci - Ce)\ V/m$$

The adsorption percentages of studied metal ions were calculated by using the fol-lowing equation:

$$\%\ \text{adsorption} = \frac{(C_i - C_e)}{C_i} \times 100$$

where Ci and Ce are the initial and equilibrium concentrations (mg/L) of the metal ions, V is sample volume (L) and m is weight of adsorbent (g).

Results and Discussion

The concentrations of the heavy metal ions present corresponding to effluent samples and Fodder/ Milk sampleswith reference to their point source of collection are listed in tables 3 and 4 in comparison with WHO standards before adsorption process.

From the table, it is obvious that the presence of Cr(VI) and Pb(II) is more common in the effluents of tannery / plating units and in one sample out of 6 samples collected from dyeing units. Ni(II) and Zn(II) ions are present in plating units alone.

The declining order of the heavy metals is: Cr(VI) > Ni(II)> Zn(II) > Pb(II) (Table 3). All the recorded concentrations areobserved to exceed the permissible limits suggested by WHO in many folds; in particular Cr(VI) concentrations are

≈ 4000 times higher than that of the WHO standards (0.1mg/L) observed to be ex- tremely alarming in tannery effluents. This may be due to the utilization of chromium salts in the protection of hides from decay to make them more vulnerable against mois- ture and ageing. Concentrations of lead/nickel/zinc in tannery and plating units are 400/40/700 folds higher than WHO permissible limit. Table 4 lists out the concentra- tions of the analyzed metal ions in fodder and milk samples.

The recorded values of Cr(VI), Pb(II) and Ni(II) in fodder samples grown in specific effluent discharged lands imply a maximum of 7: 5: 2 times the concentrations of the respective ions in comparison to WHO standards.

Increased Pb(II) ion concentrations in milk samples attributed to the grazed fodder from the tannery effluents' discharged areas indicate the bioaccumulation of lead ions to the fodder samples and reciprocate in the collected milk samples. This can be at- tributed to the fact that the Ca^{2+} ions present in milk can be replaced by Pb(II) ions.[11]

Batch Studies: Effluent samples treated as per the preparation methods possessing in-tolerant concentrations of analysed heavy metals were subjected to batch equilibration studies employing *Prosopis juliflora* Bark, a plant waste subjected to suitable chemicalmodification and activation. The results pertaining to sorption experiments are listed inthe table 5.

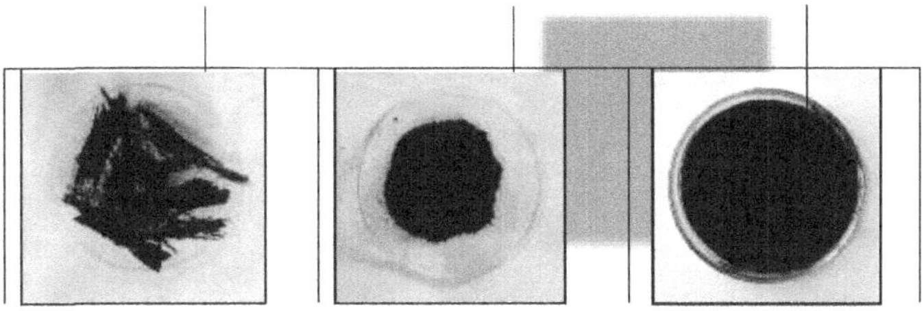

PJB	TPJB	PJBAC

Fig. 5: Adsorbent Materials

Table 3
Heavy Metal Concentrations in Effluent samples

Type of Sample	pH	Cr (mg/L)	Pb(II) (mg/L)	Ni(II) (mg/L)	Zn(II) (mg/L)
Permissible Limits (WHO*)		0.1	0.1	2.0	0.1
Dye 6	3.93	408.83	20.3	-	-
Tannery 1	3.02	433.23	18.47	-	-
Tannery 2	3.22	493.34	39.4	-	-
Plating 1	2.64	397.52	-	-	-
Plating 2	5.72	-	-	105.62	-
Plating 3	5.68	-	-	-	69.2

* World Health Organization

Table 4
Heavy Metal Concentrations in Fodder/Milk Samples

Metal Ion	WHO Standards (mg/L)		Fodder Sample (mg/L)			Milk Sample (mg/L)		
	Fodder	Milk	Dye	Tannery	Plating	Dye	Tannery	Plating
Cr	1.3	1.3	7.59	20.4	40.36	1.36	3.05	1.0
Ni	1.0	0.2	-	-	2.10	-	-	-
Pb	2.0	0.2	10.75	41.13	-	0.12	40.2	-

Table 5
Heavy Metal Removal by Batch Mode

Type of Sample	Percentage Removal (%)											
	Cr (0.1*)			Pb (0.1*)			Ni (2.0*)			Zn (0.1*)		
	Initial (mg/L)	TPJB	PJBAC	Initial (mg/L)	TPJB	PJBAC	Initial (mg/L)	TPJB	PJBAC	Initial (mg/L)	TPJB	P
D 6	**408.9**	80	77.6	**20.3**	100	100	-	-	-	-	-	
T 1	**433.2**	84	84	**18.5**	100	100	-	-	-	-	-	
T 2	**496.4**	81	82	**39.4**	73.5	100	-	-	-	-	-	
P 1	**397.5**	90	95	-	-	-	-	-	-	-	-	
P 2	-	-	-	-	-	-	**105.6**	60	83.4	-	-	
P 3	-	-	-	-	-	-	-	-	-	**69.2**	66.6	

*Permissible limits *(WHO Standards)*

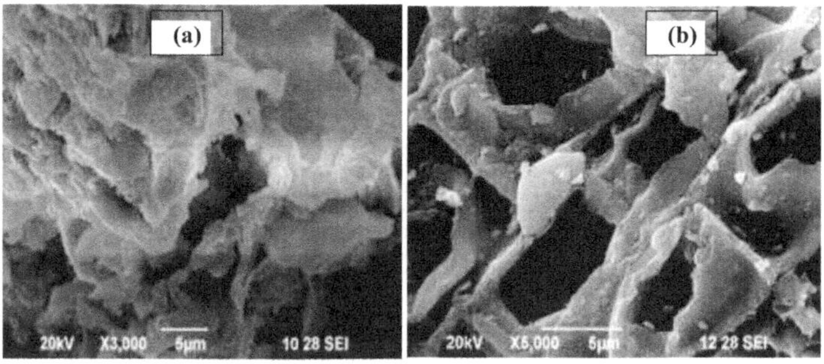

Fig. 6: SEM images (a) TPJB (b) PJBAC

Tabulated values reveal an average of 85% chromium removal except for plating efflu- ent where the removal has been excellent in case of PJBAC than TPJB. The effluent samples containing lead as pollutant have registered 100% abstraction excluding T2- TPJB system. In the cases of nickel and zinc pollutants as referred in P2 and P3 samples respectively, the activated carbon recorded about 20% excess sorption of the corre- sponding metals in relation to TPJB.

Conclusion

The present study focused on the assessment of the extent of pollution created through bioaccumulation in food chain. In this context, a total of 11 samples of effluent, fodder and milk sources were collected from various dyeing/tannery/plating industries located at Coimbatore, Erode and Tirupur districts. The concentrations of toxic metal ions pre- sent in the effluent and fodder/ milk samples were analyzed using AAS after specific di- lutions and wet digestions respectively. The recorded values revealed the occurrence of Cr, Pb, Ni and Zn in selected samples with exceeding level of concentrations as per WHO standards. Bioaccumulation of Cr, Pb and Ni are obvious wherein extended lead concentration is reported in milk samples. Suggestive reclamation studies were concentrated viaadsorp- tion methods employing P*rosopis julifora* Bark (PJB), a naturally occurring litter materi- al. PJB was modified / activated to prepare corresponding TPJB / PJBAC sorbent mate- rials subjected to characterization studies. The experimental conditions were optimized following the batchequilibration method to study the efficacy of the sorbents.

PJBAC exhibited better sorption characteristics than TPJB due to enlarged surface area supported by BET analysis. From the study design, the observations concluded that the toxicity entering into the water streams and soils from polluting elements not only leads to water/land pollution, but also reflects in bioaccumulation of these metal ions into the food chain in an alarming manner. It is the need of the hour to curtail heavy metal con- centrations through new insight in treatment technologies paving way towards retarda- tion of bioaccumulation as a suggestive measure.

References

1. Al-Busaidi M., Yesudhason P., Al-Mughairi S., Al-Rahbi W.A.K., Al-Harthy K.S. and Al-Mazrooei N.A., Toxic metals incommercial marine fish in Oman with refer- ence to national andinternational standards, *Chemosphere*, **85(1)**, 67-73 **(2011)**

2. Andal N.M., Charulatha S., Anuradha J. and Gayathri N.S., Adsorption Dynamics of Ecofriendly Litter Wastes For Zn(II)/ Cr(VI) From Electroplating Effluents - Continu- ous Column Method, *SOJ Mater Sci Eng*, **4(1)**, 1-5 **(2016)**

3. Emamverdian A., Ding Y., Mokhberdoran F. and Xie Y., Heavy Metal Stress and Some Mechanisms of Plant Defense Response, *Sci World J.*, **2015**, 1-18 **(2015)**

4. Finberg L., How Good a Food for Humans Is Cow's Milk from Hyperbole Up to Hy- perbole Down, *Am J Dis Child*, **146(12)**, 1432 **(1992)**

5. Kumar M. and Tamilarasan R., Kinetics, equilibrium data and modeling studies for the sorption of chromium by *Prosopis juliflora* bark carbon, *Arab. J. Chem*, **10(2)**, S1567-S1577 **(2017)**

6. Pathak C., Chopra A.K., Srivastava S. and Thakur D., Translocation and enrichment of heavy metals in Brassica juncea grown in Paper mill effluent irrigated soil, *J. Appl. Nat. Sci*, **5(2)**, 510-515 **(2013)**

7. Qaisar M., Audil R., Sheikh S.A., Muhammad R.A. and Muhammad B., Current Status of Toxic Metals Addition to Environment and Its Consequences, In Book Anjun N.A., Ah- mad I., Pereira E., Duarte A.,

Umar S. and Khan N., The plant family Brassicaceae, 35– 65 **(2012)**

8. Rieuwerts J.S., Thornton I., Farago M.E. and Ashmore M.R., Factors influencing metal bioavailability in soils: preliminary investigations for the development of a critical loads approach formetals, *Chem Spec Bioavailab*, **10(2)**, 61-75 **(1998)**

9. Nazir Ruqia, Elemental Assessment of Various Milk Packs Collected From KPK, Paki- stan, *Am.-Eurasian J. Toxicol. Sci.*, **7(3)**, 157-161 **(2015)**

10. Sharma S.K. and Sharma M.C., Remediation Modeling of Metal Contaminated Water and Soil Using Water Soluble Polymers – An Overview, *Int. J. Adv. Res. Phys. Sci.*, **2(2)**, 27-33 **(2015)**

11. Simons T.J.B., Lead-Calcium Interactions and Lead Toxicity, In Baker P.F., eds., Calcium in Drug Actions. Handbook of Experimental Pharmacology, Springer, Berlin, Heidelberg**(1988)**.

A Comparison Of Raw And Treated Plant Debris I The Chelation Of Anion From Aqueous Media

N.Muthulakshmi Andal* and K. Vivithabharathi
PSGR Krishnammal College for Women, Department of Chemistry, Peelamedu,Coimbatore – 641 004
**Corresponding author, Email : muthulakshmiandal@psgrkcw.ac.in*

Utilization of tea plant stems, *Camellia sinensis* (CSS), discarded as litter, col- lected from Ooty was employed for phosphate removal. The material was broken into small pieces, washed, dried, pulverized into different mesh sizes using scien- tific test molecular sieves, labelled as raw *Camellia sinensis* dust (RCSSD). Sorption efficiency of the categorized sizes was tested by applying the batch ver- ification technique, where 85 BSS exhibits better sorptive nature. The particle size of 85 BSS was determined (0.18 mm) using binocular microscope (Optika make), treated with 0.1N H_2SO_4, washed, dried, named as treated *Camellia sinensis* dust (TCSSD). Characterization studies are supported by FTIR, SEM and EDAX methods. Sorption efficiency of TCSSD was experimentally verified under varying adsorption parameters. Absorbance values were recorded using UV/VIS spectrophotometer (LABINDIA – UV 3000+) by molybdenum blue complexation method for phosphate ions. Maximum removal was registered as 48.3% and 99.7% for RCSSD and TCSSD, respectively under under optimized batch conditions of 0.18 mm particle size, 250 mg dose, 10 mg/L initial concen- tration, 9 min agitation time and pH 5 at room temperature. Experimental data were validated

using Langmuir and Freundlich isotherms wherein Freundlich plots recorded a better linear fit. Results imply that the selected material possess- es excellent anion removal capability from aqueous media.
Keywords –Adsorption , Phosphate , Plant waste, Batch process, Aqueous me- dia, Isotherms

1. INTRODUCTION

Phosphate is an essential nutrient for the growth of microorganisms in most eco- systems but in excess plays a key role in causing eutrophication [1]. Eutrophica-

tion is the rapid growth of algae due to excess nutrients in the water which co- vers the top of the water and prevents the availability of sunlight to reach aquatic plants. This leads to their death and results in a decline of dissolved oxygen lev- els. The source of excess phosphate leaching by the discharge of nutrient rich wastewater into the natural reservoirs. The point sources from which phosphate ion accumulate include natural sources (for example precipitation, dry fall, de- composition of organic materials, soil erosion) and artificial sources (for exam- ple use of fertilizers, detergents, human and animal wastes). Employment of biowaste materials to treat contamination with suitable modifications exhibit ex- cellent adsorption capacity than conventional treatment methods. The main ob- jective of this work focusses on the assessment of phosphate removal from aque- ous solutions using raw and treated tea plant stem.

2. MATERIAL AND METHOD

2.1 Collection of material

Tea plant stem of *Camellia sinensis* (CSS) discarded as litter during the winter season are collected from Ooty, were washed with distilled water in order to remove the im- purities and sundried for a period of nearly ten days. The dried material was then pul- verized using an electrical mixer (RCSSD) and sieved to various mesh sizes, namely, 85 BSS, 72 BSS, 52 BSS, 36 BSS and 22 BSS using Scientific Test Molecular sieves. The sieved material was treated with 0.1N H_2SO_4 for three hours, washed and dried (TCSSD). Figure 1 represents the cleaned tree stem, raw sieved material and its treat- ed counterpart of 85 BSS mesh size.

2.2 Microscopic analysis

Particle size determination was carried out using a binocular microscope (Optika make) for the respective sieved mesh sizes and are shown in Figure 2 for 0.18 mm particle size. Pore size diameter of TCSSD was observed to be mesoporous (13 nm) ensures better chelating property.

2.3 Preparation of sorbate

A stock solution of 1000 mg/L of phosphate ions was prepared by dissolving 1.4329
g of potassium dihydrogen phosphate.

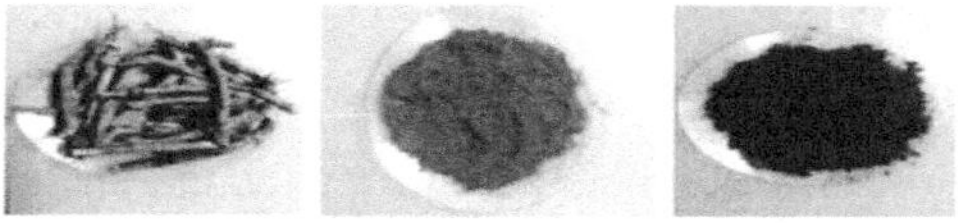

(a) CSS (b) RCSSD (c) TCSSD

Figure 1. *Clean tree stem, raw sieved material and its treated counterpart*

(a) RCSSD (b) TCSSD

Figure 2. *Sieved mesh sizes of RCSSD and TCSSD*

A standard of 100 mg/L was prepared from the stock solution and further dilutions were made as working aliquots as per the experimental requirements.

2.4 Batch experimental studies

Adsorption experiments were carried out varying the particle size (0.18, 0.21, 0.30,0.42 and 0.71 mm), agitation time intervals (3- 30 min), initial phosphate con-
centration (2,4,6,8,10,12 and 14mg/L), dosage (100, 150, 200, 250 and 300 mg), pH of the medium and co-ions (10 mg/L for NO_3^-, SO_4^{2-}, F^- and Cl^-) to optimize the best suitable condition for maximum removal by RCSSD and TCSSD. 50 ml of phosphate samples were added into the 250 ml Erlenmeyer flasks, with respective doses and particle size of raw and treated sample. The contents of the flasks were agitated in a mechanical shaker at a speed of 140 rpm. Absorbance values of initial and final read- ings were analyzed using UV/VIS spectrophotometer by molybdenum blue complex- ation method. [2]

3. RESULT AND DISCUSSION

3.1 FT IR spectral studies

FT IR spectra corresponding to TCSSD and its loaded sample recorded a shift in wavenumber of prominent peaks. These significant shifts in the wavelength may at- tribute to the phosphate binding action occurring on the material surface, 1648 /cm and 3372.66/cm corresponds to carboxyl and carbonyl group (Figure 3) [3].

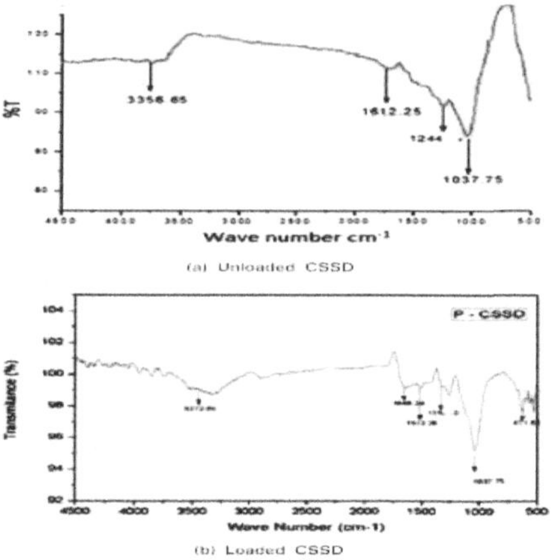

Figure 3. *FTIR images of unloaded and loaded TCSSD*

3.2 SEM/EDAX analysis

Porous nature exhibited by TCSSD is supported by the surface morphological changes as depicted in the SEM images being confirmed by the EDAX spectral peak at 1.73 keV (Figures 4 and 5).

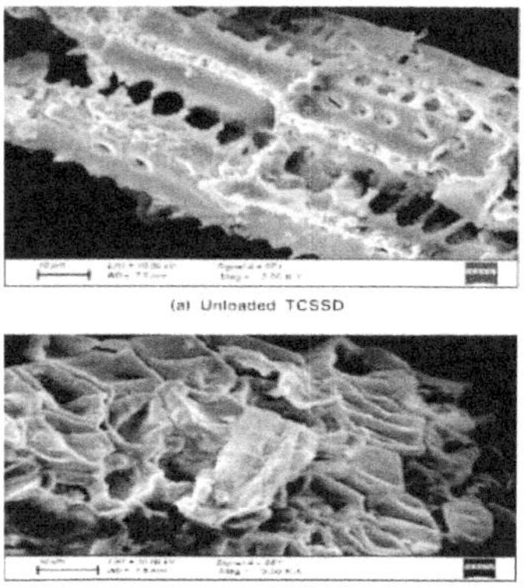

Figure 4. *SEM images of unloaded and loaded TCSSD*

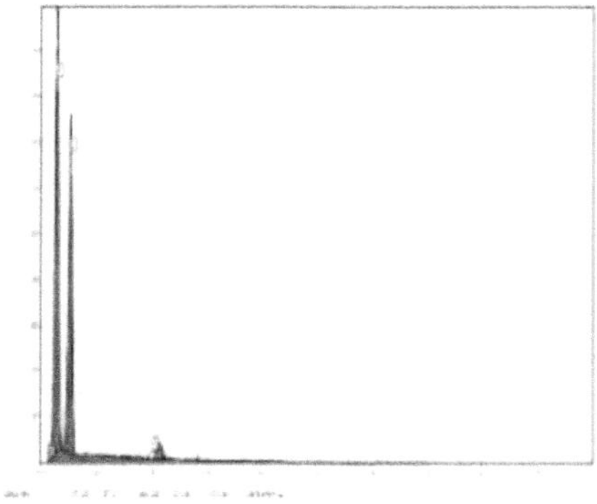

Figure 5. *EDAX spectral of loaded TCSSD*

3.3 Effect of particle size

The sorption capacity of any sorbent material is suggested as a function of particle size, probably indicating a better surface sorption phenomenon. The bar chart is indicative of almost 100% phosphate removal by TCSSD against RCSSD for 0.18 mm particle size being the most significant, is fixed for the forthcoming experi-

ments (Figure 6).

3.4 Effect of dosage

Phosphate removal was found to increase with dosages (100-300mg: 50 mg inter- val) is evidenced from figure 6 wherein the sorption rate of RCSSD/TCSSD is 48.3% and 99.7%, respectively, for 250 mg/50 mL. After the dose of 250 mg, there is no further significant increase in adsorption efficiency might be due to the ag- glomeration. Thenceforth, a dose of 250 mg was optimized for further sorption studies.

3.5 Effect of initial phosphate concentration

Figure 6 registers a maximum phosphate ions removal of 48.3% and 99.7% for raw and treated samples respectively and at an initial concentration of 10 mg/L. Further increase in the concentration resulted in waning of removal efficiency which shall be due to the buffering properties of phosphate compounds, therefore 10 mg/L was fixed for forthcoming experiments.

3.6 Effect of agitation time

The minimum period of agitation time required for the maximum removal of phos- phate was recorded as 9 min (figure 6). Declination in the curve was observed at later time frames which shall be due to the very rapid initial uptake rate and over- lapping of adsorption sites after the equilibrium point.

3.7 Effect of pH

The results pertaining to variable pH environments for the studied systems are rep- resented in figure 6. Adsorption is observed to be maximum at pH 5 for both raw and treated systems. Further increase in pH led to a decline in the sorption curves as indicated by the sloping down of the curves. The reduced sorption rate at alka- line pH ranges may be due to repulsion between similar ionic binding sites.

3.8 Effect of co-existing ions

Marked inhibition in phosphate removal is recorded in case of Cl^- as obvious from figure 6. F^-ion exhibits lesser influence and SO_4^{2-} has minimal inhibitive property, this shall be due to the electronegative character of the specific anion as reported in

the article [4].

3.9 Adsorption isotherms

Experimental data predicted using Langmuir and Freundlich isotherms are depicted in figure 7. Better linearity was exhibited by Freundlich plots in favour of multi- layer adsorption onto the surface of raw and treated materials [5].

4. CONCLUSION

Experiments pertaining to the sorption of phosphate ion onto the raw and treated tea plant stem dust were designed

FT IR spectra revealed significant peak shifts in anion laden

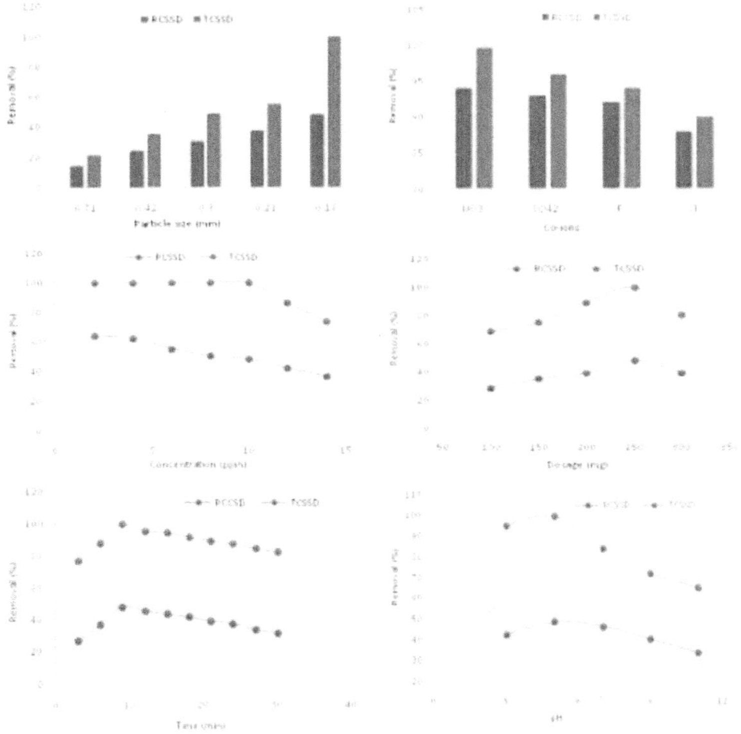

Figure 6. Percent removal of phosphate by RCSSD and TCSSD with varying particle size, initial phosphate concentration, dosage, agitation time, pH and presence of co-ions

materials due to the probable binding of phosphate ion onto the sorbent surface. SEM/EDAX images

registered a morphological change in porosity structure and the appearance of new peak, specific phosphate ion confirmed the sorption process. TCSSD exhibited al- most 100% removal than RCSSD (48.3%), being two-fold times, under optimized batch conditions of 0.18 mm particle size, 250 mg dose, 10 mg/L initial concentra- tion, 9 min agitation time and pH 5. Marked inhibition of monovalent anion was observed for phosphate uptake by the sorbent due to the charge density characteris- tics. Isothermal studies revealed the equilibrium data satisfied the linearity of Freundlich plots. (multilayer adsorption) in comparison with the Langmuir model for both the systems. The concluding remarks highlight TCSSD is potential sorbent for possible remediation of phosphate ions. Future scope focuses on the phosphate elimination from field level leachates.

Employment of Surface Modified Biomaterial in Sequestering Divalent Metal Ion

N. Muthulakshmi Andal* and J. Anuradha
Department of Chemistry, PSGR Krishnammal College for Women, Coimbatore- 641 004,Tamil Nadu, India

*Corresponding Author: muthulakshmiandal@psgrkcw.ac.in
Received: May 18, 2018; Revised: October 22, 2018; Accepted: December 9, 2018

Abstract

The study is designed to investigate the potential applicability of Nano sized Treated Goat Teeth (NTGT) as adsorbent for Ni(II) re- moval from metal laden solutions. The desized novel nanomaterial (NTGT) is characterized by AFM and TGA- DTA techniques. Batch Equilibration study is carried out for Ni(II)- NTGT system under varied influential parameters viz., dosage, agitation time intervals, optimized initial metal ion concentration (250 mg/L), pH (5.5) and temperature (303K). The results of the experiment registered a two fold increase in the removal of Ni(II) from aqueous media with one half amount of required dose and agitation time interval in compari- son with that of the counterpart TGT. High performance of NTGT in trapping Ni(II) ions is extended to field trials by treating electro- plating effluent samples collected from Coimbatore industrial belt, the results suggesting the derived NTGT as an efficient bioseques- trant

Keywords: Nickel ions; Characterization; Batch studies; Sorption

1. Introduction

The directory of "Agency for Toxic Substances and Disease Registry Priority List of Hazardous Substances" offer top 10 positions for heavy metal ions due to its toxicity and potential exposure from a wide range of sources

(Daneshfozouna *et al.*, 2017). Indiscriminate disposal of toxic metal ions are still continuingand even increasing in developing countries. Coimbatore city is well known for small and medium scale electroplating industries. Nickel plating is the most common process in the studied area and high concentrated effluent in huge volume is discharged per day, even at small units. Nickel is considered a moderately toxicelement (< 0.5 mg/L) comparing with other transition metals.

Exposure to nickel is associated with respiratory cancer, derma- titis, gastrointestinal distress, chronic bronchitis, cardiovascular, kidney diseases and a skin disorder knownas nickel-eczema (Li *et al.*, 2018). Thence, efficient removal of Ni(II) ions from industrial wastewaters are the prime need for concern. Adsorption is among the most common methods employed for heavy metal ion trap- ping due to its simplicity, feasibility and effectiveness. Biosorbents ensures maximum sorption capacity and the best alternative for non-degradable activated carbons. In the present work, Goat Teeth (GT) has been chosenas the biosorbent. GT is a shear litter, causing solid waste disposal problem, other than being helpful to determine the age of a goat.

Herein, chemically modified and nano- sized GT are prepared, characterized and equilibrium parameters are experimentally inves- tigated. To the best of our knowledge, Ni(II) ion removal employing goat teeth havenot been explored so far.

2. Experimental Section

2.1 Materials and Modification

Goat Teeth were collected from local butcher shop, detached from jaw bones, washed well with running water to eliminate colloi- dal impurities, later sun-dried for removalof moisture content. Goat Teeth consist of high percentage of mineral contents, especially exchangeable Ca^{2+} ions and organic/ bio-chemical materials such as collagen, proteoglycan (chondroitin sulfate, keratin sulfate), lipids. Therefore, the washed GT were subjected to chemical treatment by soaking in 0.1N HCl after pulverization into 0.18 mm particle size. The treated goat teeth (TGT) was washed and air dried (Anuradha *et al.*, 2017).

2.2 *Preparation of Nano- biosorbent*

40 grams of Treated Goat Teeth TGT was weighed and desized to nano level by a top-down approach. This was carried out ina Planetary Ball Mill (Model- VBCC/PM/24- 13/14) equipped with Tungsten Carbide bowland 2 mm balls. The above procedure wasfollowed by milling for 15 hours at 260 rpm. The resulted pow- der agglomerates was stored in an airtight container and referred as NTGT. A stock solution of 1000 mg/L Nickel nitrate was prepared, from which aliquots of varying concentrations were derived.

2.3 *Characterization*

2.3.1 *Atomic Force Microscopic (AFM)Analysis*

AFM analysis offers visualization in three dimensions and surface roughness. Individual particles and groups of particles can be re- solved finer, compared to other microscopictechniques. The top and 3-dimensional view (20μm x 20μm x 300 nm) of the nanoparticles is depicted in figures 1, 2 respectively. The thickness of the nanoparti- cle is measured usingthe data obtained from line profiles on the inter- facial region and the size was found to be in the range of 50–90 nm. The attainment of nano particles' size was confirmed by the Atomic Force Microscope histograms (fig c) and the appearance of a clear peak around 60.75 nm inconfirms the existence of a maximum num- berof the nanoparticle.

2.3.2 *TG- DTA Technique*

Thermo Gravimetric /Differential Thermal Analysis (TG/DTA) measurements were obtained using Simultaneous DTA- TGA Apparatus, (DTG- 60, Shimatzu- Japan). The small amount ~ 1 wt% of TGT was placedin the platinum thermo balance crucible for analysis. The samples were heated from room temperature to 900°C, at a heating rate of 20°C/ min; the measures were carried out ina constantAir atmosphere.

2.4 Batch Studies- Aqueous Solutions and Industrial Effluents

Batch equilibration experiments were verified for the operating factors viz., doses of TGT/ NTGT (50-200 mg : 50 mg), agitation time intervals (5-20 min: 5 min) at optimized initial metal ion con- centration (250 mg/L), pH 5.5 and temperature (303K). The efficiencies of the nano particles were also tested with effluentsample [Ni(II)- 220 mg/L, pH 5] collected from electroplating industry at Coimba- tore. After the contact period, the supernatant was filtered and the initial and residual Ni(II) ion concentrations were recorded using Atomic Absorption Spectrophotometer (Shimadzu AA 6200) at a wavelength of 232 nm.

The percentage values and the amounts of metal ion adsorbed from aqueous solutionswere calculated using the equation,

$$q = \frac{V(C_i - C_e)}{W}$$

$$\%\, adsorption = \frac{(C_i - C_e)}{C_i} \times 100$$

V - volume of the solution (L)m - mass of the adsorbent (g)C_i / C_e - initial / equilibrium metal concentrations(mg/L)

3. Results and Discussion

AFM provides us microscopic informationon the surface structure and plots the surface topography of the adsorbent. Fig 1(a), 1(b) shows 2D/3D AFM images from sorbent surface and fig 1(c) clear- cut the size of the nano material. NTGT tended to aggregate, thus contributing to some bright areas in AFM height images. According to fig 1(c), the grainsize of the crushed sample is 60.75 nm. The col- or range shows the morphological changes of the surface, which in- dicate the adsorption sites on the surface of nano TGT (Khorzughy *etal.*, 2015).

Thermo Gravimetric- Differential ThermalAnalyses for NTGT was studied to evaluate the effect of functionalization on their thermal degradation behavior under air atmosphere. Fig 2(a) indicative of

TG curve pertaining to thermal stability of NTGT, shows a deflec- tion point at 100°C corresponding to the removal of water. A typical degradation behavior of NTGT envisaged at 270°C, [Thermal De- composition Point (T_d)] confirmed the derived sample from TGT through HCl treatment (Rana *et al.*, 2017). Again, a dip in the curve at 500°C refers to the deterioration of organic matter (listed under Materials and Modification). This fact is supported by the absence of significant weight loss occurring between 600°C - 800°C, emphasiz- ing the total removal of organic matter.

An endothermic peak in the DTA curve [fig 2(b)], at 66.3°C infer the moisture loss followed by an exothermic coun- terpart at 154.8°C corresponding to the thermal decomposition / pyrolysis of organic content(Teaford *et al.*, 2007).

Experimentally verified graphical representation for pH in- fluence is depicted in fig 3. A peak height at pH 5.5, imply the maximum amounts of Ni(II) ions beingadsorbed, that too, enhanced in the case of NTGT. In view of this, pH 5.5 was fixed as the opti- mum pH for the sorption systems. Fig 4 and 5 refer to the effect of dosage at different

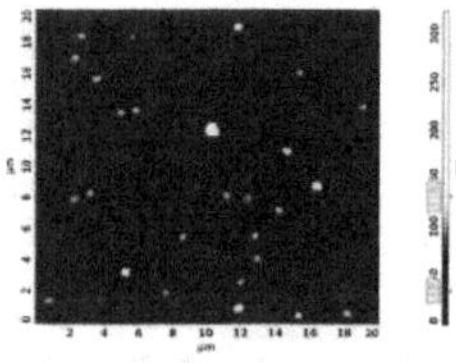

(a) 2D Surface Topography

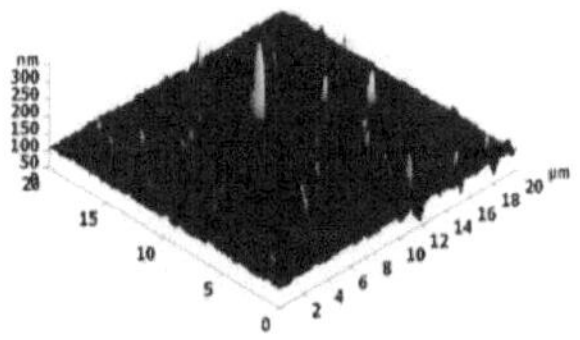

(b) 3D Surface Topography

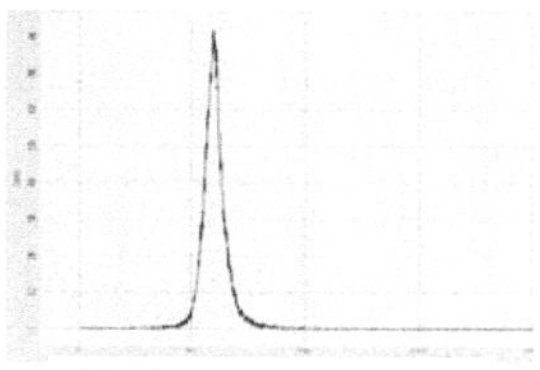

(c) Histogram (60.75 nm)

Figure 1. AFM Images - NTGT

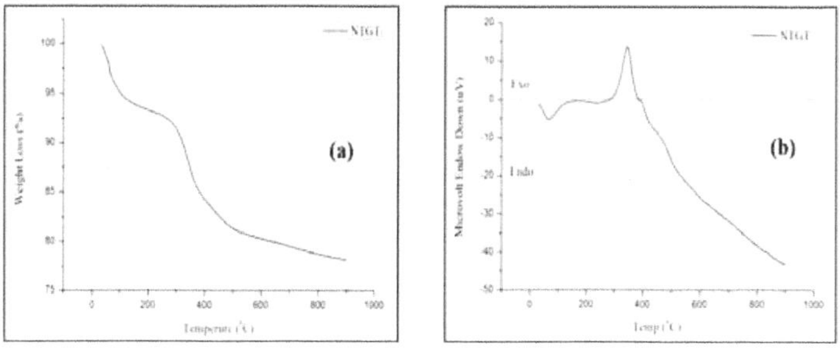

Figure 2. TG- DTA Analyses

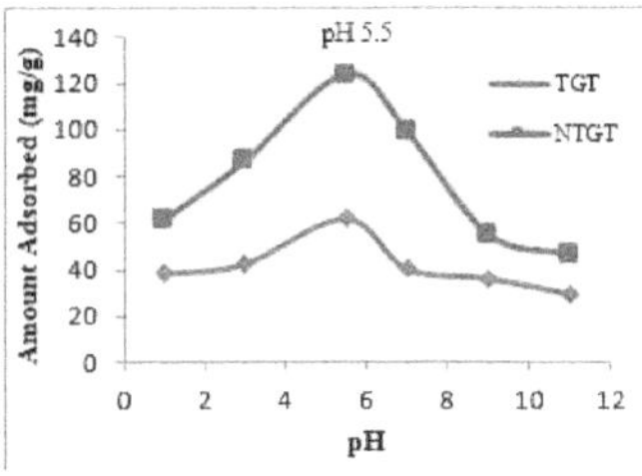

Figure 3. Effect of pH

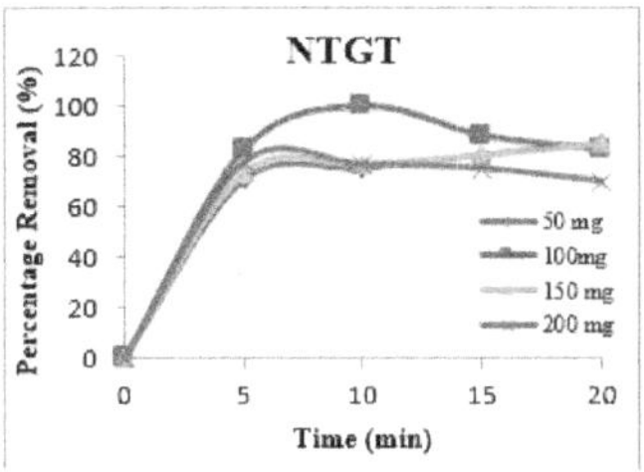

Figure 4. Effect of Dosage: Ni(II)- NTGT

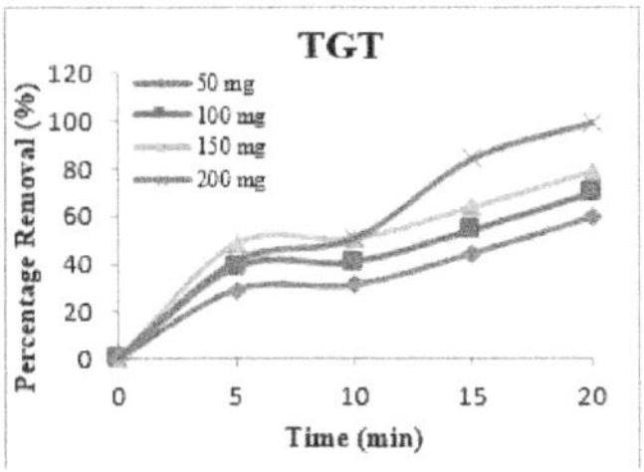

Figure 5. Effect of Dosage: Ni(II)- TGT

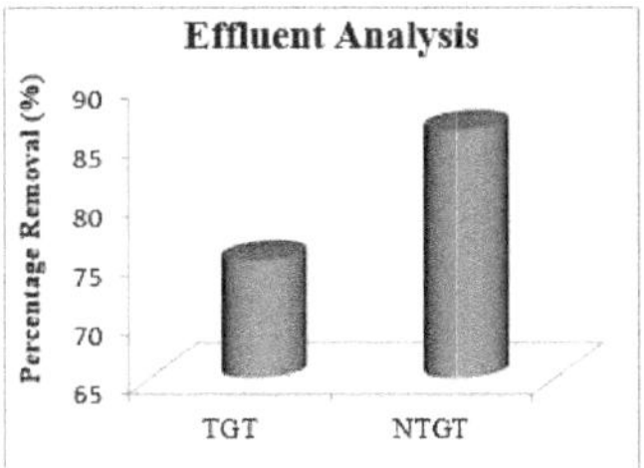

Figure 6. Nickel Plating Effluent

time periods and percentage removal of Ni(II) ions. Results pro- nounced the enhanced sorptive ability of nano particle against their modified counterpart in case of both aqueous solution and plating ef-fluent.

An agitation time/adsorbent dose of only 10 min/100 mg NTGT was sufficient against 20 min/200 mg TGT, to trap 99.8 % and 86 % Ni(II) ions from both aqueous / effluent samples (Fig 6). A decrease in adsorption rate was registered at higher time period and dosages, in- variably in all cases which may be due to the very rapidinitial uptake rate and overlapping /aggregationof adsorption sites as a result of overcrowdingof sorbent particles after equilibrium state.

Maximum adsorption capacity of 123.68 mg/g at 10 min time in- terval, 100 mg dosage of NTGT was registered, which was twofold time maximized value than the chemically modified counterpart (62.09 mg/g).

The percentage removal of the metal ions from effluent sample was observed in fig 6, to be less than the corresponding aqueous solu- tions, which shall be due to the presence of interfering ions in the caseof former.

4. Conclusion

Experiments pertaining to the adsorption of Ni(II) ions from wastewaters onto the modifications/ size reduction of chosen slaughterhouse waste were designed. Almost 100% removal of diva- lent ion was achieved under specified laboratory environment using low-cost adsorbents' viz., 250 mg/L initial metal ion concentration, pH 5.5 and different dosage/ contact time [TGT- 200 mg, 20 min; NTGT-100 mg, 10 min]. The results of the present investigations would be useful for the fabrication and designing of wastewater treatment plants, utilizing such novel bio/ nano materials in an effec- tive manner. Our future study aims at the establishment of perfor- mance oriented approach of TGT/ NTGT for other toxic heavy metal ions. This can be accomplished by developing suitable modification procedures in the sorbents viz. encapsulation and immobilization, thereby ensuring the safe disposal of the metal loaded sorbents, which in turn does not pose a great threat to the environment.

References

1. Anuradha J, Muthulakshmi Andal N, Gayathri N.S. Utilization of a Material Derived fromAnimal Waste in the Biosorption of Pb(II)Ions – Isothermal/ Thermodynamic, Kinetics and Statistical Studies. Cur- rent World Environment 2017; 12 (2): 445-455.
2. Daneshfozouna S, Abdullahb M.A, Abdullaha B. Preparation and characterization ofmagnetic biosorbent based on oil palm empty fruit bunch fibers, cellulose and Ceiba pentandra for heavy metal ions re- moval. Industrial Crops & Products 2017; 105: 93–103.
3. Mark F. Teaford, Moya Meredith Smith, Mark,W. J. Ferguson. De- velopment, Function and Evolution ofTeeth. Science. 2007.
4. Rana M, Akhtar N, Rahman S, Jamil HM, Asaduzzaman SM. Ex- traction of Hydroxyapatite from Bovine and Human Cortical Bone by Thermal Decomposition and Effect of Gamma Radiation: AComparative Study. International Journal of Complement Alterna- tive Medicine 2017; 8(3): 00263.
5. Sara Haddadi Khorzughy, Teymur Eslamkish, Faramarz Doulati Ar- dejani, Mohammad Reza Heydartaemeh. Cadmium removal from aqueous solutions by pumice and nano-pumice, Korean Journal Chemical Enggineering 2015; 32(1) : 88-96.
6. Yiran Li, Jian Zhang, Hai Liu. In-situ modification of activated carbon with ethylenediaminetetraacetic acid disodium salt during phosphoric acid activation for enhancement of nickel removal. Pow- derTechnology 2018; 325: 113–120.

Sorption Capacity of *Bivalve molluscs* Shell in the Removal of Divalent ION from Aqueous Solutions

N.Muthulakshmi Andal[1*] ,S.Charulatha[1]

[1,1*]Assistant Professor, Department of Chemistry, PSGR Krishnammal College forWomen, Coimbatore-4, India

Abstract : Environmental protection emphasizes the use of ecofriendly materials in- stead of chemicals to minimize pollution. The present work deals with the utilization of acid treated Mussel shell powder, an mollusc shell waste for the adsorption of Pb(II) from aqueous solutions. TMSP is characterized using Scanning Electron Mi- croscopy (SEM), Energy Dispersive X-ray Analysis (EDAX), Brunauer-Emmet- Teller (BET) and Barrett-Joyner-Halenda (BJH) analyses to study the surface mor- phology, elemental constitution, determination of surface area and pore structures. Batch equilibration studies are verified for the operating factors viz., particle sizes/ doses of the sorbent material upon a range of initial aqueous concentrations of Pb(II) at different temperatures, agitation time and pH of the Pb(II) -TMSP system to assessthe sorption capacity which is recorded as 83.6 mg/g implying its efficiency to be three fold times more than the reported values for varied sorbents by other research- ers. The applicability of Langmuir and Freundlich isotherms at various initial con- centrations are plotted for Pb(II) -TMSP system wherein the best straight line is well fit for Langmuir model indicating monolayer adsorption. Continuous column runningfor quantitative estimation of Pb(II) removal from the bulk of the solution is carried out by the chosen sorbent.

Keywords: bivalve molluscs shell, adsorption, lead ion, aqueous solution, column.

1. Introduction

Water pollution is of great global concern and its crisis is acute in India requiring evaluation and revision of water resource policies from time to time at all levels. Due to the rapid industrialization, toxic metal ions have been excessively released into the en- vironment. Cadmium, zinc, copper, nickel, lead, mercury and chromium are often de- tected in industrial wastewaters, which originate from metal plating, mining activities, smelting, battery/ paint/ pigment manufactures, tanneries, petroleum refining, pesti- cides, printing and photographic industries, etc[1]. The present work is focused on lead pollution, the reason being that it is lethal even at very small doses, to living beings. The observed ailments related to health issues due to lead consumption at varying levels

are headache, dizziness, nausea, vommitting, chest pain, shortness of breath, tightness of chest, rapid respiration, dry cough, nephritis and extreme weakness at a concentrationexceeding therecommended limit[2].

Several technologies such as membrane separation, electrochemical precipitation, ion exchange, preconcentration, fertilization and adsorption[3] have been reported for trapping toxic ions. Of these, asorption is a promising technique due to its simplicity, high efficiency, easy handling, technical feasibility, economic viability and social ac- ceptability[4].

Ecofriendly materials are found to possess excellent sorptive nature. The present work deals with the removal of Pb(II) from aqueous solutions using animal waste mate-rial ie; *Bivalve molluscs*Shell being litter wastes and also research work has not been reported elsewhere in analyzing its sorption potential.

2. Materials and Methods

2.1 Adsorbent preparation

Bivalve molluscs shells (Mussel Shells) were collected from the coastal areas of Tamil Nadu, India. The shells were cleaned, reduced to small sizes in a mixer blender followed by several washings in order to remove sand particles and finally sun dried for 24 hrs. The dried shells were soaked in 0.1N HCl for 4 hours to neutralize calcium car- bonate present, later the dried shells were washed many times with double distilled wa- ter, air dried and segregated into different particle sizes (85 BSS, 72 BSS, 52 BSS, 36BSS and 22 BSS) using standard test molecular sieves (JAYANT Scientific Instru- ments Co., Mumbai). The Treated Mussel Shell Powder (TMSP) of varying sizes were stored in an air tight containers.

2.2 Particle Size Determination

Varied mesh sizes of TMSP were subjected to image microscopic analysis Bin- ocular Microscope (OLYMPUS make, Model- CX21I) in order to determine the parti- cle sizes. Fifteen different granular particles of each mesh size were chosen for the measurement of lengths and breadths to arrive at an average, as no two single particles are alike.

By applying multiplication factors, the particle sizes were calculated. The calcu- lated particle sizes corresponding to the mesh sizes of 85BSS, 72BSS, 52BSS, 36BSS and 22BSS are 0.18 mm, 0.21 mm, 0.30 mm, 0.42 mm and 0.71 mm respectively. The microscopic structure of TMSP (0.18 mm) is depictedin

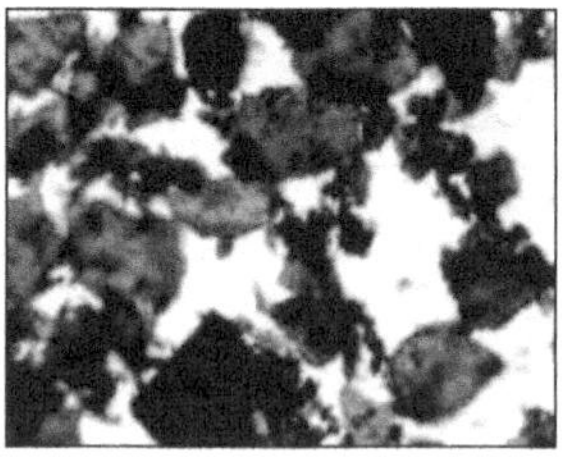

Fig 1:Optical Microscopic Image

2.3 Surface Characterization

Surface morphology of TMSP was analyzed using Scanning Electron Microscope(SEM) coupled with energy-dispersive X-ray spectroscope (EDX) (***JEOL JFM- 6390***). The surface area and volume of the pores were determined with Micromeritics ***BEL, Japan, Inc*** Surface Area Analyzer using the BET method by N_2physisorption at 77 K.

2.4 Batch adsorption experiments

The influence of varied operating parameters viz., particle sizes, (0.71 mm, 0.42 mm,

0.30 mm, 0.21 mm and 0.18 mm) dosages (200 mg, 300 mg, 400 mg and 500 mg), ini- tial concentrations of the aqueous Pb(II) solutions, (100-1000 ppm: 100 ppm intervals) predetermined time intervals between the sorbate and sorbent species, (30 min, 60 min, 120 min) pH of the medium (3, 5, 7, 9 and 11) and temperature of the system, (293K- 333 K: 10 K intervals) was experimentally verified. 50 ml of the adsorbate species tak- en in 250 ml Erlenmeyer flasks with the sorbent material agitated under the planned conditions in a mechanical shaker. The initial and residual Pb(II) concentrations were analyzed using Atomic Absorption Spectrophotometer : Shimadzu (AA 6200) model.

2.5 Data Evaluation

The percentage adsorption of metal ions from aqueous solutions is estimated us-ing the followingequation[5]

$$\% \, adsorption = \frac{(C_i - C_e)}{C_i} \times 100$$

$$q = \frac{V(C_i - C_e)}{W}$$

The amounts of metal ions adsorbed (q) are calculated using the mass balance equation

where, V is the volume of the solution (L), m is the mass of the adsorbent (g), $C_i$and C_e are the initial andequilibrium metal concentrations (mg/L) respectively.

3. Results and Discussion

3.1 SEM Analysis

The surface morphologies of the unloaded and Pb(II) loaded TMSP are depicted in figures 2a & 2b. Figures 2a represents a highly porous nature of the surface with coarse textured pores of different shapes. The smooth covered pores on the surface as observed in fig 2b against fig 2a, indicate the binding of Pb(II) ions onto the surface ac- tive sites leading to diminished ruggedness of TMSP (fig 2a). This can be evidenced from the visible surface pores as depicted in figure 2a are not distinct in fig 2b.

Fig 2a: Unloaded TMSP **Fig 2b: Pb(II) loaded TMSP**

3.2 EDAX Analysis

The EDAX spectra were recorded to analyze the elemental constitution of TMSP qualitatively.

The corresponding EDAX spectra of TMSP are depicted in figures 3a and 3b, in- dicative of the presence of O, C, Ca and Cl, theprinciple elements of any adsorbent[6]. The appearance of new peak at an energy range of 9-15 KeV (fig 2b), confirms the ad- sorption of Pb(II) by TMSP.

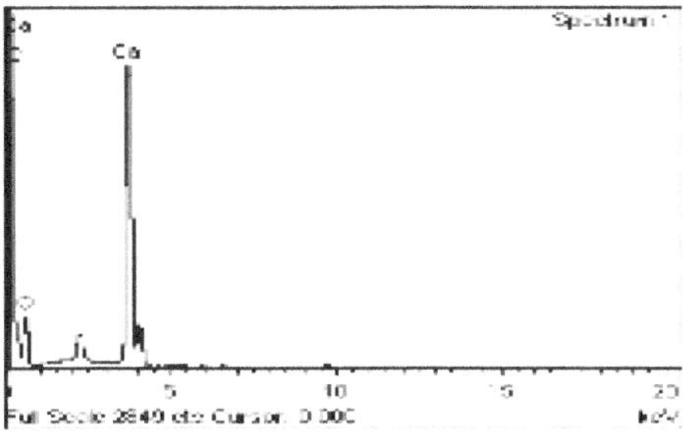

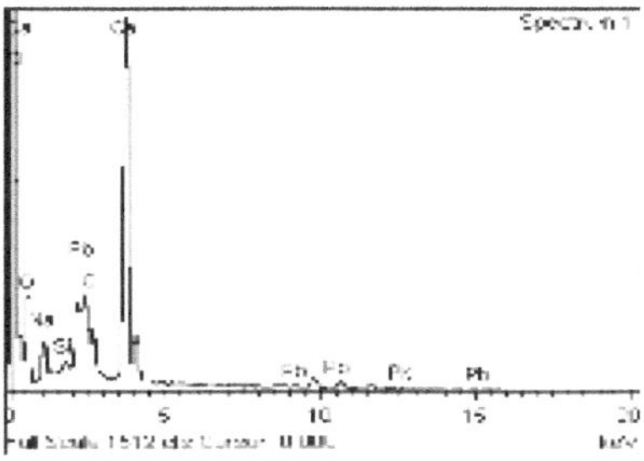

Fig 3a: Unloaded TMSP **Fig3b: Pb(II) loaded TMSP**

3.3 Effect of Particle sizes

The system response under the influence of varying particle sizes viz., 0.18mm, 0.32mm and 0.41mm for Pb(II)- TMSP system were experimentally verified and are illus- trated in figure 4. The inverted parabolic curve suggests an increase in amount adsorbed from 29.2mg/g to 83.64 mg/g, against the particle size from 0.71mm to 0.18mm. This maybe due the fact that the increase in sorption depended on the large external surface area forsmaller particle sizes. 0.18mm being the size where maximum amount had been adsorbed, is fixed as the optimumparticle size for further experiments.

3.4 Effect of Dosage

Figure 5 represents the experimental studies performed in order to optimize the dosage of TMSP required where maximum adsorption efficiency is found to be 83.6mg/g for a dose 1000 mg TMSP. Thenceforth, 1000 mg is fixed as the optimum dose for maximum removal of Pb(II) ions.

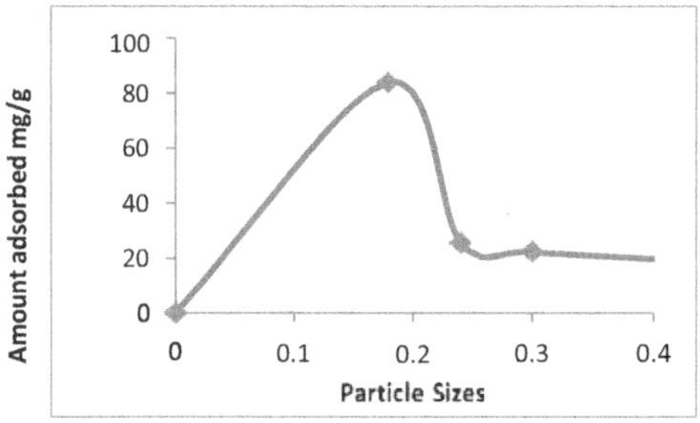

Fig 4: Influence of Particle Size

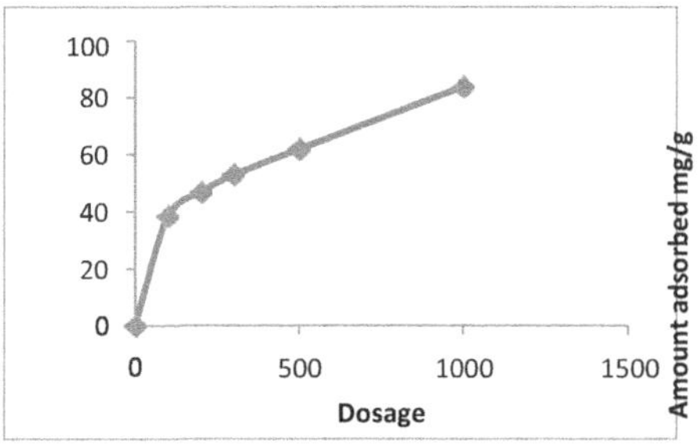

Fig 5: Influence of Dosage

3.5 Effect of Initial Concentration and Agitation Time

It is noted from the bar chart that the amount adsorbed at different initial concen- trations of Pb(II) (100- 1000ppm: 100ppm intervals) relates the direct proportion of Pb(II) adsorbed against increasing concentration of aqueous environments (23.81 mg/g to 83.6mg/g at 30 mins of contact time). Also, further increase in the contact time be- yond 30 mins, registered a negligible effect on the adsorption rate. The mechanism of solute transfer to the solid may be due to diffusion of the fluid film around the adsor- bent particle and later through the pores into the internal adsorption sites[7]. The dimin- ished rate of adsorption at further delayed agitation time shall be probably due to the slow pore diffusion of the solute into the bulk of the adsorbent.

3.6 Effect of pH

pH plays an important role in the adsorption process. The amount of Pb(II) ions adsorbed under the optimized conditions (0.18 mm & 1000 mg of TMSP, 1000mg/L of Pb(II) solutions) at varying pH environments indicated a maximum adsorption capacity to occur at pH 5.5. Mostly researchers have reported pH 5-7 to be an optimum pH for trapping heavy metal ions[8,9]. The reduced removal of Pb(II) at acidic pH 3 shall be attributed to the fact of preferential protonation of H^+ onto the sorbent surface. Similar- ly, Pb(II) undergo hydroxide precipitation at alkaline pH ranges (7 & 9) which hinders its availability to get adsorbed onto TMSP.

3.7 Adsorption Isotherms

A series of solutions containing different initial concentrations of Pb(II) were prepared and the batch equilibration studies were carried out to assess the applicability of the adsorption isotherms under the specified conditions, followed by further analysis to study the fit in of the isothermal plots based on the calculations of equilibrium con- centrations and amount adsorbed.

3.7.1 Langmuir model

The monolayer sorption model with the linearization plot of q_e vs C_e was ex- pected to give a intercept of $1/q_{max}$ and slope $1/q_m$to find the constants q_{max}

and K_a.

(C_e/q_e) = $1/(K_a$ Ce/Qe

$q_{max})$ + $C_e/$

$q_{max}$where,

q_e = the equilibrium metal ion concentration on the sorbent (mg/g), C_e = the equi-librium metal ion concentration in solution (mg/L), q_m = the maximum monolayer ad- sorption capacity of the sorbent (mg/g) and K_a= the Langmuir sorption constant (L/mg) related to the free energy of sorption.

b and q_m are Langmuir constants related to the energy and maximum adsorption capacity respectively. Langmuir plot of Ce/qe vs Ce depicted in fig.8 shows a straight line. The Langmuir constants, b and qm were calculated from the slope (1/qmax) and intercept ($1/q_{max}.K_c$) respectively. The R^2 value of 0.9919 from the plot suggests maxi- mum linearity and fit in of the Pb(II)- TMSP system on Langmuir adsorption isotherm.

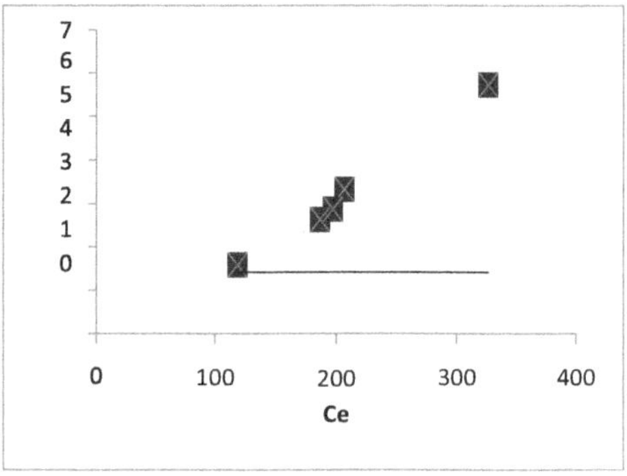

Figure 8: Langmuir Isotherm

3.7.2 Separation factor (RL)

The linearity of the Langmuir isotherm can be used to predict whether a sorp- tion system is favourable or unfavourable. Accordingly, the essential characteristics of Langmuir isotherm can be expressed in terms of a dimensionless parameter called the separation factor or equilibrium parameter R_L, which is defined by the following rela- tionship[10].The R_L values for all the system is between 0.006 and 0.096 (Table 1) indi-cating favourable adsorption.

Table 1: Equilibrium Parameter R_L for Pb(II) system

Conc. of Metal Ion (mg/L)	Pb(II)-TMSP
100	0.096
250	0.006
500	0.017
750	0.006
1000	0.016

$$R_L = \frac{1}{(1 + b C_i)}$$

where, C_i is the initial Pb(II) concentration (mg/L). The parameter R_L indicates theshape of the isotherm andthe nature of the sorption process[11] shall be $R_L > 1$ Unfavourable isotherm, $R_L = 1$ Linearisotherm $R_L = 0$ Irreversible isotherm, $0 < R_L < 1$ Favourable isotherm

3.7.3 Freundlich model

The Freundlich isotherm is the most widely used non-linear sorption model. This model proposes a multilayer sorption with a heterogeneous energetic distribution of ac- tive sites, accompanied by interaction between adsorbed molecules. The Freundlich equation is expressed linearly as

$$\log Q_e = \log K_f + 1/n \log C_e$$

The values of K_f and 1/n were obtained from the intercept and slope of a plot of log qe versus log C_e.$K_f$and 1/n are Freundlich constants related to adsorption capacity and adsorption intensity, respectively[12]. From the slope and intercept of the plot log $q_e$vs log C_e (figure 9), values of K_f and R^2 were determined (Table 2). The constants andthe correlation coefficient values (table 2) imply that Langmuir isotherm (R^2= 0.9919)is obeyed by the system less effectively as compared to Freundlich isotherm (R^2=0.9765).

Table 2: Langmuir, Freundlich and Tempkin constants data

Adsor-bent	Langmuir Constants			Freundlich Constants			Tempkin Constants		
	q_m	B	R2	K_f	1/n	R2	B_t	A_t	R2
TMSP	22.987	25.329	0.9919	10.593	0.476	0.9765	12.069	8.654	0.9721

3.7.4 Tempkin model

Tempkin isotherm equation contains a factor that explicitly takes into account ad- sorbent-adsorbate interaactions. It assumes that the heat of adsorption of all the mole- cules in the layer decreases linearly with coverage due to adsorbate-adsorbate repul- sions and the adsorption is a uniform distribution of maximum binding energy[13]. The linearized form of isotherm is [14]$q_e = B_T \ln A_T + B_T \ln C_e$

where, B_T= RT/b_T, T is the absolute temperature in Kelvin and R is the universal gas constant, 8.314 J/mol K. The constant b_T is related to the heat of sorption. The con- stant A_T is the equilibrium binding constant corresponding to the maximum binding en- ergy. The Tempkin constants A_T and b_T were obtained from the intercept and slope of the linear plots of q_e vs ln C_e as shown in figure 10.

The constants and the correlation coefficient values (table 2) imply that

Tempkin (R^2= 0.9721) and Freundlich (R^2= 0.9765) isotherms are obeyed by the system less effec- tively when compared to Langmuir model (R^2= 0.9919).

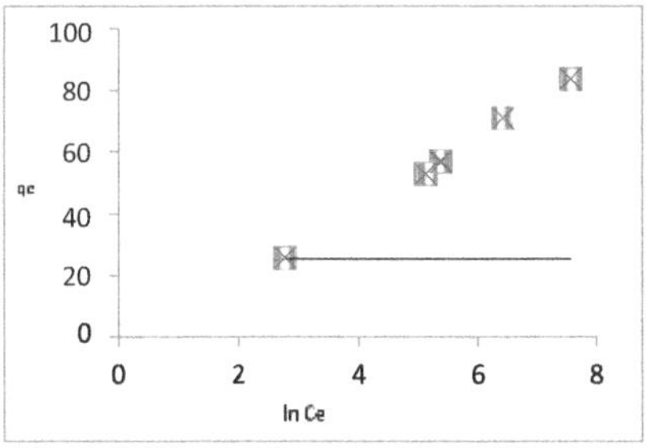

Fig 10: Tempkin Isotherm

3.8 Column Method

The results pertaining to the influential nature of variable parameters, on the basisof batch mode assessment insist the feasibility and compatibility of TMSP in trapping Pb(II) ions. Based on these observations,column experiments were performed to quanti-fy the sorbent characteristics through the continuous column run. The inlet solutions for the column being aqueous Pb(II) ions under optimized conditions of 0.18 mm particle size of TMSP at a Pb(II) concentration of 1000 mg/L followed by extension to electro- plating effluent wastewaters containing excess Pb(II) concentration.

3.8.1 Column Packing

The column is made up of cylindrical glass tube, the inner diameter being 2.5 cm and the height is 30 cm.

It was packed with TMSP between two supporting layers of glass wool, spread with the glass beads at the top of the already packed glass wool layer placed at the bot- tom. The step by step packing is as: Glass wool layer(3 cm thickness), glass beads (2cmthickness), TMSP (50 grams: 6 cm height), glass wool (1 cm thickness). These materi- als were loaded from the top of the column and allowed to settle by gravity force.

500 ml of 1000 mg/L of Pb(II) solution was poured into the column from the top and the flow rate was adjusted as 100 ml collection of the outlet sample at a rate of 20 seconds time interval. At the end of passing 1 litre of the solution, a maximum efficien- cy of 100% of Pb(II) removal was registered. The percentage removal was observed to decline upto 90% after passing 15 litre inlet Pb(II) solution. Further decline was envis- aged upto

20 litres of inlet solution and then the column went exhausted. The packed column with the experimentalsolutions are represented in fig 11.

Fig 11: Column Experiment

3.8.2 Desorption/Regeneration Studies

After the elution of Pb(II) ions, at quantities of 20 liters approximately, the col- umns packed with TMSP were washed with double distilled water followed by 0.01 M H_2SO_4 runoff, later with DD water, the collection rate being fixed as 100 ml/15mins. This desorption was carried out to assess the regeneration capacities [15,16] wherein theefficiency of TMSP was assessed as 90% recovery back against the exhausted columns.

4. Conclusion

Mussel Shell Powder (MSP) an eco-friendly and cost effective material was em- ployed for the removal of Pb(II) ions from aqueous solutions. The unloaded and metal loaded materials were subjected to SEM, EDAX and BET studies in order to determine the surface morphological differences, elemental constitutions and surface area which supported the characteristic changes that had occurred on the loaded Mussel Shell Pow- der. Parametric conditions were optimized in tune of maximum percentage removal (98.1%) for Pb(II)- TMSP system by Batch mode: 0.18mm particle size, 1000 mg dos- age, 30 minutes agitation time, 1000 mg/L initial concentration of Pb(II) ions, pH 5.5 ofthe solution medium at room temperature. The calculated correlation coefficient values from three isothermal applications viz., Langmuir, Freundlich and Tempkin indicated that Langmuir isotherm had best fit than the Freundlich and Tempkin models. Column studies were experimentally verified to quantify the sorbent characteristics and to assessTMSP efficiency in chelating Pb(II) ions from bulk of the solution. The exhausted TMSP material of the column was desorbed using 0.01M H_2SO_4 medium. The effective application of TMSP as an adsorbent to treat aqueous solution is the outcome of the pre-sent study.

Financial Support

The authors submit their due acknowledgment to Defence Research & Develop- ment Organisation (DRDO), New Delhi for the timely Financial Support in execution ofthe research work

References

1. F.D. Olivera, A.C. Soares, O.M. Freitas and S.A. Figueiredo, Global

NESTJournal, 12(2) (2010) 206-214.

2. T.O. Jimoh, Y.A. Iyaka, M.M. Nubaye, American Journal of Chemistry, 2(3), (2012)165- 170.
3. Upendra Kumar, Scientific Research and Essay 12 (2006) 033-037.
4. Sandhya Babel, Tonni Agustiono Kurniawan, J. Haz. Mat., B97 (2003) 219–243.
5. Madhavarao.M., Ramesh. A., Aparna Chandra Rao G., Seshaiah. K., J. Haz. Mat.,B129, (2006), 123–129
6. T Santhi, AL Prasad, S Manonmani, Arab.J. Che, 7 (4).(2014), 494-503
7. Mehmet EminArgun, SukruDursun, Celalettin Ozdemir, Mustafa Karatas, J. Haz.Mat., 141, (2007), 77-85.
8. Muhammad Riaz, Raziya Nadeem, Muhammad AsifHanif, Tariq Mehmood An-sari, Khalil-ur-Rehman,J. Haz. Mat., 161(2008), 88-94.
9. Nameni. M., AlaviMoghadam. M. R., Arami. M., Inter.J. Env. Sci.& Tech., 5(2008),161-168.
10. Nemr A.E., J. Haz. Mat.,161,(2009), 132- 141.
11. Ogali. R. E., Owhonda. N. K., Iyobu-Oki. S. T., Sci. Afric.,9 (1) (2010), 55-63
12. Raimundo N.P. Teixeira, Vicente O. Sousa Neto, Juliene T. Oliveira, Thalles

    C. Oliveira, Deigo Q. Melo, Marcos A.A., Silva and Ronaldo F. Nascimento,Bio. Res., 8(3) (2013), 3556-3573.
13. Sekhar. K. P.C., Vishnu Babu. R. V., Rohini. T., Ravindhranath. K., Inter. J.App. Bio. &Pharma.Tech.., 3(2), (2012), 115-125
14. Sekhar. K. P.C., Vishnu Babu. R. V., Srividhya. D., Ravindhranath. K., DerPharma Chem., 4 (2),(2012), 664-673.

15. Singh. S.R., Singh. A.P., Inter. J. Env.Res., 6(4), (2012), 917-924.

16. Stella Triyantafylou, Eirini Christoudolou, Paraskevi Neou-Singouna, Cla.&Cla.Min., 47, (1999), 567-572.

IJEP 38(10): 795-805 (2018)

Facile Synthesis, Characterization and Application of Magnetic Nano Compositesin Sequestration of Textile Dye

N. Muthulakshmi Andal* & K.S.Thangamani[1]

* *Assistant Professor, Department of Chemistry, PSGR Krishnammal College forWomen, Department of Chemistry, Peelamedu, Coimbatore – 641 004*

**Corresponding author, Email : muthulakshmiandal@psgrkcw.ac.in*

[1] *Assistant Professor, Department of Chemistry, Dr.N.G.P . Institute of Technology, Coimbatore – 641 048*

Prosopis juliflora bark activated carbon- cobalt ferrite magnetic composite (PJBAC- CFC) was synthesized by the auto-combustion method and employed for the removalof direct brown 2 (DB2) from aqueous solutions. The prepared composite was charac- terized by powder XRD, SEM, EDAX, AFM, N_2 adsorption- desorption isotherm and magnetization measurement. Operating factors influencing the rate adsorption, namely initial dye concentrations, doses of sorbent materials, preset time intervals, variable pH and temperature environments are substantiated by batch equilibration method. Adsorp-tion is observed to be more pronounced at pH 2. The magnetized composite material exhibited enhanced sorption characteristics. The isothermal adsorption data fit well with Freundlich isotherms. Kinetic studies favour the system to follow the second or- der. The outcome of the present work indicates that *Prosopis juliflora* bark activated carbon- cobalt ferrite magnetic composite can be effectively used as a cost-effective po-tential adsorbent for trapping the dye molecules.

KEYWORDS

Prosopis juliflora bark activated carbon, Cobalt ferrite, Magnetic nanocomposite, Ad- sorption, Direct brown 2, Dye removal

INTRODUCTION

Dyes are coloured organic compounds applied onto various substrates, like pa- per, leather, fur, hair, drugs, food, cosmetics, plastics and textile materials to impart colour (Samanta and Agarwal, 2009). The untreated effluents containing dyes released from various industries cause a drastic reduction in the quality of water. These dye molecules are reported to be mutagenic, teratogenic or carcinogenic and can cause se- vere damage to human beings by affecting the kidneys, reproductive system, liver,brain and central nervous system (Anne et al., 1986; Atef Alzaydien et al., 2009). Also, haemorrhage, ulceration of the skin and the mucous membrane and irritation of the res- piratory tract have been traced out (Jain et al., 1977). Therefore, it is necessary to give treatment prior to their discharge into the receiving water. Several treatment methods have been proposed. Among all, adsorption of dye molecules onto activated carbon is found to be very effective, low-cost method of colour removal (Bhatnagar and Mino- cha, 2006).

For the past three decades, activated carbon prepared from natural wastes play a vital role in water pollution control due to their high surface area, porous structure, easy operationand economic viability. But, it would be really hard to sort them from the water after satura- tion and thence have been discarded into the environment resulting in secondary pollution (Zhang et al.,2007). Under such conditions, the application of magnetic composites, being one of the most promising techniques for wastewater treatment has been focused in this pa- per to treat effluents, over a short period without producing any secondary pollutant (Luo and Zhang, 2009).

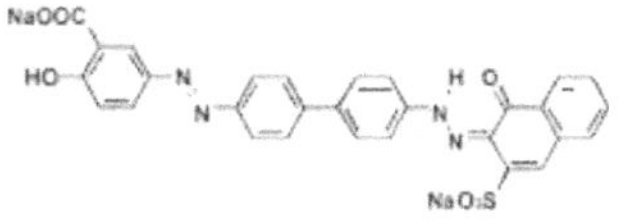

Figure 1. Structure of direct brown 2

In this method, the harmful ingredients together with the magnetic particles can be eliminated from the polluted system by a simple magnetic field. Following that, the harmful components can easily be removed from the magnetic particles and the recovered magnetic particles can be reused (Jing Hu et al., 2009). Adsorption process combined with magnetic separation has been used extensively in water treatment and environmental applications.

Nanotechnology is currently used in a powerful platform to enhance the envi- ronmental quality and sustainability through pollution prevention, treatment and re- mediation. Thereby, magnetic nanocomposites have contributed to the sequestration of dyes due to their astounding significance in accelerating separation speed and im- provising the efficiency of water treatment (Qadri et al., 2009). The viability of mag-netic Fe_3O_4 / activated carbon nanocomposite particles to trap methylene blue from aqueous solutions has been reported (Yang et al., 2008). Exploration of magnetic multi-wall carbon nanotubes for the removal of cationic dyes has been found to be efficient (Gong et al., 2009). Rocher et al., (2009) have prepared magnetic alginate beads containing magnetic nanoparticles and activated carbon for the removal of or- ganic dyes.

Wu et al., (2004) reported that $MnFe_2O_4$ and $CuFe_2O_4$ show the outstanding sorption properties with the effectual recovery of the magnetic separation technique and it has been utilized for the removal of contaminants in wastewater. Among fer- rites, $CoFe_2O_4$, a well known soft magnetic material with the cubic spinel structure, has a relatively high surface area, saturation magnetization, excellent chemical sta- bility and mechanical hardness (Liu et al., 2007). Therefore, the activated carbon prepared from tree bark of *Prosopis juliflora* and magnetic $CoFe_2O_4$ composite, a promising novel adsorbent combining the advantages of desirable absorptivity and effective magnetic separability, could be utilized for the removal of direct brown 2 (DB2) from aqueous solutions.

EXPERIMENTAL SETUP

Material

All the reagents employed were of analytical grade and used without further purifica- tion. Cobalt nitrate [$Co(NO_3)_2$. $6H_2O$], ferric nitrate [$Fe(NO_3)_3$. H_2O], urea, phos- phoric acid (H_3PO_4), sodium hydroxide (NaOH) and hydrochloric acid (HCl). The dye chosen for the present study is direct brown 2 (Figure 1): Molecular formula – $C_{29}H_{19}N_5Na_2O_7S$, molecular weight: 627.54. λ_{max} : 420 nm was purchased from S.S Brand Colours, Tirupur, Tamil Nadu. Stock and standard solutions were prepared us-ing double distilled water.

Preparation of activated carbon from Prosopis juliflora bark (PJBAC)

Prosopis juliflora bark collected from various localities in Coimbatore, Tamil Nadu, was sun dried for 10 days, crushed and soaked in a boiling solution of 30% phos- phoric acid for 24 hr. Later, the charred material was air dried and carbonized in a muffle furnace (YORCO) at a temperature of 500°C for 1 hr. The carbonized materi- al was powdered and subjected to activation for 10 min at 800°C. Activated carbon was washed several times with distilled to attain neutral pH and later dried. *Prosopisjuliflora* bark activated carbon (PJBAC) was sieved to a fixed mesh size of 85 BSS using scientific test molecular sieve (manufactured by Jayant Scientific Instruments Co., Mumbai) and stored in an airtight container.

Synthesis of *Prosopis juliflora* bark activated carbon - $CoFe_2O_4$ composite material Appropriate doses of ferric nitrate and cobalt nitrate were dissolved in 50 ml of distilledwater followed by addition of 28g urea in solution. The entire mixture was subjected toauto-combustion, later the content was heated in a magnetic plate ensuring constant stir-ring. 11g of *Prosopis juliflora* bark activated carbon was added to the hot bath and theheating was continued.

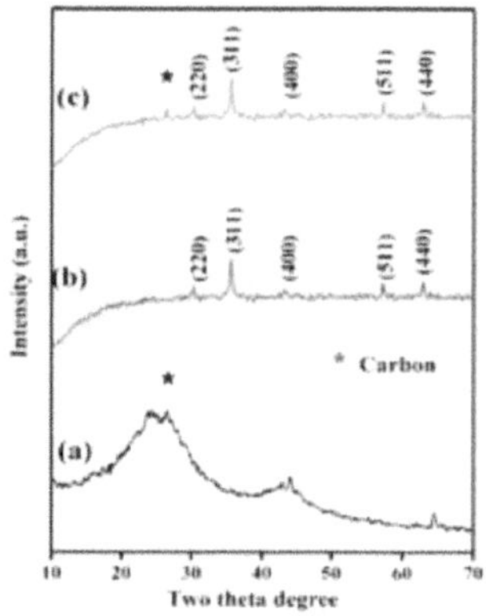

Figure 2. *XRD patterns of . (a) Prosopis juliflora bark activated carbon, (b) $CoFe_2O_4$ and (c) Prosopis juliflora bark activated carbon-cobalt ferrite magnetic composite*

The whole mixture was ignited resulting in the formation of dried gel in a burnt state through self -propagating combustion. The dry powder was ground well using mortar and pestle, stored and referred to as Prosopis juliflora bark activated carbon -cobalt ferrite magnetic composite (Thangamani et al., 2017)

Characterization studies

The surface morphology of the prepared composite was studied using scanning electron microscope (JEOL model JSM-6390 LV). The crystalline nature of *Prosopis juliflora* barkactivated carbon -cobalt ferrite magnetic composite was analysed with advance x- ray dif-fractometer (XPERT – PRO, Bruker AXS D8). The particle size of the composite was vis- ualized through atomic force microscope, NTMDT (NTEGRA PRIMA) model. BET sur- face area, pore diameter and pore volume were measured through the nitrogen adsorption at77K using a surface area/ pore analyzer (Coulter OMNISORP). The specific magnetic satu-ration was assessed using vibrating sample magnetometer (Lakeshore VSM 7410). SolutionpH values and dye concentrations for all the experiments were registered using pH meter(LABTRONICS) and UV- Visible spectrophotometer (Lab India UV 3000^{+}), respectively. **Batch adsorption experiment**

Batch equilibration studies of the DB2/ *Prosopis juliflora* bark activated carbon - cobalt fer-rite magnetic composite system were experimentally verified with the conditions: 50 ml volume of adsorbate solution at varying initial dye concentrations (25 to 150 mg/L), agitat- ed at preset time intervals (10 to 120 min : 10 min interval) *Prosopis juliflora* bark activat- ed carbon -cobalt ferrite magnetic composite doses (25 – 100 mg) under pH (2 -12). The adsorbent and adsorbate species were separated by centrifugation at 2000 rpm for 10 min, where initial and residual dye concentrations were determined spectrophotometrically at λ_{max} values of 420 nm. Duplicate experiments were performed to ensure the reproducibility values within + 0.1%. The amounts of dye adsorbed was calculated as:

$$qe=C_o- C_e/V*W \qquad ---- \qquad (1)$$

where C_o and C_e are initial and equilibrium concentrations (mg/L) of the dyes, respectively.V is the volume of the solution and W is the weight of the adsorbent used.

RESULT AND DISCUSSION

Characterization of *Prosopis juliflora* bark activated carbon -cobalt ferrite magneticcomposite

Figure 2 comprises of the X-ray diffraction patterns for *Prosopis Juliflora* bark activated carbon, $CoFe_2O_4$ and *Prosopis Juliflora* bark activated carbon -cobalt ferrite magnetic composite.

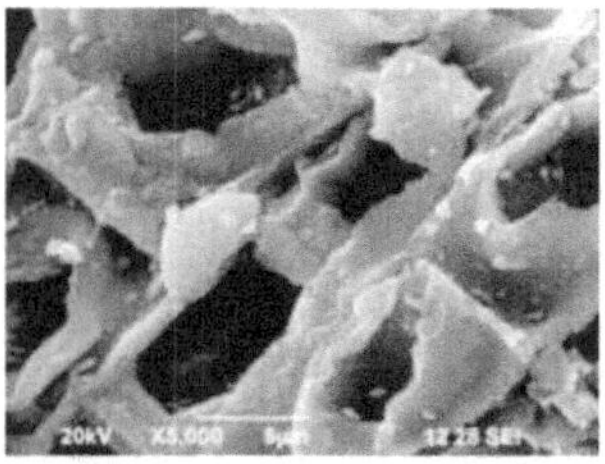

Figure 3a. *Scanning electron microscope image of Prosopis juliflora bark activated carbon*

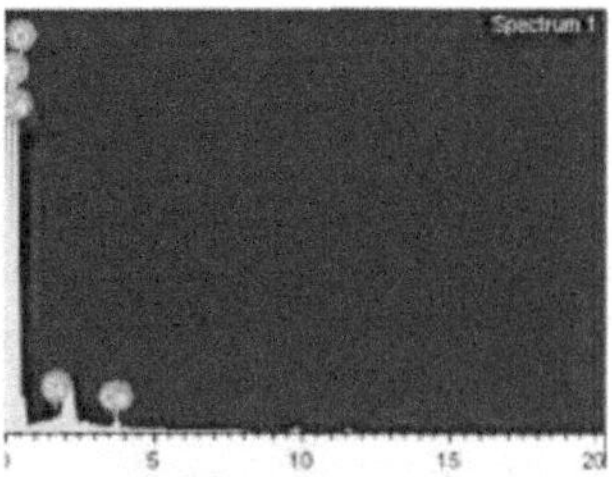

Figure 3c. *EDAX spectrum of Prosopis juliflora bark activated carbon*

Figure 3b. *Scanning electron microscope image of Prosopis juliflora bark activated carbon cobalt ferrite magnetic composite*

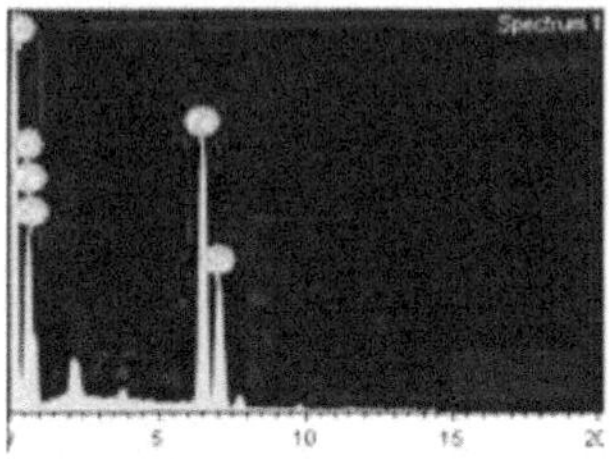

Figure 3d. *EDAX spectrum of Prosopis juliflora bark activated carbon cobalt ferrite magnetic composite*

The diffraction peaks indexed at (220) , (311), (400), (511), (440) in figure 2c perfectly matches the cubic spinel structure (JCPDS card no. 22-1086) of cobalt ferrite. The absence of characteristic impurities as obvious from figure 2c supports the formation of single- phase spinel ($CoFe_2O_4$). The weak peak appearing at $2\theta = 26.5^o$ in figure 2a corresponds to *Prosopis Juliflora* bark activated carbon, whose pattern is obviously different from that of figure 2b and 2c. The average crystalline size of *Prosopis juliflora* bark activated carbon - cobalt ferrite magnetic composite particles was calculated using Scherrer's equation:

$$D = 0.9\lambda / \beta \cos\theta \text{ .... } (2)$$

where D is an average crystalline diameter, k is a constant (0.9 for Cu-kα), λ is an x- ray wavelength (0.15405 nm for Cu-kα), β is the peak width of half maximum in XRD diffraction lines and θ is Bragg's diffraction angle in degree. The averagegrain size measured was 20 nm.

The surface morphological changes of *Prosopis juliflora* bark activated carbon and

Prosopis juliflora bark activated carbon -cobalt ferrite magnetic composite

Figure 4a. *Surface topography of Prosopis juliflora bark activated carbon cobalt ferrite magnetic composite*

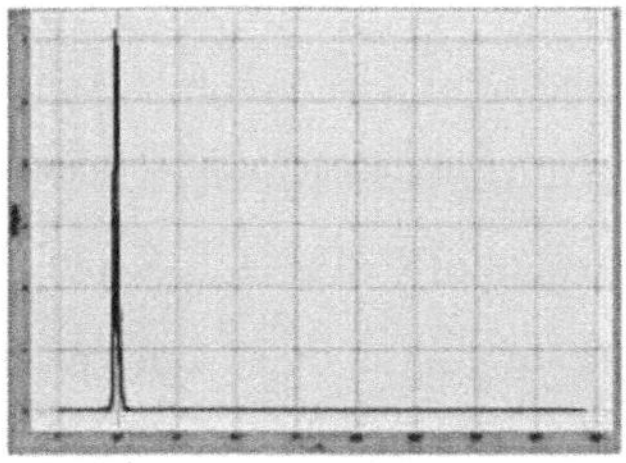

Figure 4b. *Histogram of Prosopis juliflora bark activated carbon cobalt ferrite magnetic composite*

were studied using SEM/EDAX and AFM techniques. Figure 3a reveals many poreswhich percolate the surface and inner parts of *Prosopis juliflora* bark activated car- bon. The pores may be formed by decomposition and carbonization of organics at high temperature (Shuangshuang et al., 2009). Figure 3b is indicative of the smooth surface of $CoFe_2O_4$ and even distribution of holes on the surface of activated carbon with a uniform diameter. This structural morphology of nano-sized $CoFe_2O_4$ parti- cles will enhance the adsorption capacity of the nanocomposite. EDAX spectra of *Prosopis juliflora* bark activated carbon and *Prosopis juliflora* bark activated carbon

-cobalt ferrite magnetic composite in figures 3c and 3d, show the presence of Co, Feand O in the synthesized composite.

Table 1. *Surface characterization*

Sample	BET-surface area (m²/g)	Total pore volume (cm³/g)	Saturation magneti-zation (emu/g)
PJBAC	185.2	0.132	0
PJBAC/CFC	108.4	0.035	32.1

Figures 41 and 4b register the 3D topography and histogram, registered with clarity and peak appearance at 2 nm confirming the prevalence of nano-sized composite material.

BET surface area, pore volumes and surface magnetization measurements of the prepared activated carbon and the derived composite listed in table 1, imply that sur-face area and total pore volume of the composite diminished obviously, compared tothat of the activated carbon. The reason can be due to the relatively small surface of $CoFe_2O_4$, partially covering the activated carbon surface. Similar results have been reported in the literature for the magnetic oxide supported activated carbon (Yang etatl., 2008; Castro et al., 2009).

Magnetic hysteresis loop of nano composite, *Prosopis juliflora* bark activated carbon -cobalt ferrite magnetic composite (Figure 5), demonstrates the values of saturation magnetization (Ms) and remnant magnetization (Mr) as 22.1 emu/g and
1.3 emu/g, respectively. It is observed that there is a decrease in Ms value of com- posite than bare $CoFe_2O_4$, indicating that the metal oxide is embedded with activat- ed carbon. Magnetic separability nature of the prepared composite was confirmed by placing a magnet near a glass bottle containing dispersed composite in water.The dispersed black particles were attracted towards the magnet and the solution turned crystal clear which was easily separated by filtration (Figure 5). This sup- ports the fact that *Prosopis juliflora* bark activated carbon -cobalt ferrite magnetic composite can be

effectively used as a magnetic adsorbent to uptake dyes from the aqueous solutions.

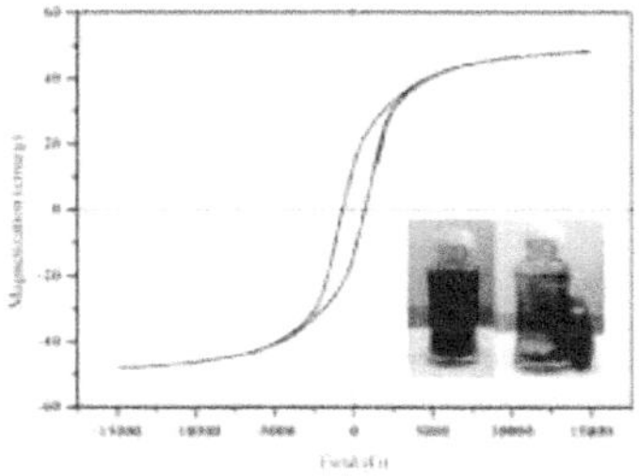

Figure 5. Magnetic hysteresis loops of Prosopis juliflora bark activated carbon-cobalt ferrite magnetic composite the inset of figure shows the well-dispersed solution by magnetic separation under an external magnetic field

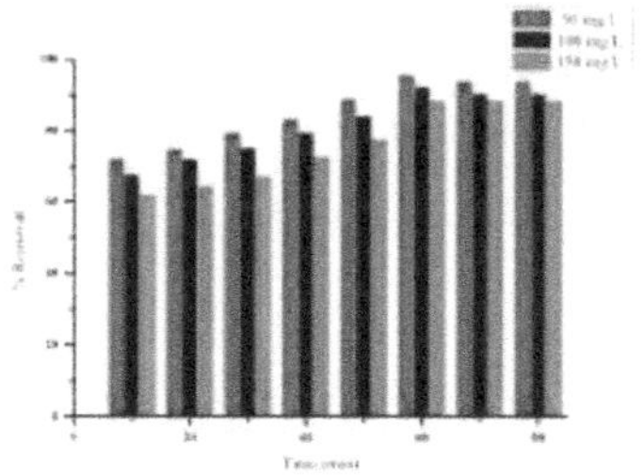

Figure 6. Effect of initial dye concentration and contact time

Adsorption studies

Effect of initial dye concentration and contact time: Varying ranging of initial dye con- centration (25 mg/L – 150 mg/L : 25 mg/L interval) registered a gradual decline in percent- age dye removal (95.6 to 88.4) with the actual amount of dye adsorbed rising from 31.8mg/g to 88.4 mg/g. This may be due to the saturation of activated sites on the sorbent surfaceat higher initial dye concentration.

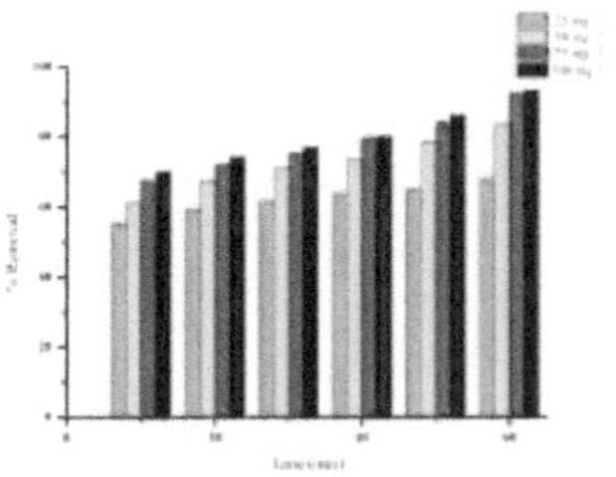

Figure 7. Effect of dosage

Employability of pine leaves (Yagub et al., 2012), saw dust (Kannan and Sundaram 2001) and rice husk (Ramaraju et al., 2013) for dye removal support our findings as shown in fig- ure 6.

As far as agitation time interval is concerned, more than 60% removal of the dye oc- curred in the first 10 min and thereafter adsorption attained a saturation rate slowly. The rap-id adsorption at the initial contact time is due to the availability of more positively charged surface which later declined in the uptake, was probably due to the electrostatic repulsion between the negatively charged species onto the adsorbent surface and the existing anionic sorbate species in solution (Heibati et al., 2009). The system reached equilibrium at nearly 50 min where the maximum uptake of dye molecules obviously had occurred.

Effect of sorbent dose: Figure 7 illustrates the influence of the different dosages (0.025 g-

0.01 g *Prosopis juliflora* bark activated carbon: 0.025 g interval) on the adsorption pattern ofdirect brown 2. Dye removal was observed to be directly proportional to doses initially dueto the availability of more active sites and later attained saturation beyond 0.01g.

Effect of pH: The adsorption capacity of an adsorbent is mainly influenced by the textural properties and its chemical nature of the surface (Hadi et al., 2015).

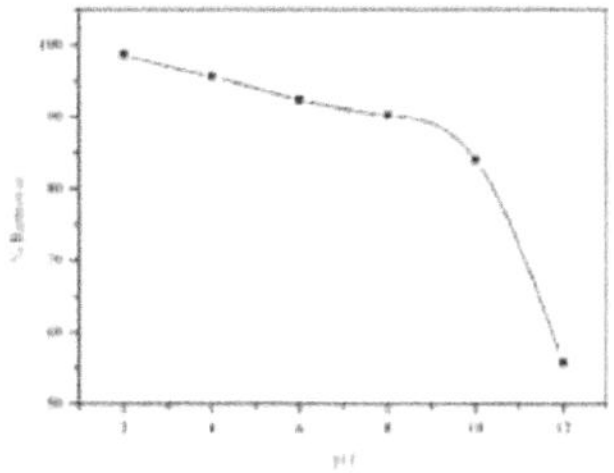

Figure 8. *Effect of pH*

One of the most important factors affecting the capacity of adsorbent is solution pH as it al- ters the chemical characteristics of the adsorbent surface. Furthermore, the pH value can af- fect the ionization degree of the adsorbate and metal ion speciation in solution (Xi and Guo 2013). The dye removal efficiency decreased from 98.8% to 55.8% with an increase in pH (2
– 12) and maximum direct brown 2 removal was observed at pH 2 (Figure 8). This may be due to the chemical and electrostatic interactions between the adsorbent surface and the dye molecules. At acidic pH, an enhanced number of positively charged sites favours the adsorp-tion of the anions due to electrostatic attraction. At higher pH, the increase in negatively charged binding sites leads to least sorption of anionic dye due to electrostatic repulsion of the negatively charged surface (Heibati etal., 2015).

Adsorption isotherms

Langmuir isotherm: In 1916, Irving Langmuir proposed an adsorption isotherm, which de- picted the relationship between number of active sites on the sorbent and pressure of the sys- tem. The Langmuir isotherm is based on the following assumptions (Langmuir, 1918):

1. Availability of fixed vacant sites
2. Uniformity in the size of the sites
3. The capability of the site to hold one sorbate molecule during the adsorption

process

4. Lack of interaction between sorbate molecules and neighbouring sites

5. Monolayer adsorption

The Langmuir equation maybe written as:

$$\frac{C_e}{q_e} = \frac{1}{K_L Q_m} + \frac{C_e}{q_m} \quad ......(3)$$

where q_e corresponds equilibrium amount of solute adsorbed per unit weight of adsorbent (mg/g); Ce is the equilibrium concentration (mg/L); q_m is the maximum amount adsorbed per unit mass of adsorbent (mg/g); b is a constant related to the affinity of the binding sites and the energy of adsorption (L/mg). The values of q_m and b for the system were calculated from the slope and intercept of plot Ce/q_e vs Ce (Figure 9a), respectively. The greater value of b confirmed the affinity of *Prosopis juliflora* bark activated carbon -cobalt ferrite magnet-ic composite for direct brown 2. The separation factor (R_L) was calculated to confirm the fa- vourability of the

$$R_L = \frac{1}{(1+bC_i)} \quad ...(4)$$

adsorption process using the following equation:

Where, C is the initial dye concentration (mg/L). The separation parameter R_L indicates theshape of the isotherm and the nature of the sorption process. If R_L>1 unfavourable isotherm,R_L=1 linear isotherm, R_L=0 irreversible isotherm, 0< R_L< 1 favourable isotherm. The $R_L$values obtained were lying between 0 and 1 (0.11, 0.63, 0.43 for 50, 100 and 150 mg/L, re-spectively) indicative of favourable adsorption process (Shahwan and Erten, 2004) **Freundlich isotherm**: Freundlich isotherm is based on the assumption that adsorption ofsorbate occurs on a sorbent surface with multilayer adsorption and the amount of sorbate ad-sorbed increases infinitely with the rise in concentration. A linear form Freundlich equationis written as:

$$\log q_e = \log K_F + 1/n \log C_e \quad \ldots(5)$$

where, q_e and Ce indicate the amount of solute adsorbed per unit weight of adsorbent (mg/g) and equilibrium concentration of the solute in the bulk solution (mg/L), K_F is Freundlich constant, which implies the relative adsorption capacity of the adsorbent [mg/g $(mg/L)^{-1/n}$] and 1/n is an adsorption intensity. The values of 1/n lying between 0-1 support favourable nature of adsorption process (Table 2). The Freundlich constants K_F and n were determined from the intercept and slope values of the plot log q_e versus log Ce, respectively (Figure 9 b).

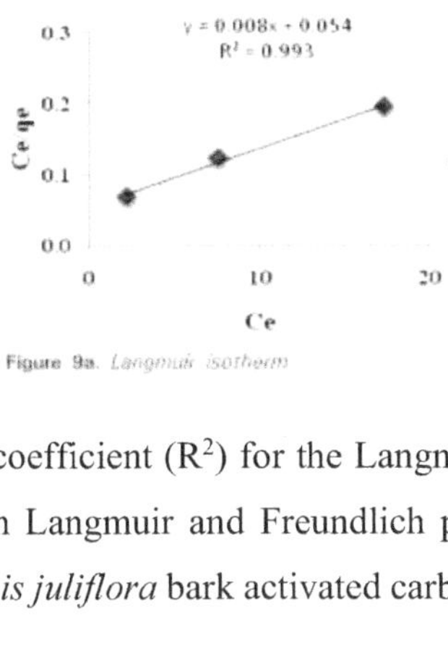

Figure 9a. *Langmuir isotherm*

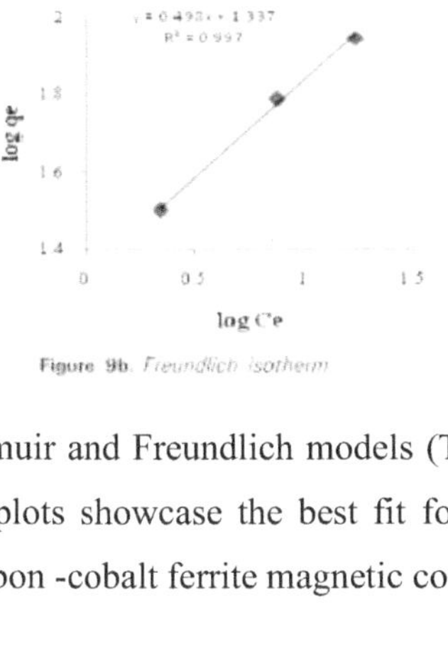

Figure 9b. *Freundlich isotherm*

The correlation coefficient (R^2) for the Langmuir and Freundlich models (Table 3) implies that both Langmuir and Freundlich plots showcase the best fit for direct brown 2- *Prosopis juliflora* bark activated carbon -cobalt ferrite magnetic composite system.

Adsorption kinetics

Adsorption kinetics studies describe the uptake rate of ions of interest by the adsorbent, and the reaction pathways of adsorbent and adsorbate interaction until it attains equilibrium. Pseudo first order and pseudo second order kinetic models were applied to the system. Pseu-do first order kinetic model rate equation is as follows:

$$\ln(q_e - q_t) = \ln q_e - k_1 t \quad \ldots(6)$$

Where q_e (mg/g) – the amount of dye adsorbed at equilibrium, q_t (mg/g)- the amount of dye adsorbed at time t and k_1 (min^{-1}) – pseudo first order rate constant. A straight line plot of ln (q_e - q_t) vs t (Figure 10a) was obtained, where q_e and k_1 were calculated from the intercept and slope of the plot, respectively.

Table 2. *Isotherm parameter*

Dye	Langmuir			Freundlich			
	q_m (mg/g)	b (L/mg)	R^2	R_L	1/n	K_F [(mg/g)(mg/L)$^{1/n}$]	R^2
DB2	125.0	0.1481	0.993	0.0632	0.498	101.8	0.997

Table 3. *Kinetic parameter*

Dye	Conc., mg/L	q_e exp (mg/g)	First order kinetics			Second order kinetics		
			k_1(min^{-1})	q_e cal (mg/g)	R^2	k_2x10^{-2} (g/mg/min)	q_e cal (mg/g)	R^2
DB2	50	31.9	0.066	14.0	0.904	0.004	34.5	0.986
	100	61.6	0.034	22.1	0.952	0.002	66.7	0.983
	150	88.5	0.023	35.2	0.963	0.001	99.2	0.966

The equation for Pseudo second order kinetic model is given below:

$$\frac{t}{q_t} = \frac{1}{K_2 q_e^2} + \frac{t}{q_e} \qquad \text{...(7)}$$

K2 (g/mg/min) second order rate constant. Plots of (t/qt vs t) (Figure 10b) were found to be linear. From the slope and intercept values, q_e and K_2 were calculated, respectively. Table 3 lists the constant values calculated from pseudo first order and pseudo second order kinetic models. The correlation coefficients R^2 for the pseudo first order kinetic model obtained is lower than that of the pseudo second order kinetic model.

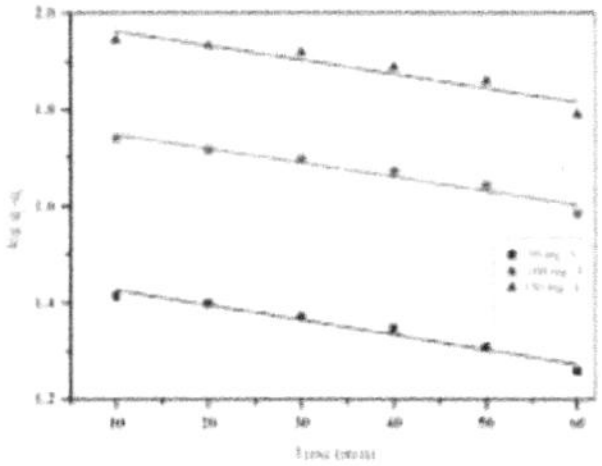

Figure 10a. *Pseudo first order*

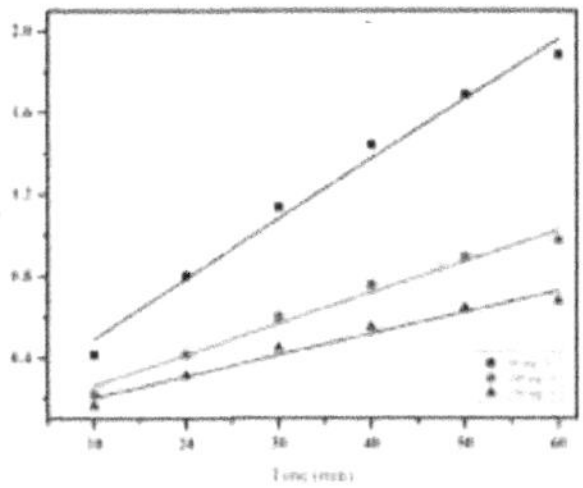

Figure 10b. *Pseudo second order*

The calculated and experimental values of qe for pseudo first order significantly varied from each other, but larger deviations were found insignificant in the case of pseudo second order kinetic model (Sivrikaya et al., 2013). The above statements favour the better description forthe adsorption system to follow pseudo second order model.

CONCLUSION

In this study, *Prosopis juliflora* bark activated carbon -cobalt ferrite magnetic composite (PJBAC - CFC) was productively synthesized by a facile one – step up combustion method. XRD, SEM, AFM, BET surface area and magnetic measurements were carried out to char- acterize the prepared composite. The brown 2 onto *Prosopis juliflora* bark activated carbon -cobalt ferrite magnetic composite were investigated based on the optimized variable parame-ters. The equilibrium of concentration of dark brown 2 onto *Prosopis juliflora* bark activatedcarbon -cobalt ferrite magnetic composite was suitably described by both Langmuir and Freundlich models with maximum adsorption capacity of 125 mg/g. The process of adsorp-

tion was relatively rapid and was best described by the pseudo second order kinetic model. The magnetic properties of prepared adsorbent allowed their separation from water in pres- ence of magnetic field, indicating *Prosopis juliflora* bark activated carbon -cobalt ferrite magnetic composite could be a promising and effective adsorbent for the removal of dyes from water.

ACKNOWLEDGEMENT

The authors would like to thank the Department of Civil Engineering, Dr. N.G.P. Institute ofTechnology, Coimbatore for rendering the laboratory facilities.

REFERENCES

Anne, J., *et al.*, 1986. Rabbit lung after inhalation of hexa and trivalent chromium. *Env. Res.*, 41: 110-119.

Atef Alzaydien, S. 2009. Adsorption of methylene blue from aqueous solution onto a low- cost natural Jordanian Tripoli. *Am. J. Env. Sci.*, 5(3): 197-208.

Bhatnagar, A. and A.K. Minocha. 2006. Conventional non-conventional adsorbents for removal of pollutants from water-A review. *Indian J. Chem. Tech.*, 13: 203-217.

Castro, C.S. 2009. Activated carbon/iron oxide composites for the removal of atrazine from aqueous medium. *J. Hazard. Mater.*, 164: 609-614.

Gong, J.L., *et al.* 2009. Removal of cationic dyes from aqueous solution using magnetic multi-wall carbon nanotube nanocomposite as adsorbent. *J. Hazard. Mater.*, 164 : 1517-1522.

Hadi, Pejman, *et al.* 2015. A critical review on preparation, characterization and uitlization of sludge-derived activated carbons for wastewater treatment. *Chem. Eng. J.*, 260 : 895-906.

Heibati, Behzad, *et al.* 2009. Removal of direct N blue-106 from artificial textile dye effluent using activated carbon from orange peel. Adsorption isotherm and kinetic studies. *J. Hazard. Mater.*, 165:100-110.

Heibati, Behzad, *et al.* 2015. Kinetics and thermodynamics of enhanced adsorption of the dye AR 18 using activated carbons prepared from walnut and poplar woods. *J.Mol.Liq.*, 208:99-105.

Jain, K.K., *et al.* 1977. Chromium VI removal from aqueous solution by wollastonite. *J.Inst.Public. Health Eng.*, 4:101-109.

Jing Hu, Irene, M.C. Lo and Guohua Chen. 2007. Comparative study of various mag-

Liu, X.M., and S.Y. Fu and L.P. Zhu. 2007. High-yield synthesis and characterization of mono disperse sub microsized $CoFe_2O_4$ octahedra. *J. Solid State Chem.*, 180 : 461-466.

Luo, X. and L. Zhang. 2009. High effective adsorption of organic dyes on magnetic cellulose beads entrapping activated carbon. *J. Hazard. Mater.*, 171: 340-347.

Qadri, S., A. Ganoe and Y. Haik. 2009. Removal and recovery of acridine orange from solutions by use of magnetic nanoparticles. *J. Hazard. Mater.*, 169:318-323.

Ramaraju, B., P. Manoj Kumar Reddy, C. Subrahmanyam. 2013. Low cost adsorbents from agricultural waste for removal of dyes. *Env. Prog. Sust. Energ.*, 13: 1-9.

Rocher, V., *et al.* 2008. Removal of organic dyes by magnetic alginate beads. *Water Res.*, 42: 1290-1298.

Samanta, A.K. and P. Agarwal, 2009. Application of natural dyes on textiles. *Indian J.Fibre Text. Res.*, 34:384-399.

Shahwan, T. and H.N. Erten. 2004. Temperature effects in barium sorption on natural kaolinite and chlorite-illite clays. *J. Radio Anal. Nucl. Chem.*, 260 : 43-48.

Shuangshuang. Lv, *et al.* 2009. Rice hull/ $MnFe_2O_4$ composite: Preparation, characterization and its rapid microwave - assisted COD removal for organic wastewater. *J. Hazard. Mater.*, 171: 634-639.

Sivrikaya, S., *et al.* 2013. 1,3,5 triazine triethylene tetramine polymer: An efficient adsorbent for removal of Cr(VI) ions from aqueous solution. *Toxicol. Env. Chem.*, 95: 899-908.

Thangamani, K.S., *et al.* 2017. Utilization of magnetic nano cobalt ferrite doped Capra aegagrus hircus dung activated carbon composite for the adsorption of anionic dyes.

J. Env. Chem. Eng., 5: 2820-2829.

Wu, R. and J. Qu. 2005. Removal of water-soluble azo dye by the magnetic material $MnFe_2O_4$. *J. Chem. Tech. Biotech.*, 80 : 20-27.

Wu, R., *et al.* 2004. Removal of azo-dye acid red B (ARB) by adsorption and catalytic combustion using magnetic $CuFe_2O_4$ powder. *Appl. Catal.*, B 48: 49-56.

Xi, X. and X. Guo. 2013. Preparation of bio-charcoal from sewage sludge and its performance on removal of Cr(VI) from aqueous solutions. *J. Mol. Liq.*, 183: 26-30.

Yagub, M.T., *et al.* 2012. Equilibrium kinetics and thermodynamics of methylene blue adsorption by pine tree leaves. *Water Air Soil Poll.*, 223(8): 5267-5782.

Yang, N., *et al.* 2008. Synthesis and properties of magnetic Fe_3O_4- activated carbon nanocomposite particles for dye removal. *Mater. Lett.*, 62: 645-647.

Zhang, G., *et al.* 2007. $CuFe_2O_4$/activated carbon composite: A novel magnetic adsorbent for the removal of acid orange II and catalytic regeneration. *Chemosphere.* 68 : 1058-1066.

AUTHORS

1. K.S. Thangaman, Assistant Professor, Department of Chemistry, PSGR Krishnammal College for Women, Coimbatore - 641 004.

2*. N. Muthulakshmi Andal, Assistant Professor, Department of Chemistry, Dr. N.G.P. Institute of Technology, Coimbatore-641 048.

3. E. Ranjith Kumar, Assistant Professor, Department of Physics, Dr. N.G.P. Institute of Technology, Coimbatore-641 048.

4. V. Suba, Department of Chemistry, Dr. N.G.P. Institute of Technology, Coimbatore-641 048.

EFFICACY OF MODIFIED TREE BARK IN THE SCREENING OF DIVALENT IONS IN AQUEOUS MEDIA: CHARACTERIZATION AND ISOTHERMAL STUDIES

N. Muthulakshmi Andal[1],Gayathri.N.S[2], J. Anuradha[3]

[1]Assistant Professor, [2,3]Research Scholar Department of Chemistry, PSGR Krishnammal College for Women, Coimbatore-4.

Abstract

The present work aims at examining the potentiality of the identified no-cost sorbent *Prosopis juliflora* barks to remove Pb(II) and Cd(II) ions from aqueous solutions. *Prosopis juliflora* barks are collected from Coimbatore, washed with doubly distilled water, dried, pulverized and treated using 0.1NHCl. The characteristics and functional groups present in the treated barks are analyzed by SEM,EDAX and FTIR techniques. Variable parameters in- fluencing the adsorption of divalent ions onto the treated barks are verified by Batch Equilibration method. Effective experimental results are derived under the following optimized conditions: 15 min agitation time, pH 5 envi- ronment for the trapping of Pb(II) ions (93% removal) and 20 min agitation time, pH 7 environment for trapping Cd(II) ions (81% removal) at a parti- cle size of 0.18mm, 300 mg dosage and an initial metal concentration of 100 mg/ L. Isothermal analysis of Langmuir and Freundlich models are studied where the best linear fit is exhibited by Langmuir model for both the sys- tems. The obtained results indicate that TPJB is a promising material in chelating Pb(II) in comparison to Cd(II) ions. This shall be attributed to the fact of enlarged ionic radii of Pb(II) ions than Cd(II) ions favoring better sorption in case of the former. Extension of the study is driven towards the applicability of TPJB to sequestrate the heavy metals from industrial out-lets.

Keywords: *Prosopis juliflora* bark, characterization, divalent ions, biosorp-

tion

Introduction

Pollution of the environment with toxic substances in wastewater effluents is a major concern for public health and environmental quality, with heavy metals being the most dangerous entity[1]. Lead and Cadmium are introduced into water bodies from smelting, metal plating, batteries, mining, pigments, stabilizers, al- loy industries and sewage sludge. Lead, one of the three most toxic heavy met- als, has long-term potential negative impacts like anemia, encephalopathy, hepatitis and nephritic syndrome[2]. The harmful effects of cadmium include number of acute and chronic disorders such as "itai–itai" disease, renal damage, emphysema, hypertension and testicular atrophy[3]. The removal of these toxic heavy metals from wastewaters is of great interest and important in the field of water pollution control, since metals are persistent and have potential to be bio- accumulated by a range of organisms[4]. Conventional treatment methods have been found to be difficult owing to high capital/ operational costs, resulting in the generation of chemical sludge waste that must be treated before disposal as it poses hazards and pollution risks to the atmosphere[5]. Adsorption process in- volves use of low-cost, naturally occurring and readily available agricultural/ animal waste materials as adsorbents[6]. In this current study, *Prosopis juliflora* bark an agro-waste has been investigated for the remediation of industrial leachates containing Pb(II) and Cd(II) ion through adsorption process.

Materials and Methods

Prosopis juliflora bark (PJB) was collected from various localities in Coimbatore, washed thoroughly with distilled water to remove impurities, air-dried and sieved into different particle sizes viz., 22 BSS (0.71 mm), 36 BSS (0.42 mm), 52 BSS (0.30 mm), 72 BSS (0.21 mm) and 85 BSS (0.18 mm) using standard Scientific Test Molecular Sieves. In order to enhance the ac-

tive sites and metal sorption capacity, the sieved materials were treated with N HCl for 3 hours, washed several times toattain pH 7 with doubly distilled water, dried, stored in airtight containers for further experiments and re- ferred to as TPJB. Stocksolutions of Lead and Cadmium (1000 mg/ L) were prepared by dissolving Lead nitrate and Cadmium acetate salt, in 1000 ml of doubly distilled water. Working standards were preparedby progressive dilution of stock copper and cadmium solutions. The influence of various parameters like particle size (0.18mm-0.71mm), sorbent dosage (100mg- 500mg), initial concentration (50, 100, 150 and 200 mg/ L), agitation time interval (5-30
minutes) and pH (3, 5, 7, 9 and 11) were assessed by Batch Equilibration experiments at room temperature to optimize the favorable conditions for maximum uptake of the selected sorbate species. pH measurements were per- formed with a controlled pH analyzer (LABTRONICS). The agitated sam- ples were filtered and the initial/residual concentrations of Pb(II) and Cd(II) were analyzed using Atomic Absorption Spectrophotometer (AA-6200, Shi- madzuMake).

Surface characteristic Studies

The treated and metal loaded TPJB were subjected to spectral analyses (within the rangeof 400-4000 cm-1) to determine the functional groups pre- sent in TPJB using *Shimadzu Infrared Spectrophotometer*. Morphology and elemental constitutions of the unloaded and loaded adsorbent material wereidentified using Scanning Electron Microscope (SEM) and Energy Disper- sive X-ray Analysis (EDAX)(*JEOL JFM-6390*).

RESULTS AND DISCUSSION

FTIR Spectral analysis

The possibilities of functional groups present in TPJB involving sorption of

Pb(II) and Cd(II) ions determined using FT-IR Spectroscopic studies are de- picted in fig 1a – 1d. Fig 1(b) shows the appearance of peaks at 3335.79 cm^{-1} and 1746.62 cm^{-1} corresponding to O-H and C=O. The disappearanceof O-H band at 3335.79 cm^{-1} in fig 1c and 1d indicate the probable formation of metal oxygen bond during sorption.

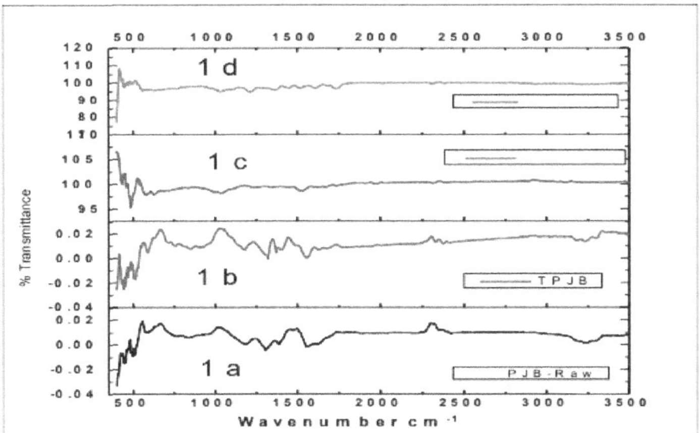

Fig 1a: Raw PJB, Fig 1b: unloaded PJB, Fig 1c: Pb(II) loaded TPJB and Fig 1d: Cd(II) loaded TPJB

Surface Morphology

SEM images of unloadedand metal laden TPJB have been depicted in figs2a, 2b& 2c. Fig 2a registers a high porousmorphology and a coarse surface tex- ture with different shapes and sizes. The morphologicalchanges as obvious from fig 2b & 2c suggest thateffective sorption of Pb(II) and Cd(II) hadoccurred on the sorbent surface.

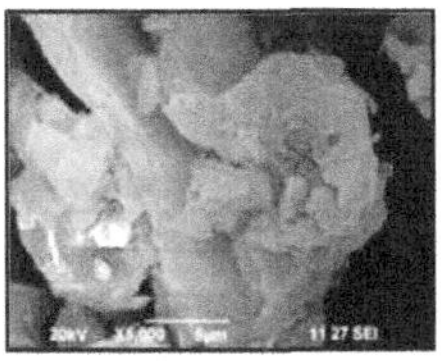

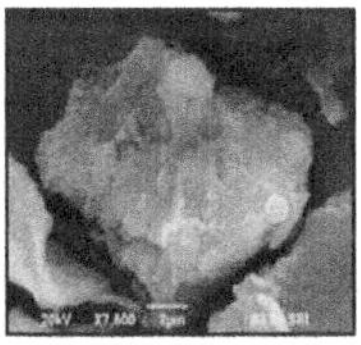

Fig 2a: Raw PJB **Fig 2b: unloaded TPJB** **Fig 2c: Pb(II)loaded TPJB**

Fig 2d: Cd(II)loaded TPJB

EDAX Analysis

The recorded EDAX spectra registers the elemental constitution qualitatively wherepercentage of O, C, Ca and Cl are 47.13, 52.2, 0.58 and 0.09 respectively in TPJB. These have been reported as the principle elements of theadsorbent. The appearance of new peak in figure 3c and 3d, confirm the presence of Pb(II) and Cd(II) in the loaded counterparts.

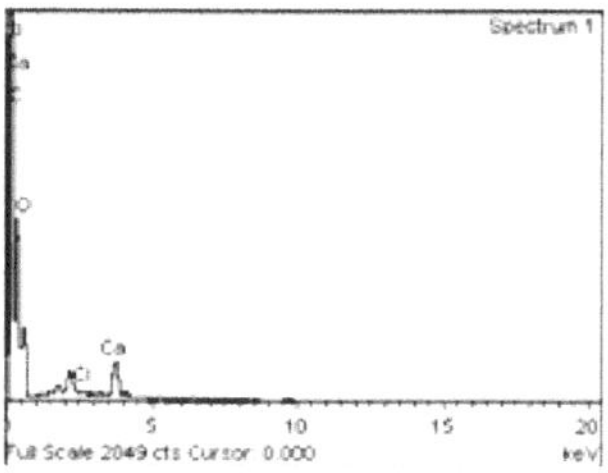

Fig 3a: Raw PJB

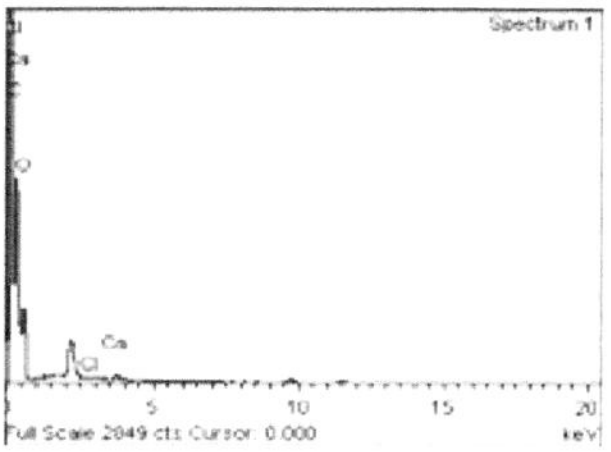

Fig 3b: unloaded TPJB

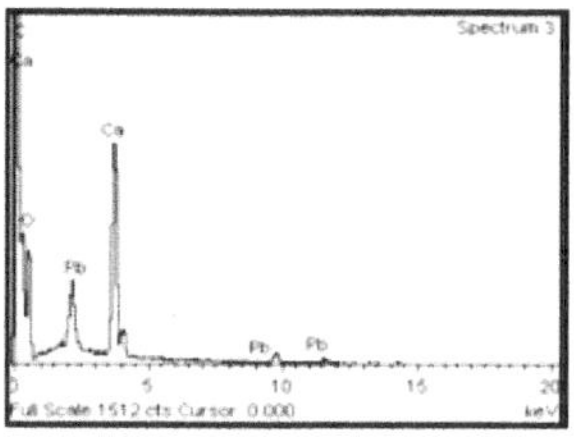

Fig 3c: Pb(II)loaded TPJB

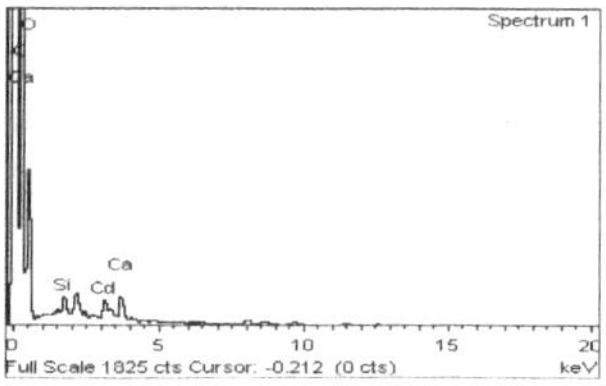

Fig 3d: Cd(II)loaded TPJB

Effects of Particle Size

Particle size plays a vital role in determination of metal sequestering capacity of TPJB from aqueous solutions. Both the systems registered a maximum sorption capacity (mg/g) against particle size

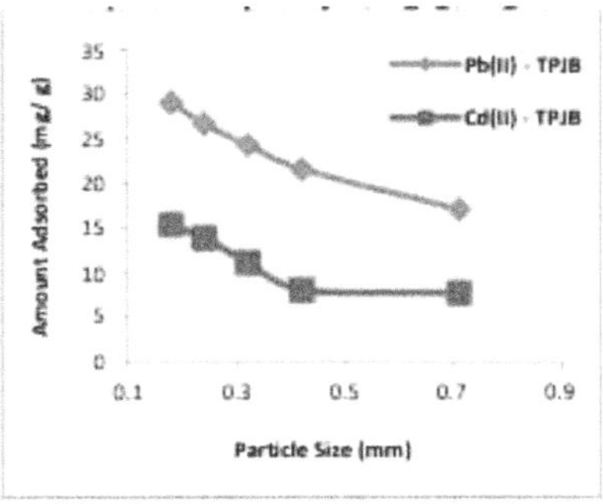

Fig 4: Effects of Particle Size

Effects of Dosage

Removal of Pb(II) and Cd(II) ions were found to be less at lower doses, but exponentially increased and reached an equilibrium point at 300 mg (figs 5). The declination in the metal uptake above equilibrium state is due to the partial aggregation among the active sites at higher doses and lack of binding site at lower doses[8] are responsible for the retardation of metal ions sorption onto the TPJB. smallest the particle size of 0.18mm amongst the chosen sizes (fig 4). This can be explained by thefact that relatively smaller particle size yields large surface area[7].

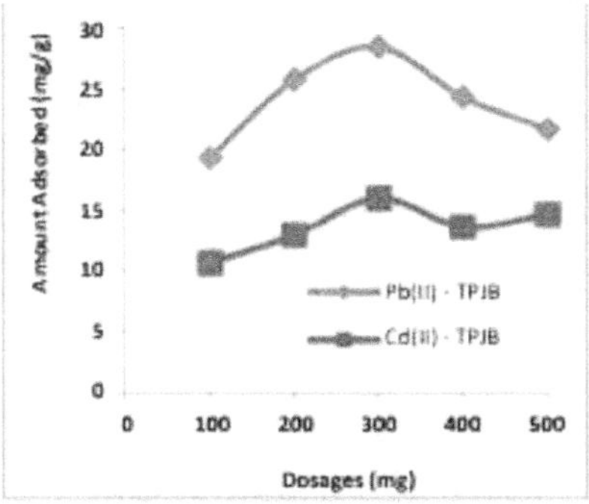

Fig 5: Effects of Dosage

Effects of Initial Concentration and Agitation Time

Pb(II) and Cd(II) ion removal as a function of initial metal ion concentration andagitation time are given in fig 6 and 7. An inclination in the adsorption curve from 50 - 100 mg/ L is obvious in fig 6 which may be due to abundant availability of active sites on thesorbent surface to trap these metal ions and the declining curves indicates the saturation of the sorption sites for complexation of metal ions at higher concentration[9]. The maximum removal observed for Pb(II) -TPJB and Cd(II) -TPJB are 93% (15 min) and 81% (20 min) respec- tively. The highest uptake value of Pb(II) corresponds to the lower hydrated ionic radius of Pb(II) ions (2.61 Å), compared with Cd(II) (2.75 Å), also higher charge density favours attractiveelectrostatic interactions between the TPJB andPb(II) ions[10].

Effects of pH

pH dependance for maximum removal of metal ions onto TPJB was studiedat varying pH

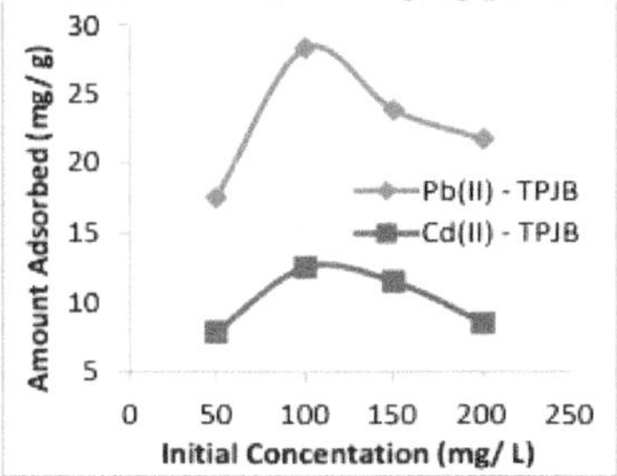

Fig 6: Effects of Initial Concentration

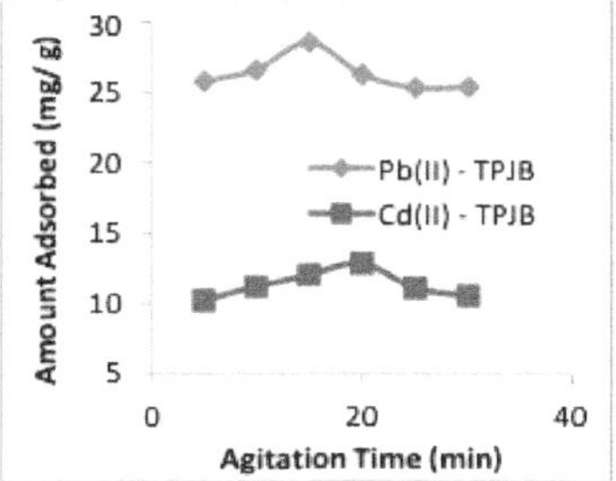

Fig 7: Effects of Agitation Time

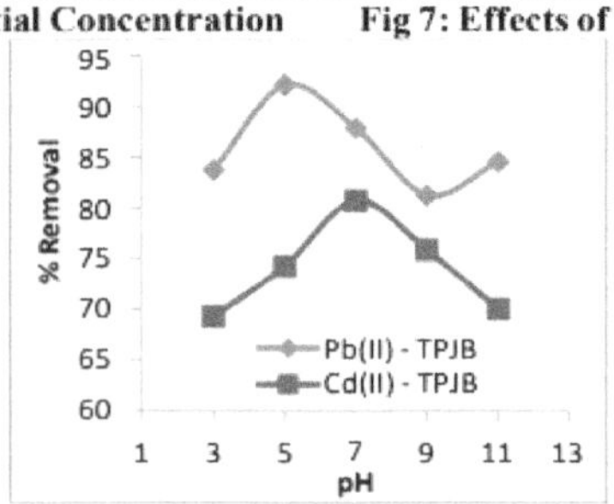

Fig 8: Effects of pH

environments viz., 3-11. It is evident from Fig. 8 that maxima adsorption had oc- curred at pH 5 for Pb(II) and pH 7 in case of Cd(II) ions as indicated by the inverted parabolic curves. Thereduced sorption rate at highly acidic and alka- line pH ranges may be attributed to the protonation of the sorption sites and preferentialhydroxyl complex formation of the metal ion[11].

Isotherm studies

The equilibrium data were modeled using Langmuir, Freundlich iso- therms and Table 1, lists out the isothermal constants with correlation coeffi- cients. The relation between sorbent and sorbate is provided by Langmuir con-stant *b*.[10] Higher *qm*, b and R^2 value were observed for Pb(II)- TPJB system than Cd(II)-TPJB system (fig 9).

Freundlich isotherm [fig 10] presents lower correlation coefficient (R^2) values when compared to Langmuir values. Kf is an indicator of the adsorp- tion capacity and 1/n is the adsorption intensity. The magnitude of the expo-nent 1/n, shows the favorability of adsorption.

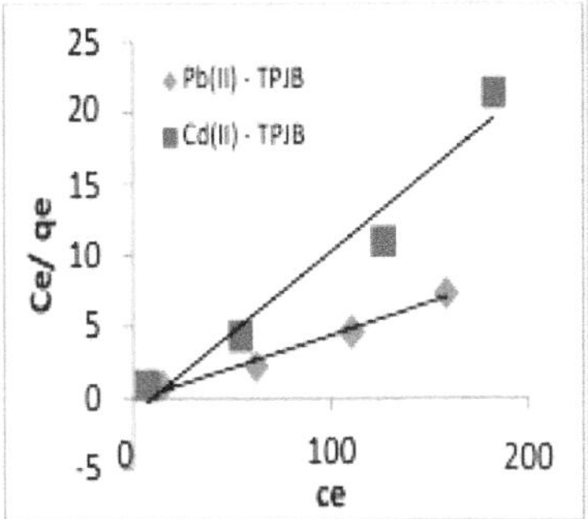

Fig 9: Langmuir Isotherm

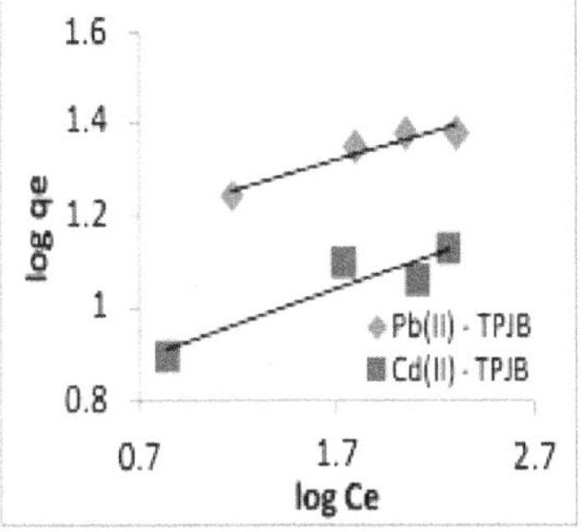

Fig 10: Freundlich Isotherm

Table 1: Isothermal Constants

System	Langmuir			Freundlich		
	Qm	b	R^2	n	Kf	R^2
Pb(II) - TPJB	27.4725	5.1621	0.9818	7.9428	12.8026	0.9178
Cd(II) - TPJB	11.0011	12.7304	0.9574	0.7833	6.0715	0.8596

Conclusion

Tree barks of *Prosopis juliflora* was an effective biosorbent for the adsorption of Pb(II) and Cd(II) ions from aqueous solutions. The FTIR spectra showed that O-H functional groups are responsible binding sites of the metal ions in both the systems. SEM and EDAX analyses of unloaded / loaded material, registered a significant changes in porosity structure andappearance of Pb(II) and Cd(II) peaks supporting the adsorption positively. Batch equilibration ex- periments revealed maximum percentage removal of 92% of Pb(II) – TPJB and 81% of Cd(II) – TPJB systems under the optimized conditions of 0.18 mm particle size, 300 mg dosage, 100 mg/ L initial concentration at 15 min(pH5) and 20 mins (pH 7) time intervals respectively. The isothermal data fa- voured the monolayer adsorption of Pb(II) and Cd(II) ions. The concluding remarks promises TPJB as potential adsorbent for possible remediation of wastewater contaminated with lead andcadmium ions.

References

1. Tatah VS, Otitoju O, Ezeonu CS, Onwurah INE, Ibrahim KLC, "Characterization and Adsorption Isotherm Studies of Cd (II) And Pb (II) Ions Bio-remediation from Aqueous Solution Using Unmodified Sorghum

Husk", J Appl Biotechnol Bioeng, vol. 2(3), March 2017, pp. 1-9, DOI: 10.15406/jabb.2017.02.00034

2. Mr. R. P. Suresh Jeyakumar and Dr. V. Chandrasekaran, "Desorption and Reutilization Studies of Copper and Lead Lons by Activated Carbons Prepared fromUlva Fasciata sp", Volume : 2, Issue : 6, June 2013, pp. 60-63, DOI : 10.15373/22778179
3. M. Madhava Rao, A. Ramesh, G. Purna Chandra Rao, K. Seshaiah, "Re-moval of copper and cadmium from the aqueous solutions by activatedcarbon derived fromCeiba pentandra hulls", Journal of Hazardous Mate-rials vol. B129 September,(2006) pp. 123–129, doi:10.1016/j.jhazmat.2005.08.018.
4. Lawala,O.S., Sannia,A.R., Ajayib,I.A.and Rabiua,O.O., "Equilibrium, thermodynamic and kinetic studies for the biosorption of aqueous lead(II)ions onto the seed husk of Calophyllum inophyllum'', J. Hazard. Mat., vol. 177, 2010, pp. 829–835.
5. N.S. Gayathri, N. Muthulakshmi Andal and J. Anuradha, "An Investigation Approach on the Sequestration of DivalentMetal Ions Employing Animal Waste", *Orient. J. Chem.,* Vol. **33**(3),March 2017, pp. 1406- 1413, doi.org/10.13005/ojc/330342
6. Oboh I, Aluyor E, Audu T (2009) "Biosorption of heavy metal ions from aqueous solutions using a biomaterial", Leonardo Journal of Sciences vol.(14) May2009, pp. 58-65.
7. Zainul Aknar Zakaria, Marlini Suratman, Nurfadilah Mohammed, Wan Azlina Ahmad, "Chromium(VI) removal from aqueous solution by un-treated rubber wood sawdust", *Desal.* Vol. 244, Aug 2009, pp. 109-121, doi: 10.1016/j.desal.2008.05.018
8. Pungam Bark and CAC, T. Shanthi, V. M. Selvarajan, "Removal of Cu (II) Ions from Aqueous Solution by Carbon Prepared from Henna Leaves",

J Chem. Pharm. Res., vol.4(9), 2012, pp. 4296-4306.

9. J. Anandkumar and B. Mandal, "Removal of Cr(VI) from aqueous solution using Bael fruit (Aegle marmelos correa) shell as an adsorbent", *J.Hazard. Mater.* vol. 168, Sep 2009,pp. 633-640, 10.1016/j.jhazmat.2009.02.136
10. Roshanak Khandanlou, Mansor B. Ahmad, Hamid Reza Fard Masoumi, Kamyar Shameli, Mahiran Basri, Katayoon Kalantari, "Rapid Adsorption of Copper(II) and Lead(II) by Rice Straw/Fe3O4 Nanocomposite: Optimi- zation, Equilibrium vol. 10(3), March 2015, pp. 1- 99, doi:10.1371/journal.pone.0120264
11. Jeyaseelan Aravind, Shanmugaprakash Muthusamy, Sangeetha HubbathalaiSunderraj, Lenin Chandran, Kanmanipod as a novel eco-friendly bio- sorbent: a study on equilibrium and kinetics of Ni(II) biosorption", Int J Industrial Chem., vol. 4, March 2013, pp. 1- 25, doi: 10.1186/2228-5547-4-25

APPLICATION OF CAPRA AEGAGRUS HIRCUS DUNG ACTIVATEDCARBON FOR THE REMOVAL OF ANIONICDYES- ISOTHERMAL STUDIES

K.S.Thangamani[2,3] N. Muthulakshmi Andal[1*]

[1*]Assistant Professor, Department of Chemistry, PSGR Krishnammal Col-lege for Women, Coimbatore-4

[2]Assistant Professor, Department of Chemistry, Dr. N.G.P. Institute of Technology, Coimbatore-48

[3]Research Scholar, Department of Chemistry, PSGR Krishnammal Collegefor Women, Coimbatore-4

Abstract

A novel activated carbon prepared from profusely available bio-waste material is employed for the removal of anionic dyes viz., Direct Brown 2 (DB2) and Reactive Red 152(RR152) from aqueous solutions, followed byspecific characterization studies in support of the system. Batch equilibra-tion experiments pertaining to the influence of operating factors viz., vary- ing initial dye concentrations, preset time intervals, doses of sorbent mate-rial, variable pH and temperature environments are substantiated to evaluate the sorptivenature. Adsorption capacities are observed to be 35.7mg/g and 17.8 mg/g for the concentrations of 100 mg/L of DB2 and RR152respectively. Adsorption is more predominant in acidic pH. Isothermal ad- sorption data fitswell with Freundlich isotherm. The aftermath of the pre-sent work implies that the derived activated carbon can be effectively employedas potential scavenger for trapping dye molecules.

Key Words: Activated carbon, Adsorption, Anionic dyes, Capra Aega-grus Hircus Dung

1. INTRODUCTION

Textile, cosmetic, paper, leather, pharmaceutical and food industries discharge huge quantities of dyes as wastewaters. It is estimated that 10–15
% of the dyes employed inthe textile processing enter the effluent streams[1]. It is very difficult to degrade them due to their complex structure and synthet- ic origin [2]. The presence of even very low concentrations of dyes can cause severe health hazards. Many ofthese dyes are carcinogenic, mutagenic, and teratogenic and also toxic to human beings, fish species and microorganisms. Hence, their removal from aquatic wastewater has become inevitable [3]. Currently, there are several treatment methods available for the removal of dyes from wastewaters viz., electrochemical treatment, coagulation, precipi- tation, froth flotation, solvent extraction, membrane filtration, adsorption and advanced oxidation process [4]. Amongst, the adsorption process is found to bemost effective over the other methods and employing activated carbon s a widely followedprocedure for dye removal in effect to large surface area, lowcost and sludge free clean operation. The present work deals with the utiliza- tion of Activated Carbon prepared from Capra Aegagrus Hircus Dung for theuptake of Direct Brown 2 and Reactive Red 152 dyes from aqueous solutions,wherein least reports are available in literature.

2. MATERIALS AND METHODS

2.1. Preparation of Goat Dung Activated Carbon

Capra Aegagrus Hircus Dung - Goat Dung (low cost, ecofriendly) collected from various localitiesin Coimbatore, Tamil Nadu, India, was sun dried, crushedand soaked in 1:1 sulphuric acid for 24 hours. Later, air dried and carbonizedin Muffle furnace (YORCO) for 60 minutes. The carbonized material was powdered and activated for 10 minutes, washed several times with distilled water to attain neutral pH and dried. The prepared Goat Dung Activated Carbon (GDAC) was categorized to a fixed particle size (0.18mm)and stored in air tight containers [5].

2.2. Preparation of Aqueous Dye Solutions

Dyes used in the present study wereselected on the basis of their vivid presence frequently in textile leachates. These dyes belonging to 'azo' (-N=N-
) group were procured from S.S Brand colours at Tirupur, Tamil Nadu, India. The colour index name, properties and thecorresponding applications of chosen dyes are listed in Table.1 and their chemical structures are represented in Fig.1. All the reagents employed viz., Hydrochloric acid, Sodium hydroxide and Sulphuric acid were of an Analytical grade purchased from Sigma Aldrich Company and used without further purification. Stock solutions with a concen- tration of 1000 mg/L of DB2 and RR152 were prepared using doubly distilled waterand the aliquots of required concentrations weremade.

Batch Adsorption Studies

A series of Erlenmeyer flasks of 500 mL capacity with 50mL of two aqueous dye solutions of varying concentrations (25 – 150 mg/L) and doses of activated carbon samples (50mg to 250mg) were agitated in a temperaturecontrolled shaker (30°C to 60°C) at 180 rpm to a tune of preset time frames (10-100 min) with different pHenvironments (2 – 12). Concentrations of the dye molecules in the supernatant solutions were estimated by recording the ab- sorbance values at respective λ max in UV-Visible spectrophotometer (Lab In- dia UV 3000^{+}). The amount of dye adsorbed was calculated as

Amount adsorbed (qe) = $\frac{Co-Ce}{m}$ V ... (1)

where Co = initial concentration of dye solution (mg/L), Ce = equilibrium concentration of dye solution (mg/L), m = mass of the adsorbent (g),V = Volume of test solution (L).

Table 1: Properties of Dyes

C.I Name	**Direct brown 2**	**Reactive Red 152**
Molecular formula	C29H19N5Na2O7S	C52H30Cl2N14Na6O20S6
Molecular weight	627.54	1752.11
λmax	540 nm	420nm
Application	dyeing cotton fibres	dyeing cotton, wool, silk and nylon fibres

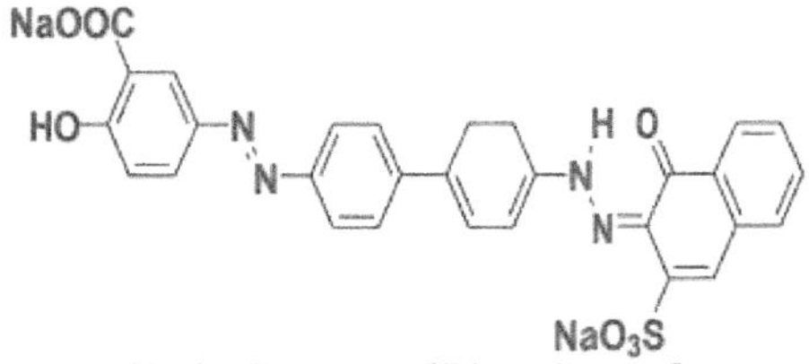

Fig 1a: Structure of Direct Brown 2

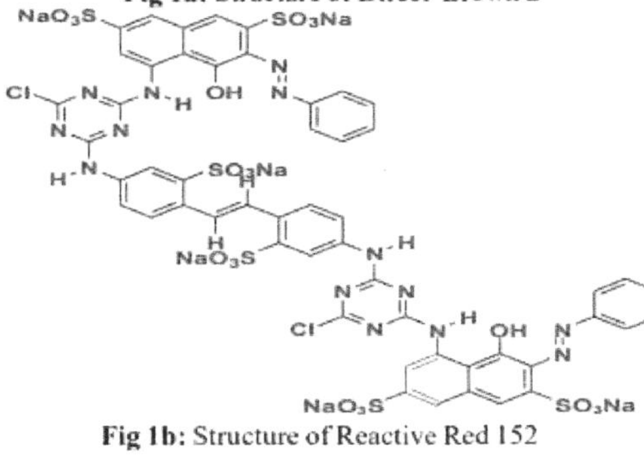

Fig 1b: Structure of Reactive Red 152

2.3. SURFACE CHARACTERISTICSTUDIES

Surface morphology of the unloaded and loadedadsorbent material were identified using Scanning Electron Microscope (SEM)(JEOL JFM- 6390).

Functional groups present in GDAC and DB2, RR152 dye loaded GDAC were deter- mined (within the range of 400-4000 cm-1) using Shimadzu Infrared Spec- trophotometer.

RESULTS AND DISCUSSION

4.1 Morphological Studies (SEM)

SEM images of the GDAC, loaded with specified dyes are represented in Figs 2a, 2b & 2c respectively. Fig.2a. shows a layered structure with considera- ble number of surface pores promoting rapid sorption of dye molecules ontothese pores. Well defined closed pores in Figs.2b & 2c reveal the occupationof DB2 and RR152 dye molecules respectively

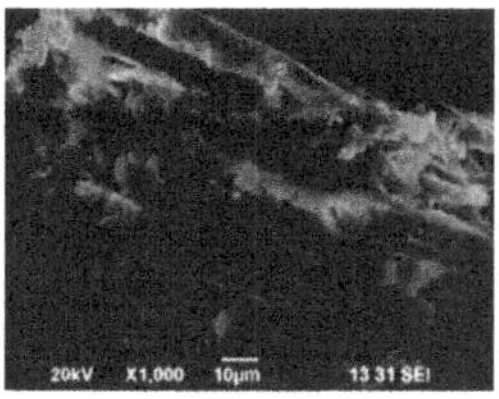

Fig 2a: GDAC **Fig 2b**: DB2 loaded GDAC **Fig2c**:RR152loadedGDAC

4.1 FTIR Spectral analysis

The FTIR spectrum of unloaded GDAC, loadedwith DB2 and RR152 is shownin Fig 3a, 3b &3c. The existence of amine, hydroxyl, carbonyl and amide groups on GDAC is obvious from the Fig 3a, which are the sorption sites for dyes. Afteradsorption of dyes, the band at 3757.3, 3452.5, 2279.8, and 1438.9 cm^{-1} (shown in Fig.3a) were shifted to higher frequency at 3768.9, 3572.1,2291.4 and 1469.7 cm^{-1} (shown in Fig.3b & 3c). The band at 1635.6 cm^{-1} disappeared after adsorption. The shift and disappearance of different functional groups on the adsorbent after adsorption indicated the possible in- teraction of those active sites for uptake of DB2 and RR152. Similar results were observed for the removal Pb (II) using activated carbon prepared from goat dung [5,6].

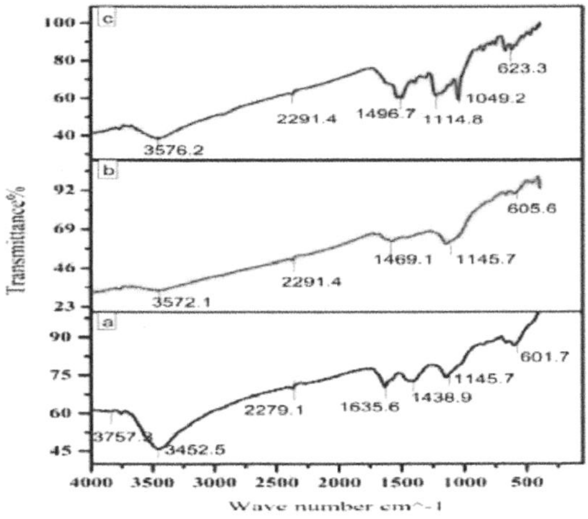

Fig 3a: GDAC, **Fig 3b**: DB2 loaded GDAC, **Fig 3c**: RR152 loaded GDAC

4.2 Effect of Initial Dye Concentration and Contact Time

The initial dye concentration provides necessary driving force to over- come the resistance of the mass transfer between theaqueous and solid phase [7]. Figs 4a & 4bevidenced that percentage of dye removal declined with risein initial dye concentration (25 – 150 mg/L). However, at equilibrium time of 60 min (DB2) and 90 min (RR152), all concentrations registered more or less approximate percent removal. Thence, an initial concentration of 100 mg/L was fixed for further experiments where an appreciable dye removal of 92% (DB2) and 86% (RR152) is recoded.

Initially, dye removal registered a steep increase with contact time, later ap- proached almost a constant value further which least difference was observed, indicative of attainment of equilibrium, thereby restricting the batch experi- ments to 60 minutes and 90 minutes for DB2 and RR152 respectively.

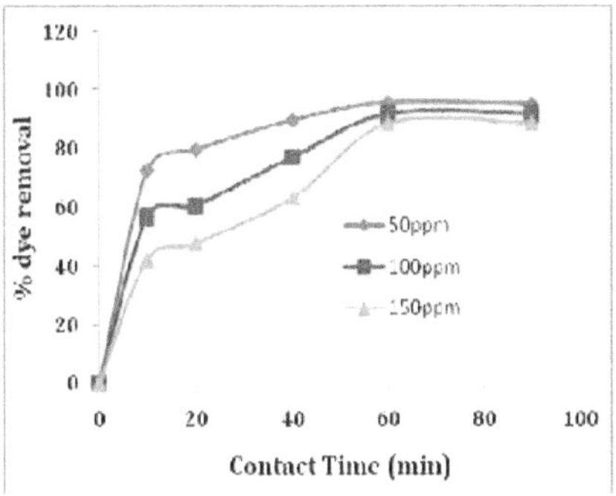

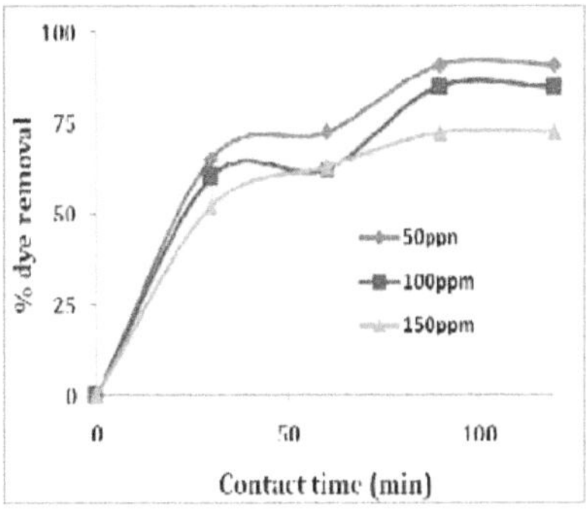

Fig 4a: Effect of Initial Concentration andContact Time (RR152-GDAC)

Fig4b: Effect of conc &Contact Time (DB2-GDAC)

4.3 Effects of Dosage

The amount of respective dye moleculesadsorbed(mg/g) against the var-ied carbon dosesshows a maximum requirement of 150 and 250 mg for effec- tive sorption of DB2 and RR152 dye respectively, above which minimal up- take is evident. These dosage values correspond to 35.7 mg/g and 17.8 mg/g amount adsorbed as perFig.5. The increase in the adsorption efficiency with the adsorbent dosage could be due to the increase of the adsorbent surface area, as the mass increases it implies towards the presence of more available adsorp- tion sites [8]. Thence, optimum doses of 150 and 250 mg were fixed for forthcoming experiments.

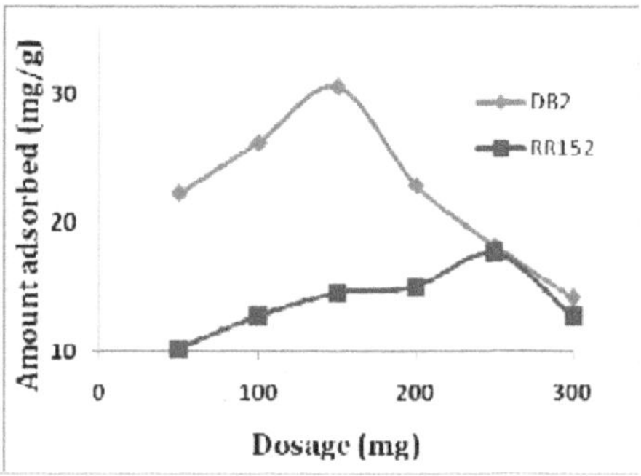

Fig 5: Effect of Dosage

4.1 Effect of pH

A maximum dye removal at pH 2 is obvious from Fig.6, which decreased along with increasing initial pH of the dye solution. This may be due to acidic pH the surface of the adsorbentbecomes positively charged, which enhances the adsorption of negatively charged dye anions through electrostatic forces of at-traction [9]. Also at alkaline pH, excess OH^- ions compete with the anionic group of the dye thereby declining the percent removal.

4. 5 Effect of Temperature

Reduction in percentage removal of dye with increasing temperature is seen in Fig.7 shall be due to weakening of the bond formation between dye moleculesand the binding sites [10].Also increase in kinetic energy of dye molecule athigh- er temperatures may cause escape of dyes from solid surface to solution phase promoting desorption in preference to adsorption. This shows that the adsorption of dyes onto GDAC is exothermic in nature and all the experi- mental results accounted so far are within the room temperature environ- ment.

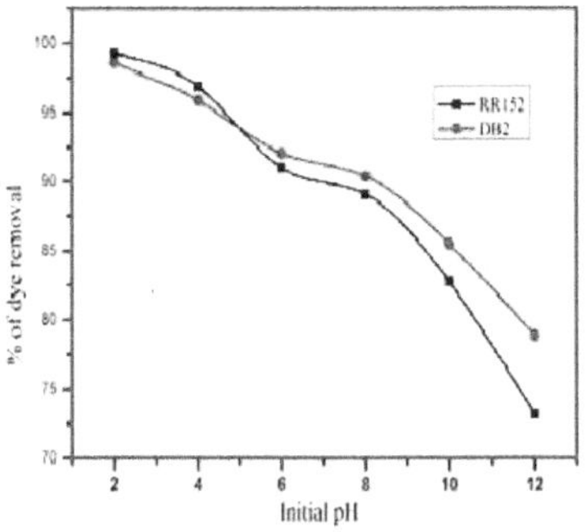

Fig 6: Effect of pH

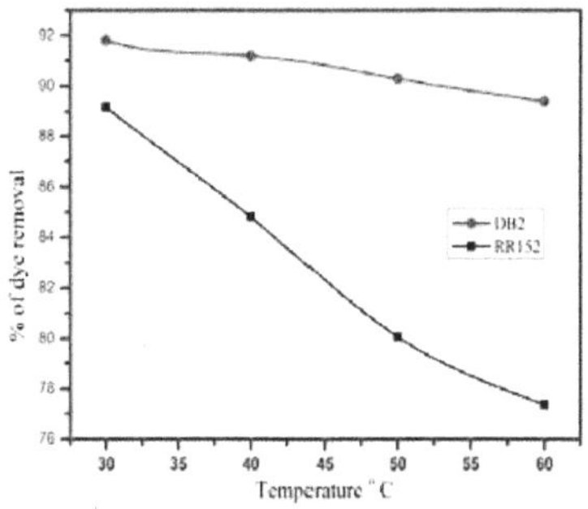

Fig 7: Effect of Temperature

4.6 Isotherm studies

The Langmuir and Freundlich isotherms was tested based on the adsorption studies for DB2 –GDAC and RR152-GDAC systems. Theapplicability of eachisotherm is evaluated by the value of isothermal constants (calculated from the slope and intercept of the specific plots) andcorrelation coefficients (Table 2 and Figs 8 & 9). qm and KF values of DB2 is observed to be two fold that of RR152, which supports the adsorptioncapacity values mentioned under influence of dosage. Also better linearity of the plots is exhibited by Freundlich model (marginally greater R^2 values) favoring multilayer adsorption

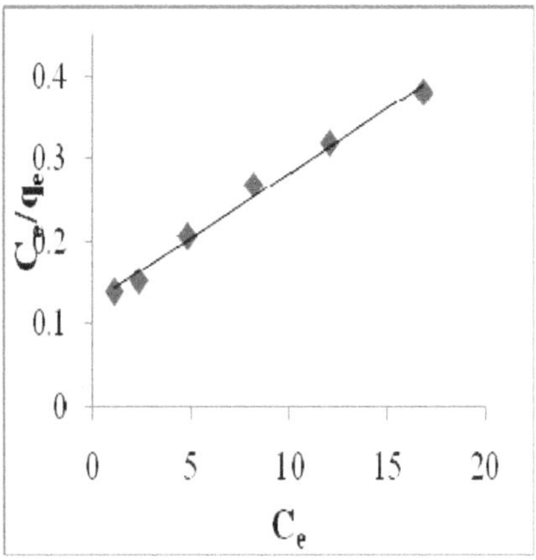

Fig 8a: Langmuir Isotherm Isotherm(DB2-GDAC)

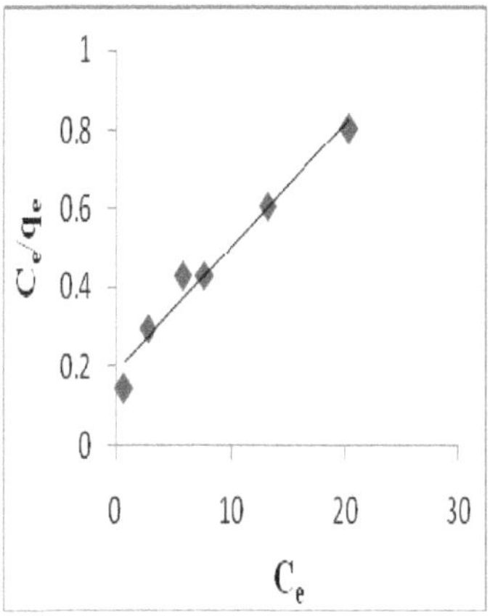

Fig 8b: Langmuir (RR152-GDAC)

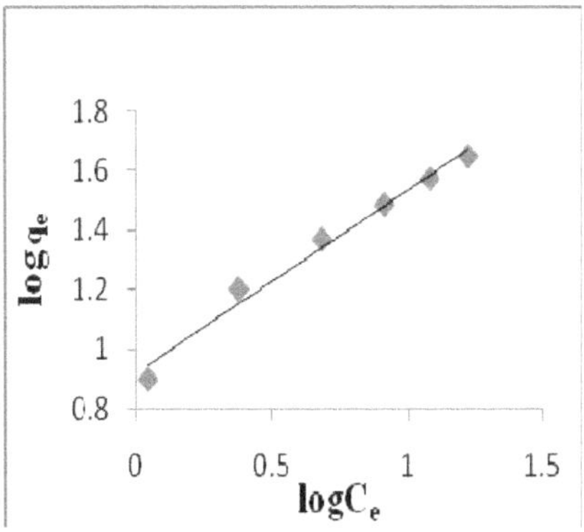

Fig 9a: Freundlich Isotherm Isotherm(DB2-GDAC)

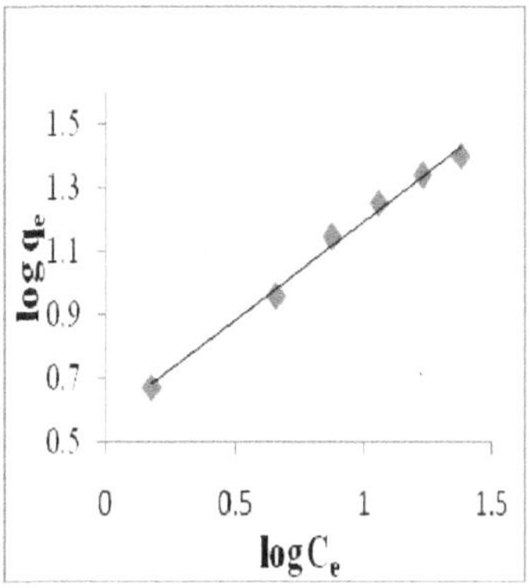

Fig 9b: Freundlich (RR152-GDAC)

Table 2: Isothermal Constants

	Langmuir				Freundlich		
Dye	qm (mg/g)	b(L/mg)	R^2	RL	1/n	KF (mg g^{-1})	R^2
DB2	66.66	0.121	0.988	0.076	0.612	8.356	0.994
RR 152	32.26	0.115	0.967	0.057	0.624	3.706	0.992

5. CONCLUSION

Adsorption of Direct Brown 2 and Reactive Red 152 onto activated car-bon derived from goat dung was investigated. GDAC and its dye laden samples were characterized using SEM, which revealed the expected morphological changes. Optimized condition of 100 mg/L initial dye concentration, 60 and 90 minutes preset time interval, 150 and 250 mg carbon doses of (DB2 and RR152), pH 2 and room temperature derived from batch results. From the experimental observation, it is concluded that GDAC proved to be a promising sorbent for dye molecules, where the further studies will proclaim its application to effluent leachates.

REFERENCES

1. V.K.Garg, R.Gupta, A.B.Yadav, R.Kumar, "Dye remov-al from aqueous solution by adsorption on treated sawdust, Bioresour. Technol". Volume: 89, September2003,pp.121-

124,Doi:org/10.1016/S0960-8524(03)00058-0

2. Ramakrishnan, K. and C. Namasivayam, "Development and characteristics of activated carbons from Jatropha husk, an agro industrial solid waste, by chemical activation methods", J.Environ. Eng. Manage. Volume 19, Issue: 3, January 2009,pp.173-178.

3. Konicki W, Sibera D, Mijowska E, Lendzion-Bielun Z, N, U.," Equilib- rium and kinetic studies on acid dye Acid Red 88 adsorption by magnetic ZnFe2O4 spinel ferrite nanoparticles", J Colloid Interf Sci. Volume 398, February 2013, pp. 152-160,Doi:10.1016/j.jcis.2013.02.021

4. Mohammad Foroughi-dahr, Hossein Abolghasemi, Mohamad Esmaieli,Ghadir Nazari, Bettina Rasem, "Experimental Study on the AdsorptiveBehaviour of Congo Red in Cationic Surfactant-Modified Tea Waste",Process Safety and Environment Protection. Volume 95, May 2015, pp226-236, Doi: 10.1016/j.psep.2015.03.005

5. K.S. Thangamani, N. Muthu-lakshmi Andal, E. Ranjith Kumar, M. Saravanabhavan,"Utilization of magnetic nano cobalt ferrite doped Capra aegagrus hircus dung activatedcarbon composite for the adsorption of anionic dyes", J. Environ. Chem.Eng. Volume 5, May 2017,2820–2829. Doi:org/10.1016/j.jece.2017.05.030

6. Mambo Moyo, George Nyamhere, EdithSebata & Upenyu Guyo, "Kineticandequilibrium modelling of lead sorptionfrom aqueous solu- tion by activated carbonfrom goat dung", Desalin. Water Treat.Volume 57, Issue :2,October 2014, pp. 1-11. Doi:org/10.1080/19443994.2014.968217.

7. Chen.X.C,Wang.Y.P,Lin.Q,Shi.J.Y,Wu. W.X, and Chen.Y.X "Biosorption of copper(II) and zinc(II) from aqueous solution by Pseudomonas putida cz1' Colloid and Surface B", Biointerfaces,Volume 46, Issue 2, 2005, pp. 101-107, Doi.org/10.1016/j.egypro.2012.05.106.

8. Namasivayam C and Yamuna R T, "Adsorption of direct red 12 B by biogas residual slurry: Equilibrium and rate processes" Environ Pollut, Volume 89, December 1995, pp 1-7, Doi:org/10.1016/0269- 7491(94)00056-J

9. Han, R., Ding,D., Xu,Y., Zou,W., Wang,Y., Li,Y., Zou,L." Use of ricehusk for the adsorption of Congo red from aqueous solution in columnmode", Bioresour.Technol, Volume 99, Issue 8:May 2008, pp2938– 2948,DOI:10.1016/j.biortech.2007.06.027.

10. Saha. P, Chowdhury, S., Gupta. S, Kuma, I." Insight into adsorption equilibrium, kinetics and thermodynamics of malachitegreen onto clayey soil of Indian origin". Chemical Engineering Journal, Volume 165, December 2010, pp. 874–882, Doi:org/10.1016/j.cej.2010.10.048

Functionalized Novel Nano Composites in Ni(II)Removal from Aqueous /Effluent Media

N. Muthulakshmi Andal[1] and G. Gohulavani[2]

[1]Assistant Professor, Department of Chemistry, , PSGR Krishnammal College for Women, Coimbatore- 04

[2]Assistant Professor, Department of Chemistry, Ethiraj college for Women, Chennai-08

Abstract: Nanohydroxy apatite – Azaridachtaindica nutshell composite material was prepared via co-precipitation method. The product was characterized by FT-IR and Atomic force microscopy. The Ni(II) removal capacity of the synthesized novel ma- terial is investigatedby employing batch mode to verify the sorption efficiency under optimized conditions viz.,11 mg/L initial concentration of Ni(II), 21 minutes agitation time between sorbent/sorbate molecules and 50mg dosage of the nanocomposite. The results of the experiment registered a significant increase in the sorption nature of nanohydroxy apatite composite of the chosen nutshell against the chemically modi- fied Azaridachtaindica nutshell where 450 mg dosage is employed under similar op- erating conditions. The successfullness of the nanomaterial in trapping Ni(II) ions is extended to field levels through the assessment of effluents collected from electroplat- ing industry suggesting that nano composited biomass of Azaridachtaindica nutshell is an efficient bioaccumulant.

Keywords: adsorption, nanocomposite, characterization, *Azaridachtaindica* nutshell, efflu-ent

1. Introduction

Water is one of the most indispensable and unique substance, because it can naturally re- new and clean itself, by allowing pollutants to settle out through the process of sedimenta- tion. Ground water is highly valued because of certain properties not possessed by surface water[1]. It is required by industries for various reasons such as cooling, steam generation, heat exchange, gas scrubbing, washing of solids, surface rinsing etc. Today, the water is polluted with hundreds of toxins, heavy metals and impurities. This is a major problem in the global context. It has been suggested that it is the leading worldwide cause of deaths and diseases which accounts for the death of more than 14000 people daily.[2]Such toxic metals are removed from the aqueous and industrial streams using various sequestration techniques. The cost-effective, being the adsorption method which employs chemically modified agriculture materials as biosorbents for the removal. Such modification of agri- cultural waste materials can extract soluble organic compounds and enhance their chelat- ing efficiency. Due to dearth of information on the adsorption of heavy metals on Nano- modified sorbents, this modification procedure was adopted and applied to the selected ag- ricultural sorbent.

2. Materials and Methods

Neem tree, botanically known as *Azadirachta indica*, belongs to the family Meliaceae.The tree is tall, fast growing and is known for its drought resistance [3].Neem seed nut shell husk were thoroughly washed well with double distilled water and dried in the air. The dried shells were powdered, sieved and the fractions measuring from 0.18 to 0.71 mm wereprepared and designated as AINS.

A one-step co-precipitation method[4] was adopted to prepare theNano- composite of 0.18mm AINS-Nanohydroxyapatite. The white nano hydroxyapatite powder obtained through themethod was washed well with double distilled water several times to bring the pH to 7 and later dried under normal conditions. The dried nano powder was then mixed with 0.18 mm of AINS in the ratio of 1:1 by adopting the one step co- precipitation meth- od. pH adjustments were done using acetic acid and NaOH solutions. The obtained nano hydroxyapatite- adsorbent composite was then dried well and utilized for the specific stud- ies. The Nanohydroxyapatite –

Azaridachta indica nut shell composite is mentioned as NAINSC.

Known weights (50mg-200mg) of the adsorbent were taken in 250 ml Erlenmeyer flasks at room temperature (250C) and equilibrated with 50 ml of Ni(II) solution of known con- centrations of 11 ppm was prepared from its sulphate salt [$Ni(NH_4)_2SO_4.9H_20$]. Fresh dilu-tions were used for each study. The pH values (3, 5, 7, 9) of the solutions were adjusted with 0.1M HCl /0.1M NaOH solutions respectively. The flasks were then agitated in an orbital mechanical shaker at 120 rpm at predetermined time intervals(3-30 min.) and the supernatants were filtered. The residual concentrations of Ni(II) ions were then determined by Atomic Absorption Spectrophotometer (Perkin Elmer AA Analyst 100 with air- acety- lene oxidizing flame).

Figure1:*Azaridchta indica* nut shell

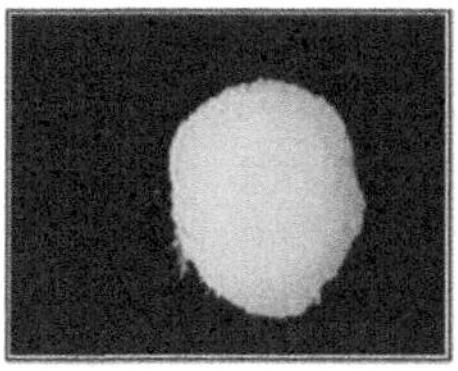

Figure2:NAINSC

1. Results and Discussion

1.1 Characterization by AFM &FTIR

Atomic Force Microscopic Analysis offers visualization and analysisin three dimensions. Using the AFM, individual particles and groups of particles can be resolved finer, com- pared to other microscopic techniques. Thus synthesised Nanohydroxyapatite-*Azaridachta indica* nut shell composite (NAINSC) was analyzed for the confirmation of its nano size, employing the AFM technique. The histogram of NAINSC (fig.3) shows a clear peak <18 nm which confirms the prevalent existence of maximum number of nanoparticles in the composite material. The three dimensional surface topography of NAINSC is also repre- sented in figure 4.The FTIR spectrum of NAINSC is presented in figure 5

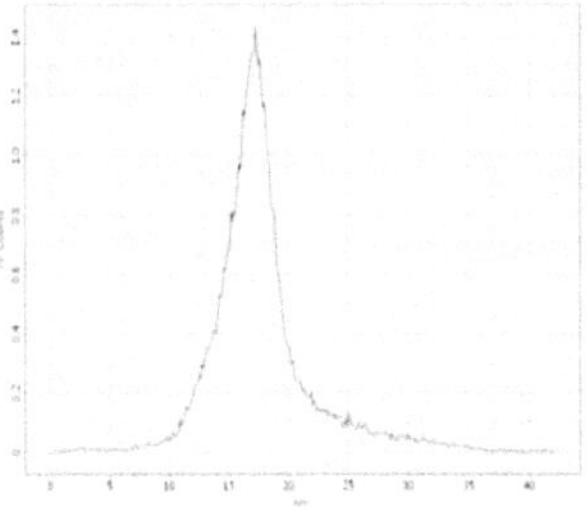

Figure 3: Histogram (NAINSC)

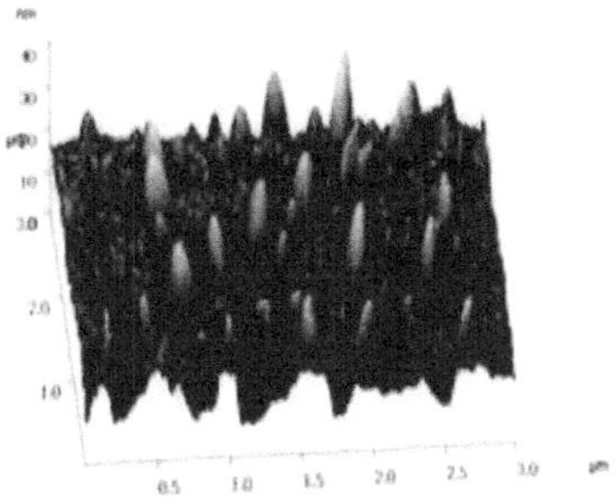

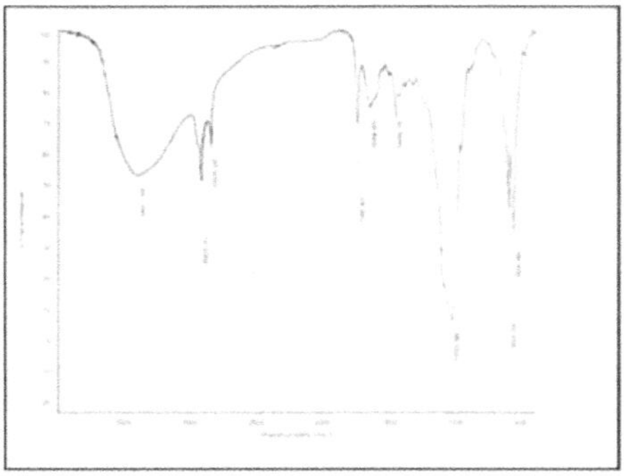

Figure 4: 3-D surface Topography (NAINSC) **Figure 5:** FTIR Spectrum of NAINSC

3.2. Effect of NAINSC dosage on Ni(II) removal

The efficiency of the synthesized NAINSC on Ni(II) was experimented by varying the composite dose(10-50 mg:10 mg interval). 11 ppm nickel solution was prepared froma stock of Ni(II) solution of 1000 mg/L was prepared by dissolving 6.7303 g of anhydrous Nickel ammonium sulphate in 1000 ml of double distilled water.Batch adsorption studies for different composite dosages with a contact time of 21 minutes for both the metal ions [Ni(II):pH 6.9; at 11 ppm initial metal ion concentration] were studied by varying the NAINSC dosage from 10 mg to 50mg. After agitation, the supernatant was filtered and analyzed using AAS.

The percentage removal of metal ions are presented in figure 7.15, the corresponding val- ues are tabulated in table 7.4. The results show that 50 mg of NAINSC was sufficient to trapthe divalent Nickel ion, where 74% of the metal has been removed, The sorption affin-ity of the metal on NAINSC was found to be good. This may be due to the solution pH, which is an important parameter in controlling composite sorption onto themetals. The physiochemical properties of n-HAP such as solubility, stability, crystallinity etc., vary depending upon the pH conditions[291]. At nearly a neutral pH of 6.9 employed for Ni(II)- NAINSC system, the composite material is probablyrelated to its lower solubility, This solubility nature of nano composites at neutral pH rangeenhances the availability of sorp- tion sites, thereby increasing the percentage removal. Thus, the higher capability of NAINSC to adsorbNi(II), may be attributed to optimum pH conditions.

3.3 Effect of NAINSC dosage on industrial effluents

The eficiency of NAINSC was tested on industrial effluents collected from chemical and electroplating industries. The percentage removal of nickel from the respective effluents was observed to increase with dosages. Table -1support the experimental values of the NAINSC- Ni(II) systemto exhibit better nickel removal capacity.

Table 1: Effect of NAINSC dosage on Industrial effluent

Weight of adsorbent(mg)	% Removal of metal ions from effluents
	Nickel
10	69.61
20	70.58
30	70.62

40	70.76
50	71.18

2. Conclusion

From the above reports, it is concluded that the low-cost NAINSC was effective, easily available, non-toxic, indigenous adsorbents for the removal of Ni(II)from aqueous solu- tions and industrial effluents. The results of the present investigations would be useful for the fabrication and designing of wastewater treatment plants, utilizing such novel nanocomposites in an effective manner. Our future study aims at the establishment of performance oriented approach of NAINSC for other toxic heavy metal ions. This can be accomplished by developing suitable modification procedures in the sorbents viz. encapsulation and immobilization, thereby ensuring the safe disposal of the metal load- ed sorbents, which in turn does not pose a great threat to the environment. Also the fur- ther work accomplishes to synthesize novel nanocomposites utilizing various agricul- tural waste materials to enhance and upgrade the sorption quality of such low cost sorbents.

References

[1] P.N.Rajankar, S.R.Gulhane, D.H. Tambekar, D.S.Ramteke, A study on the physico- chemical characteristics of the municipal solid waste of Amravati city. *Waste Elec.j.Chem.,*6(3) 905(2009)

[2] "The cost of meeting the Johannesburg targets for Drinking Water" *Water Academy Org.,*(2004)

[3] Vijayalaxmi, K Radha, K S and Vandana Shiva., 1995,

[4] Neem - A User's Manual Zhang Li,LiYubao, Yang Yiping, Preparation and Invitro investigation of chitosan/Nanohydroxyapatite composite used as bone substitute ma-terials, *J.MatSci: Materials in Medicine.,*16(2005) 213-219

Utilization of magnetic nano cobalt ferrite doped Capra aegagrus hircus dung activated carbon composite for the adsorption of anionic dyes

K.S. Thangamani[a,b], N. Muthulakshmi Andal[a,*], E. Ranjith Kumar[c], M. Saravanabhavan[b]

[a] *Department of Chemistry, PSGR Krishnammal College for Women, Coimba-tore-48, India*

[b] *Department of Chemistry, Dr. N.G.P. Institute of Technology, Coimbatore-48*

[c] *Department of Physics, Dr. N.G.P. Institute of Technology, Coimbatore-48*

A R T I C L E I N F O

Keywords:

Capra aegagrus hircus dung activated carbon Magnetic nanocomposite Adsorption Anionic dyes

A B S T R A C T

A novel activated carbon prepared from profusely available bio-waste ma- terial, goat dung is employed for an- ionic dyes removal viz., Direct Brown 2 (DB2) and Reactive Red 152 (RR152) from aqueous solutions. Goat Dung Activated Carbon-Cobalt Ferrite Magnetic Composite (GDACCFMC) is synthesized by auto-combustion method. The surface characteristics of the prepared adsorbents are analysed by SEM, EDAX, BET and BJH and the size of the synthesized GDACCFMC is characterized by AFM. From the AFM results, the synthesized GDAC-CFMC is found to be lesser than 3 nm. The magnetic property of GDAC-CFMC is characterized by using vibrating sample magnetometer (VSM) equipment which reveals its excellent magnetic performance in dye removal. The batch equilibration experiments pertaining to the influence of operating factors viz., varying initial dye concentrations, doses of sorbent materials, preset time intervals, variable pH and temperature environments are sub- stantiated to evaluate the sorptive nature. Adsorption is observed to be more pronounced at acidic pH. Sorption efficiency of Goat Dung Activat- ed Carbon (GDAC) is compared with the derived GDAC-CFMC, whose experimental results indicate better performance of the magnetized materi-al. The isothermal adsorption data fits well with Freundlich isotherms. Thethermodynamic studies reveal that the adsorption process is favourable, exothermic and spon- taneous. Adsorption kinetic studies favour the sys- tems to follow the second-order. The aftermath of the present work impliesthat both GDAC and GDAC-CFMC can be effectively used as potential adsorbents for trapping the dye molecules.

1. Introduction

Huge quantities of dyes are discharged as wastewaters from various in- dustries such as textile, cosmetic, paper, leather, pharmaceutical and foodindustries. The effluents from dyeing industries are highly co- loured/toxic to the aquatic life. It is estimated that 10–15% of the dye employed in the textile processing, enters the effluent streams [1]. Even meagre amounts of dyes can cause severe health hazards like nausea, haemorrhage, cancer, ulceration of skin and mucous membranes which pose severe damage to kidneys, liver, brain and central nervous sys- tems. A few remedial measures can be implemented to restrict further damage. Firstly dyes are to be annihilated from these wastewaters be- fore being expelled into the environment to prevent the ecosystem from being polluted. Several treatment processes available for effluent treat- ment include bio degradation [2], chemical oxidation [3], froth flotation [4], electrolysis

[5] , adsorption [6], electro-coagulation [7] and photo catalysis [8]. The adsorption technique is observed to be one of a popular physicochemical treatment for the removal of dissolved dyes from textile effluents. Many new promising materials have been reported inthe past few years viz., tea waste [9–11], fish scale [12], banana peel [13], almond shell [14], pomelo (Citrus grandis) peel [15], broad bean peel [16], peanut hull [17] and raw barley straw [18] for the dye re- moval. In addition activated carbons have been widely employed as effective sorbents for the uptake of dye molecules from aqueous solu- tions. Fabrication of nanomagnetic activated carbon has been fo- cussed in recent times to enhance the dye removal efficiency [19].The derived metal oXide doped nano sized activated carbons are con- sidered as prospective adsorbents for aqueous pollutants due to their high surface area, the unique advantage of being easy separation un- der external magnetic fields [20]. Shenmin Zhu et al. [21] have syn- thesized activated carbon/iron oXide magnetic composite for the ad- sorption of methylene blue. Lunhong Ai et al. [22] in their findings brought to light commercially available activated carbon/$CoFe_2O_4$ magnetic composite for the removal of malachite green dye. $CoFe_2O_4$ is found to be an effective fascinating magnetic material as it holds the properties of moderate saturation magnetization, excellent chemical stability and mechanical hardness [23]. The present work deals with the preparation of Goat Dung Activated Carbon and its synthesized compo- site using $CoFe_2O_4$ for the uptake of Direct Brown 2 and Reactive Red152 dyes from aqueous media. Seldom work had been reported else- where in literature using modified/synthesized goat dung for the removal of the aforesaid dyes which possess high commercial values.

2.1. *Reagents*

All the reagents employed viz., Cobalt nitrate [Co $(NO_3)_2$ $\cdot 6H_2O$],analysed by X-ray Powder Diffraction technique using Advance X-Ray Diffractometer (XPERT-PRO, Bruker AXS D8). Particle size of the com- posite was visualised using Atomic Force Microscope, NTMDT (NTEGRA PRIMA) model. Surface area of the samples was measured through the nitro- gen adsorption at 77 K using Surface area/Pore ana- lyzer (Coulter OM- NISORP). The magnetic property of the prepared composite was assessed us- ing Vibrating Sample Magnetometer (Lakeshore VSM 7410). Thermal stabilityof GDAC was evaluated by TGA-DSC (NETZSCH JUPITOR STA 449F3). Functional Solution pH values and dye concentrations for all the experiments were registered using pH meter (LABTRONICS) and UV- visible spectropho- tometer (Lab India UV 3000 $^+$) respectively.

Ferric nitrate [Fe $(NO_3)_3$ $\cdot 9H_2O$], Urea, Sodium hydroxide and Sulphuric acid were of an Analytical grade procured from Sigma Aldrichcompany and further used without purification. Direct brown 2 and Reac-tive Red 152 dyes were purchased from S.S Brand Colours at Tirupur,Tamil Nadu, India.

Direct brown 2: Molecular formula- $C_{29}H_{19}N_5Na_2O_7S$, Molecular weight: 627.54. The dye structure reveals the presence of sodium salt of sulphonic ac- ids and azo groups as substituents. These are mainly used for dyeing cotton fibres. Almost, all dyeing units employ this dye due to their colour fastness property and easy application.

Reactive Red 152: Molecular formula – $C_{52}H_{30}Cl_2N_{14}Na_6O_{20}S_6$, Mo- lecular weight: 1752.11. The principal chemical classes of reactive dyes include azo, triphendioxazine, phthalocyanine, formazan and an- thraqui- none apart from sodium salt of sulphonic acids group as substituent. These dyes are extensively utilized in dyeing cotton, wool, silk and nylon fibres.

2.5. Equilibration experiments – GDAC

Batch adsorption experiments were performed for DB2–GDAC and RR152–GDAC systems with 50 mL of aqueous solutions at varying in-itial dye concentrations (25 to 150 mg/L) agitated at 200 rpm in mechanical shaker (TECHNO) at preset time intervals (10 to 120 min: 10 min interval). Experiments were carried out under different sorption con-ditions: GDAC doses (50–250 mg), pH (2–12) and temperature (30–60°C). The initial pH was adjusted by the addition of 0.1N HCl/NaOH for acidic and alkaline environments respectively. The adsorbent and adsorbatewere separated by centrifugation at 2000 rpmfor 10min, where initial and residual dye concentrations of DB2 and RR152 were determined spectrophotometrically at λ_{max} values of 420 nm and540 nm respectively. Duplicate experiments were performed to ensure the reproducibility values within ± 0.1%.The amounts of dyes adsorbed were calculatedas:

2.2. *Preparation of Goat Dung Activated Carbon (GDAC)*

Goat Dung was collected from various localities in Coimbatore, Tamil Nadu, India, which is sun dried and crushed. The crushed material were soaked in 1:1 sulphuric acid and kept for 24 h at room temperature. Later, it was air dried and carbonized in Muffle furnace (YORCO) at temperature of 500°C for a time period of one hour. The carbonized ma- terial was powdered and is subjected to activation for 10 min at 650°C, since high surface area and high micro porosity can be obtained through chemical activation at moderate temperature [24]. After that, the pre- pared material was washed several times with distilled water to attain neutral pH and dried further. The prepared Goat Dung Activated Carbon

(GDAC) was categorized to a fixed mesh size of 85 BSS, and stored in air tight containers.

2.3. *Synthesis of GDAC composite material*

A composite of GDAC was prepared by an auto combustion method,with 2 mol of ferric nitrate and 1 mol of cobalt nitrate which was pre- paredas a mixture by dissolving appropriate doses of both salts in 50 mL of distilled water. 28 g of urea was dissolved in distilled water and added to the above solution to ensure the combustion process, followed by heating the contents in a magnetic hot plate with constant stirring using magnetic stirrer. 11 g of GDAC was added to the hot bath and the heating was con- tinued. Further, the ignition of the mixture was ensured resulting in the formation of dried gel in a burnt state through self propagating combustion. Those obtained dry loose powder was ground well using mortar and pestle and stored as done earlier. This synthesized material is referred to as GDACCFMC.

2.4. *Equipments*

The presence of carbon, hydrogen, nitrogen and sulphur content in the activated carbon was determined using Elementary Vario ELIII-Germany CHNS Analyser. The surface morphology of prepared samples was inves- tigated using Scanning Electron Microscope (JEOL model JSM-6390 LV). High purity and crystallinity of the samples werewhere, C_o and C_e are the initial and equilibrium concentrations (mg/L) of the dyes respectively. 'V'is the volume of the solution and 'W' is the weight of the adsorbent used.

3. Results and discussion

3.1. *Morphological studies (SEM/EDAX)*

The SEM images of the unloaded GDAC, loaded with DB2 and RR152 are shown in Fig. 1a, b & c. Fig. 1a exhibits appreciable pores on the unloaded surface, possessing a layered structure and had disappeared in Fig. 2b & c due to the occupation of respective dye molecules. Nu- merous pores with irregular sizes, suitable for trapping extensive dye molecules are obvious in Fig. 2a. The composite surface inspite of being covered by small aggregates of cobalt ferrite, a porous structure is still seen leftover, indicative of its high adsorbable properties (Fig. 2b& c)

The EDAX spectra of GDAC and GDACCFMC composite are depicted Fig. 3a & 3b correspondingly, where the presence of Co, Fe and O in the synthesized composite is confirmed (Fig. 3b).

3.2. *X-Ray diffraction studies (XRD)*

Elemental investigation and crystal structure identification of the pre- pared nano material were carried out using X-ray diffractometer. Fig. 4 comprises of the X-ray diffraction patterns of Goat dung activated carbon, magnetic cobalt ferrite and GDAC CFMC. The diffraction peaks indexed at (220), (311), (400), (511), (440) in Fig. 4c is perfectly matched with cubic spinel structure (JCPDS card no. 22–1086) of cobalt ferrite.

No characteristic impurity peaks were detected in Fig. 4c, which confirmed the formation of a single phase spinel ($CoFe_2O_4$). The weakpeak appearing at $2\theta = 26.5°$ in Fig. 4a corresponds to GDAC, whosepattern is obviously different from Fig. 4b & 4c. The average crystalline size of GDACCFMC particles were calculated using Scherrer's equation

Fig. 1. (a) SEM images of GAC. (b) SEM images of GDAC-

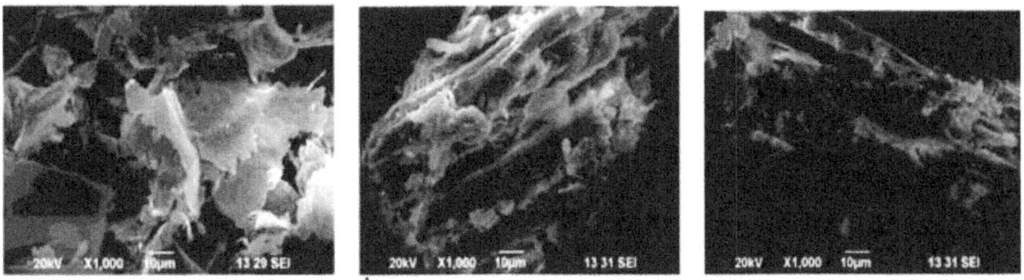

a b c

DB2. (c) SEM images of GDAC- RR152.

$$D = \frac{0.9\lambda}{\beta\cos\theta}$$

where

D = average crystalline diameter k = constant(0.9 for Cu-kα)

λ = X-ray wavelength (0.15405 nm for Cu-kα)

β = peak width of half maximum of XRD diffraction linesθ = Bragg's diffraction angle in degree.

The average grain size measured was 10 nm.

3.3. *Atomic force microscopic analysis*

Atomic Force Microscopic analysis offers visualization in three di- men- sions. Where individual particles and their groups can be resolved finer, compared to other microscopic techniques. The synthesized GDACCFMC was subjected to this analysis and its topography/histo- gram are represent- ed in Fig. 5a & b.

The 3D topography and histogram, registered clarity and peak appear- ance at 2 nm confirmed the pre- valence of nanosized composite material.

3.4. *BET surface area & magnetic properties*

BET surface areas, pore volumes and surface magnetization measurements of the prepared activated carbon and its composite are listed in Table 1, reveal approXimately a twofold decrease in the calculated values for the composite. This may be due to the relatively small surface area of $CoFe_2O_4$partially covering the activated carbon surface. Similar results have been reported in literature for the magnetic oXide- supported activated carbon [21,25].

Magnetic metal composite and magnetic metal oXide supported adsor- bents possesses a relatively high surface area, a smaller particle size and porous nature, which enhances the adsorption capacity towards organic and inorganic waste water coming out from textile and dyeing industries. Particularly, the excellent magnetic property of the metal oXide composites makes them easily recoverable by magnetic separation technology after adsorption. Commercially available

large scale processes, magnetic modification of abundantly available low cost adsorbents could lead to materials suitable for environmental applica- tions. The magnetic separation provides an enviable path for online sepa- ration, where particles with affinity to target species are fused with the heterogeneous solution. Upon merging with the solution, the particles tag the target species. An external magnetic field is then applied to separate the tagged particles from the solution [26].

The magnetic particles of sizes which are lesser than 30 nm exhibit su- perparamagnetism and it responded well to the magnetic fields without any permanent magnetization [27].The superparamagnetic properties ofthe synthesized magnetic composite was verified by the magnetization curve measured by vibrating sample magnetometer (VSM) at room tem- perature.

The result of the typical hysteresis loop relating to the magnetic behav- iour of GDACCFMC as shown in Fig. 6.

The high saturation magnetization value (26.7 emu/g), being lower than that of bare $CoFe_2O_4$ (86.1 emu/g), is because of the presence of GDAC. Therefore, this sample with higher saturation magnetisation value re- sponded to an external magnetic field, supporting the en- hancement in the adsorption capacity. The magnetic separability of prepared composite is further confirmed by placing a magnet near a glass bottle containing water

with the dispersed composite, which was turbid, black, would be changed into clear solution (Fig. 6b). From the literature studied, the saturation magnetization value of cobalt ferrite is comparatively greater than other metal ferrites ($MnFe_3O_4$- 78.4 emu/g and $NiFe_3O_4$- 58.0 emu/g), which in- creases the magnetic property and dye removing capacity of the prepared composite [28].

3.5. *TGA and DSC analysis*

Thermal gravimetric analysis (TGA-DSC) measures changes in physicaland chemical properties of matter as a function of increasing temperature at constant heating rate or as a function of time. This technique finds out selected characteristics of matter that demonstrates either mass loss or gain due to decomposition, oxidation, or loss of volatiles and assesses thermal stability of a matter in a desired

a b c

Fig. 2. (a) SEM images of GDACCFC. b SEM images ofGDAC-CFC-DB2. c SEM images of GDACeCFC RR152.

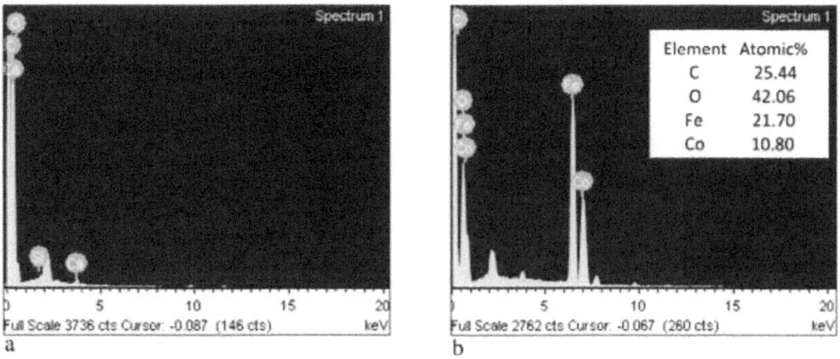

Fig. 3. a EDAX Spectrum of GDAC. b EDAX Spec-trum of GDAC-CFC.

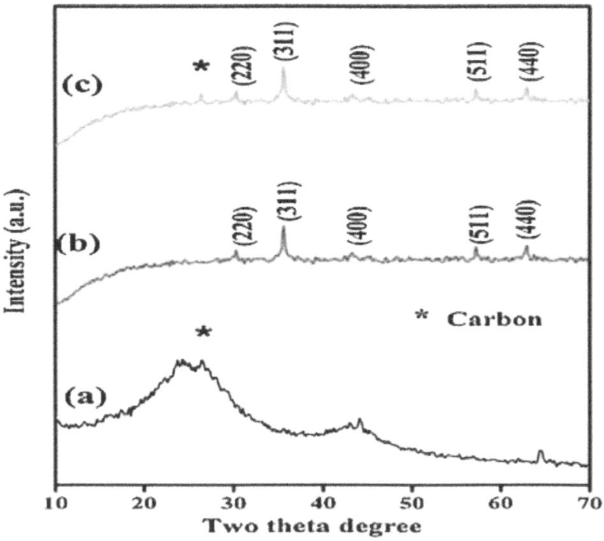

Fig. 4. XRD patterns of (a) GDAC (b) reference standard data for $CoFe_2O_4$ of the JCPDS file No. 22–1086 and (c) and GDAC/CFC.

temperature range. There is no observed mass change takes place if the ma-terial shows high thermal stability. Negligible mass loss corresponds to little or no slope in TGA trace.TGA data also gives the upper usage limit temperature of a material, beyond which material begin to de-grade/unstable. Thermal behaviour of GDAC was observed by TGA-DSC curve in N_2 gas environment to study the weight loss during chemical acti-vation (Fig. 7). The experiment was conducted at a ramping rate 20°C/min from 26°C to 1000°C. The TGA curves showed that the weight of GDAC decreased steeply from 75°C to about 150°C and the weight decreases slowly from 300°C to about 700°C. The weight lost at the beginning of the process was due to evaporation of free moisture con- tent. With further temperature increase, the gradual decrease in mass was ascribed to the evaporation of bound water and the release of vo- latile matter present in GDAC. Sharp decrease of weight evidenced at temperatures between 80°C and 1000°C was caused by thermal degradation of some structure within GDAC takes place. The appro- priate temperature for the carbonization of prepared carbon is found to be 650°C.

The DSC curve (Fig. 7) shows a little drop at a temperature of 100°C, because the sample needs heat to evaporate the remaining water. From 125°C, organic compounds started to decompose and release heat. When temperature was about 500°C, DSC arrived at valley, where the activated carbon containing organic compounds were decomposed into gases like CO, CO_2, H_2, H_2O, CH_4, SO_3, acetic acid and methanol. After 500°C, DSC curve decline slowly till 800°C due to the vaporization of the products by adsorbing large heat and there was a steep decline from 800°Cto 1000°C this may be due to the release of residual volatiles from the firststage. Pyrolysis reaction was over at 1000°C and the residue was just carbon and ash. The yield obtained was 56%.

3.6. *Proximate and ultimate analysis*

Moisture content (%) by mass, ash content (%) by mass, volatile matter and fixed carbon content were analysed as per standard procedures and listed in Table 2. The moisture content has no effect on the adsorptive power of the carbon, but its presence dilutes and increases the weight of carbon. Ash content generally gives an idea about inorganic constituents associated with carbon [29], the lesser value of ash content indicate that the prepared GDAC has lower inorganic content and higher fixed carbon. Volatile matter is due to the presence of highly porous organic compounds present in the raw material. The fixed carbon content was obtained by subtracting moisture, volatile matter and ash content from 100%.High value of ash and volatile matter reduces the quantity of fixed carbon.

The percentage (by mass) of carbon, hydrogen, nitrogen, sulphur and oxygen (by difference) present in the adsorbents were quantified by Ele- mentary Vario ELIII-Germany CHNS Analyser and listed in Table 2

3.7. *FTIR spectral analysis*

The FTIR spectra of GDAC before (Fig. 8a) and after adsorption of DB 2 (Fig. 8b) and RR152 (Fig. 8c) were carried out by Shimadzu (IR affinity-1) Spectrometer. For the unloaded GDAC, a small band locat- ed at 3757.3 cm^{-1} due to NeH stretching vibration and a wide band lo- cated at 3452.5 cm^{-1} represent OeH stretching vibration. The band at 2380.1 cm^{-1} and 2276.8 cm^{-1} indicate presence of both meth- ylene (–CH_2–) bridges and aromatic CeH stretching vibrations. The band at 1759.1 cm^{-1} indicates C▐O stretching vibration of amide, also the bands located at 1581.6 cm^{-1} and 1635.64 cm^{-1} are the existence of amide I (a CO stretching mode conjugated to a NeH de- formation mode) and amide II (the NeH deformation mode conjugated to a C]N

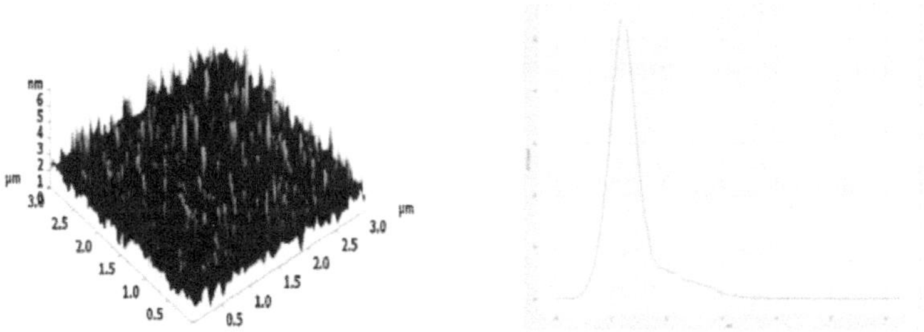

a b

Fig. 5. a Three-D surface Topography ofGDACCFC. b Histogram of GDAC-CFC.

Table 1
Surface Characterisation.

Samples	BET-surface area (m^2g^{-1})	Total pore volume (cm^3g^{-1})	Saturation magnetization ($emug^{-1}$)
GDAC	253.8	0.495	0
GDAC/ CFMC	129.1	0.283	26.7

deformation mode) respectively [30]. The band at 1438.9 cm^{-1} cor- responds to C]C vibrations of aromatic ring. The stretching vibra- tion of eCH_3 at 1396.4 cm^{-1} is related to methyl structures [31]. Small additional peak observed at 1265.3 cm^{-1} and 1145.7 cm^{-1} indicate the presence of S]O group. These FTIR results (Fig. 8a) show the ex- istence of amine, hydroxyl, carbonyl and amide groups which are the sorption sites for dyes. After adsorption of dyes, the band at 3757.3, 3452.5, 2279.8, and 1438.9 cm^{-1} (shown in Fig. 8a) were
shifted to higher frequency at 3768.9, 3572.1, 2291.4 and 1469.7

cm^{-1} (shown in Fig. 8b & c). The band at 1635.6 cm^{-1} disappeared af- ter adsorption.

The shift and disappearance of different functional groups on the ad- sorbent after adsorption indicated the possible interaction of those ac- tive sites for uptake of DB2 and RR152. Similar results were observed for the removal Pb (II) using activated carbon prepared from goat dung [32].

3.8. *Effect of initial dye concentration and contact time*

The data interpretation of DB 2 and RR 152 sorption on GDAC at vari- ous initial concentrations are shown in Fig. 9a & b. The percentage of ad- sorption efficiency was observed to decline with increasing initialutiliza- tion of fly ash and saw dust have been studied [34,33]. Contact time isthe agitation time required for the ad- sorbent–adsorbate system to attain anequilibrium point. The equilibrium attainments of the mesoporous activat- ed carbons are greater than the microporous carbons. Also, the equilibra- tion time for the adsorbents with smaller particle sizes have been reportedto be slightly lower than those with larger particle sizes [35]. The present findings indicate a maximum amount was adsorbed initially at 60 min and 90 min contact time for DB 2 and RR152 respectively being fixed as the equilibration time intervals. In Fig. 9a & b, the curves are single, smooth and continuous indicative of saturation, suggesting the possibility of monolayer coverage of the dye onto the carbon surface [36]. Least changes were recorded in the sorption capacity of GDAC after the at- tainment of equilibrium, henceforth 60 min for DB 2 and 90 min for RR 152 were fixed as an agitation time intervals.

3.9. *Effect of adsorbent dosage*

The effect of sorbent dosage was studied with initial dye concentration of 100 mg/L, pH of 6, temperature of 30°C and an agitation time of 60 min (DB 2) and 90 min (RR152) to achieve maximum dye removal.

Significant rise with doses were observed, this increasing trend of sorption against doses indicates greater availability of GDAC molecules for trapping the dyes. Saturation point of sorption takes place

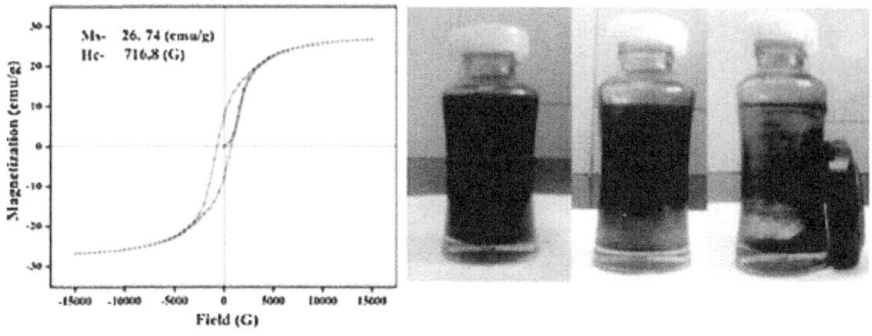

Fig. 6. Magnetic hysteresis loops of synthesized GDACCFC at room temperature; the inset of figure shows the well-dispersed solution by magnetic separation under an external magnetic field.

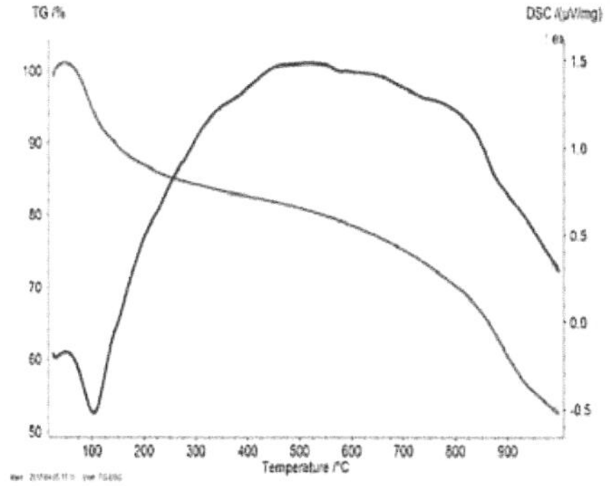

Table 2
Elemental Analysis of GDAC

Proximate analysis (%)		Ultimate analysis (%)	
Moisture content	6.5	Carbon	69.2
Volatile matter	13.2	Hydrogen	1.44
Ash content	4.1	Nitrogen	2.84
Fixed Carbon	70.2	Sulphur	4.17

at 150 mg and 250 mg for DB2 and RR152 respectively (Fig.S3- Supplimentary material). So these doses were optimized for further experiments

3.10. Effect of pH

The adsorption capacity of any material is influenced not only by its textural properties such as specific surface area, pore size and pore size distribution, but also by the chemical nature of its surface. The chemical characteristics can considerably be altered by adjusting the pH level of an adsorption system [37]. At low pH, the prevalent positive surface charges of GDAC attract the negatively charged functional groups pre- sent in the anionic dyes. An increase in pH value leads to the rise in the number of negatively charged sites, which further promotes the competi- tion among the sorption sites between the negatively charged hydroxyl ions and anionic dyes, resulting in the decline of adsorption rate [38]. The above statement matches with our experimental results where max- imum of 98.6% (DB2) and 99.3% (RR152) were observed at pH 2 (Fig. 10), followed by a dip in the curves.

3.11. *Effect of temperature*

Effect of temperature is a significant physico-chemical process parame- ter, since temperature influences adsorption capacity. A decrease in the adsorption capacity with increasing temperature has been observed for boththe systems as it is obvious from the significant decline in the straightlines (Fig. 11). This may be due to the weak bond formation at higher tem- peratures between the interacting dye mole- cules and the functional groups of the adsorbent [39].

3.12. *Adsorption isotherm*

The fit in, Langmuir and Freundlich isotherm models have been stud- ied to assess the adsorption capacity. The surface phase may be considered as a monolayer or multilayer depending upon the linearity of the plots

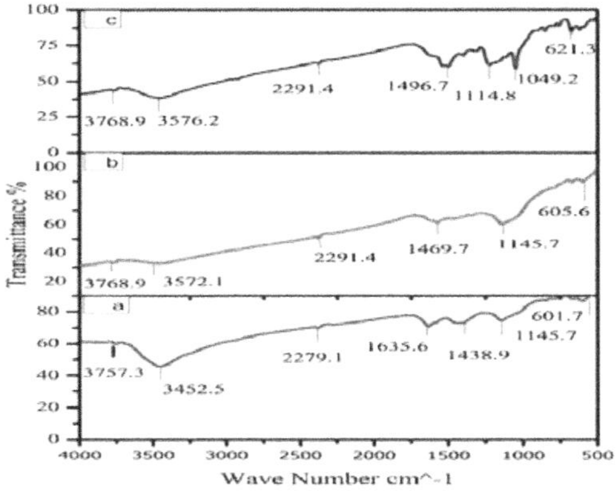

Fig. 8. FTIR Spectra of GDAC (8a) before and after adsorption ofDB2 (8b) and RR152 (8c).

3.12.1. *Langmuir isotherm*

Langmuir model can be expressed by following equation (Langmuir,1918) [40]

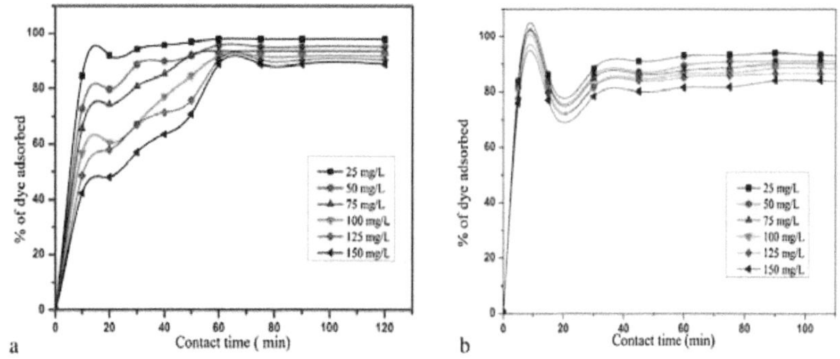

Fig. 9. a Eaffect of initial concentration and contact time(GDAC − DB2). b Effect of initial concentration and contact time (GDAC − RR152).

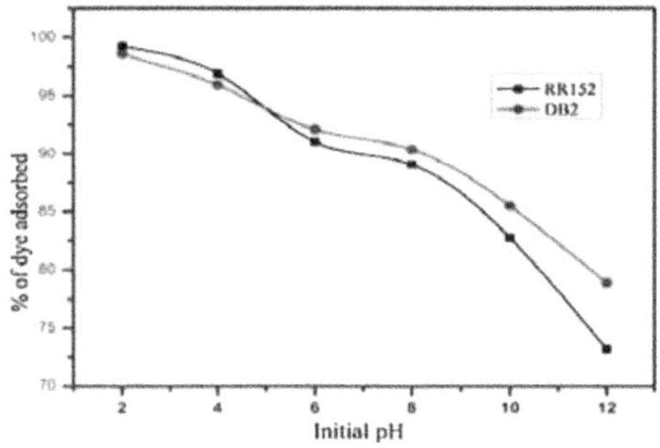

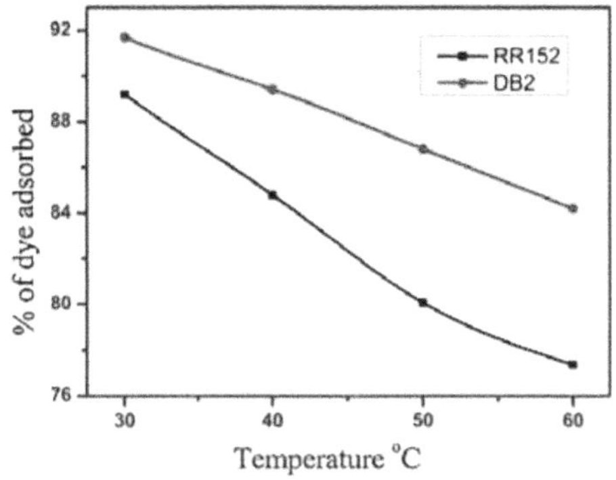

Fig. 10. Effect of pH. Fig. 11. Effect of Tempera-ture.

$$\frac{C_e}{q_e} = \frac{C_e}{q_m} + \frac{1}{bq_m}$$

where q_e is equilibrium amount of solute adsorbed per unit weight of adsorbent (mg/g); C_e is equilibrium concentration (mg/L); q_m is the maximum amount adsorbed per unit mass of adsorbent (mg/g); b is a constant related to the affinity of t he binding sites and energy of adsorption in (L/mg). A plot of C_e/q_e vs C_e (Fig.S4-Supplementary material) would give the value of q_m and b from the slope and intercept respectively.Higher the value of b is the affinity of adsorbent for the metal to be sorbed.

The linearity of the Langmuir isotherm can be used to predict whether a sorption system is favourable or unfavourable. The essential characteristics of Langmuir isotherm can be expressed in terms of a dimensionless param- eter called the separation factor or equilibrium parameter R_L, which is defined by the following relationship

where, C_i is the initial metal ion concentration (mg/L). The parameter R_L indicates the shape of the isotherm and the nature of the sorption processshall be

$R_L > 1$ Unfavourable Isotherm $R_L = 1$ Linear iso-therm

$R_L = 0$ Irreversible isotherm

$0 < R_L < 1$ Favourable isotherm

The Langmuir isotherm data for GDAC – DB2 and GDAC – RR152 systems are given in Table 2. The coefficient of determination (R^2) for both the systems studied was approximately 0.98. The R_L values calcu-lated with initial concentration range 25–150 mg/L were between ($0 < R_L < 1$) (listed in Table S7-supplimentry material), indicating favoura- ble adsorption.

3.12.2. *Freundlich isotherm*

Freundlich equation (Freundlich 1906) [41] is basically empirical, often used to describe heterogeneous systems with low to intermediate concen- tration ranges. The isotherm assumes that adsorbent surface sites have a spectrum of different binding energies and adsorption could occur via multiple layers instead of single layer [42,43]. The Freundlich equation is given by,

$$q = K_F\ C_e^{1/n} \qquad (5)$$

A linear form of above equation is used to calculate the constants'K_F' and 'n'

$$\log q_e = \log K_F + 1/n \log C_e \qquad (6)$$

whereq_e (mg/g) = amount of solute adsorbed per unit weight of ad- sorbent ,C_e (mg/L) = equilibrium concentration of the solute in the bulk
solution

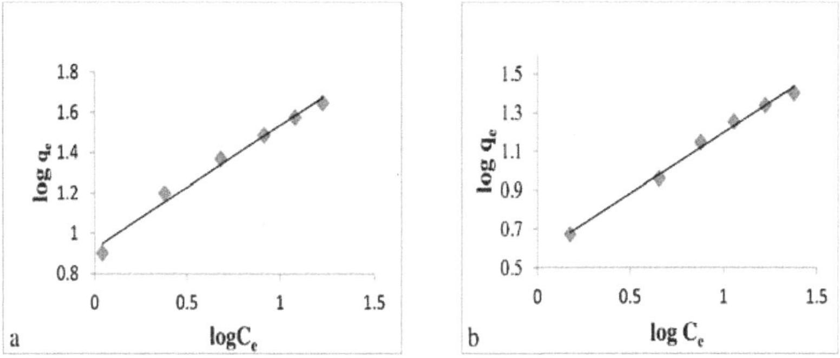

Fig. 12. a. Freundlich isotherm (GDAC-DB2) Fig. 12b Freundlich isotherm (GDAC-RR152).

Table 3
Isotherm parameters.

	Langmuir				Freundlich		
Dye	q_m (mg/g)	b(L/mg)	R^2	R_L	1/n	K_F (mg g^{-1} (mg $L^{-1})^{-1/n}$	R^2
DB2	66.66	0.1209	0.988	0.0763	0.612	8.356	0.994
RR152	32.26	0.1648	0.967	0.0571	0.6240	3.706	0.992

Table 4
Adsorption capacity – Comparison.

Adsorbent	Dye	Adsorption capacity (mg g^{-1})	Reference
Goat dung	Direct brown 2	66.6	Present study
Goat dung	Reactive red 152	32.3	Present study
Sewage sludge	Direct red	49.2	[44]
	Direct black	28.9	
Rice husk	Direct Red 31	74.1	[45]
	Direct orange 24	29.4	
Orange peel	Direct red 23	10.7	[46,47]
	Direct red 80	21.1	
Fly ash	Reactive Red 194	4.71	[34]
Pine-fruit shell	Reactive Red 194	20.9	[48]

Table 5
Kinetic parameters.

Dye	Conc Mg/L	q_e exp (mg/g)	First order Kinetics			Second order Kinetics		
			k_1 (min-1)	q_e cal (mg/g)	R^2	k_2 - 10^{-4} (g/mg min)	q_e cal (mg/g)	R^2
DB2	50	16.0	0.023	2.0	0.946	0.004	17.2	0.998
	100	30.6	0.032	7.1	0.982	0.006	35.7	0.992
	150	44.4	0.048	31.2	0.915	0.009	55.5	0.889
RR152	50	9.1	0.008	2.1	0.888	0.002	9.2	0.999
	100	17.7	0.006	11.8	0.936	0.011	17.8	0.999
	150	25.2	0.004	16.2	0.941	0.014	25.6	0.999

K_F (mg g^{-1} (mg $L^{-1})^{-1/n}$ = Freundlich constant indicative of the relative adsorption capacity of the adsorbent

1/n = adsorption intensity.

The value of 1/n lying between 0 and 1 indicates the adsorption processto be favourable. The Freundlich constants K_F and ‘n’ were determined from the intercept and slope of the plot log q_e versus log C_e respectively (Fig. 12a & b).The values of ‘n’ greater than 1 (Table 3)

Table 6
Effect of GDAC-CFMC dosage on DB2 and RR152 dyes.

Mass of adsorbent (mg)	Percentage Removal (GDAC-CFMC)		Percentage Removal (GDAC)	
	DB2	RR152	DB2	RR152
25	82.5	73.8	50.1	31.5
50	99.0	80.4	71.6	58.2
100	99.2	92.5	83.4	61.36
150	99.2	98.2	90.0	72.98

implies favourable nature of adsorption.

The values of correlation regression co-efficients obtained for GDAC – DB2 and GDAC - RR152 suggested that both the systems fitted well to the Freundlich isotherm model.

Table 4 lists the comparative study on the adsorption capacities of different materials with the chosen material for the removal of direct and reactive dyes. Goat dung is observed to possess better sorption capacity than those reported in literature.

3.13. *Adsorption Kinetics*

Pseudo first order and pseudo second order kinetic models were applied to the DB2-GDAC and RR152-GDAC systems. Pseudo first order kinetic model rate equation is as follows

$$\ln(q_e - q_t) = \ln q_e - k_1 t \qquad (9)$$

where q_e (mg/g) and q_t (mg/g) shows amount of dyes adsorbed on the GDAC at equilibrium and at time t and k_1(min^{-1}) is pseudo first order rate constant. A plot of ln (q –q) vs‘t’ gave a straight line, whereto be

$$\frac{t}{q_t} = \frac{1}{k_2 q_e^2} + \frac{t}{q_e}$$

fa- vourable, spontaneous and exothermic. Experimental results pertaining to the utilization of GDACCFMC implied better sorptive nature as compared to that of GDAC even at lower dosages. Thus it is concluded that, GDAC and GDACeCFMC are promising derived sorbents from litter goat dung, the latter possessing greater potentiality due to its nanosize and magnetic property resulting in the enhancement of dye removal due to its larger sur- face area.

where k_2 (g/mg/min) is second order rate constant. Plots of (t/qt vs t) (Fig. S6-suplimentary material) were found to be linear. From slope and inter- cept values, q_e and k_2 were calculated respectively. The Constants were
calculated from pseudo first order, pseudo second order kinetic models are given in Table 5.

The correlation coefficients R^2 for the pseudo first order kinetic model ob- tained for both dyes are lower than that of the pseudo second order kinetic model. The calculated and ex- perimental values of q_e for pseudo first ordersignificantly varied from each other, but larger deviations were found in- significant in the case of pseudo second order kinetic model [51].The abovestatement supports the better description for the adsorption system by pseudo second order model.
4. Effect of magnetic nano activated carbon-Cobalt ferrite composite on the removal of DB2 and RR152

The efficiency of the synthesized GDACCFMC for the removal of DB2 and RR152 was experimentally verified under optimized conditions, where the composite dose requirements alone were significantly different. Further it was confirmed by performing batch experiments with varying doses, wherein lower doses of 50 mg and 150 mg were sufficient enough toexhibit maximum percentage removal of DB2 and

RR152 dye molecules respectively against the doses of 150 mg and 250 mg for GDAC as ad- sorbent (Table 6).

5. Conclusion

Adsorption process of anionic dyes onto activated carbon prepared from goat dung (GDAC) using H_2SO_4 and $CoFe_2O_4$ (GDACeCFMC) were studied. Both the GDAC and GDACCFMC were characterized using SEM/EDAX and XRD. The magnetic parameter of the GDACCFMC was confirmed by Atomic Force Microscope and Vibrating Sample Magnetometer. Two dyes, one direct dye and reactive dye were selected for the present investigation (DB2 & RR152). DB2-GDAC and RR152- GDAC systems were optimized through batch equilibration experi-
ments. The operating conditions were fixed as 60 min and 90 min contact time, pH 7, and 100 mg/L initial dye concentration at room temper-ature for the systems respectively.

The adsorption equilibrium was best described by Freundlich model for DB2-GDAC and RR152- GDAC system. Kinetic studies reveal the better fit of pseudo second order kinetic equation for both the systems. The positive values of ΔS^{oo}, the negative value of ΔG^{oo} and ΔH^{oo} revealed the nature of adsorption

Appendix A. Supplementary data

Supplementary data associated with this article can be found, in the online version, at http://dx.doi.org/10.1016/j.jece.2017.05.030.

[1] I. Feddal, A. Ramdani, S. Taleb, E.M. Gaigneaux, N. Batis, N. Ghaffour, Adsorption capacity of methylene blue, an organic pol-lutant, by montmorillonite clay, Desalin. Water Treat. 52 (13–15)(2014) 2654–2661.

[2] Y.Z. Jin, Y.F. Zhang, W,Micro-electrolysis technology for in- dustrial wastewater treatment, Bioresour. Technol. 15

(2003) 334–338.
[3]W.S. Wan Ngah, L.C. Teong, M.A.K.M. Hanafiah, Adsorption of dyes and heavy metal ions by chitosan composites: a review, Carbohydr. Polym. 83 (2011) 1446–1456.
[4]G. Sami, Sorption kinetics for dye removal from aqueous solu- tion using natural clay, J. Environ. Earth Sci. (2012) 30–40 (ISSN2).
[5]M.R. Malekbala, S.M. Soltani, S.K. Yazdi, S. Hosseini, Equilibrium and kinetic stu-dies of safranine adsorption on alkali-treated mango seed integuments, Int. J. Chem. Eng. Appl. 3 (3) (2012) 160–166.
[6]M. AzharulIslama, A. Benhouriaa, M. Asif, B.H. Hameed,Methylene blue adsorption on factory rejected tea activated carbon prepared by conjunction of hydrothermal carbonization and sodium hydroXide activation processes, J. Taiwan Inst. Chem.Eng (2015) 1–8.
[7] Mohammad Foroughi-dahr, Hossein Abolghasemi, Mohamad Esmaieli,Ghadir Nazari, Bettina Rasem, EXperimental study on the adsorptive behaviour of Congo red in cationic surfactant-modified tea waste, Process Saf. Environ. Prot. 95 (2015) 226– 236.
[8] Mohammad Foroughi-Dahr, Hossein Abolghasemi, Mohammad Esmaili, Alireza Shojamoradi & Hooman Fatoorehchi, adsorption characteristics of Congo red from aqueous solution onto tea waste, Chem. Eng. Commun. 202 (2015) 181–193.
[9] F. Marrakchi, Muthanna J. Ahmed, W.A. Khanday, B.H. Hameed, Mesoporous carbonaceous material from fish scales as low-cost adsorbent for reactive orange 16 adsorption, J. Taiwan Inst. Chem. Eng. (2016) 1–8.
[10]G. Annadurai, R.S. Juang, D.J. Lee, Use of cellulose-based wastes for adsorption of dyes from aqueous solutions, J.

Hazard. Mater. B 92 (2002) 263–274.
[11]F.D. Ardejani, K. Badii, N.Y. Nimall, S.Z. Shafaei, A.R. Mirhabibi, Adsorption of direct red 80 dyes from aqueous so- lution on to almond shells: effect of pH, initial concentration and shell type, J. Hazard. Mater. 151 (2008) 730–737.
[12]B.H. Hameed, M.I. El-Khaiary, Removal of basic dye from aqueous medium using a novel agricultural waste material: pumpkin seed hull, J. Hazard.,Mater. 155 (2008) 601–609.
[13]B.H. Hameed, M.I. El-Khaiary, Sorption kinetics and isotherm studies of a cationic dye using agricultural waste: broad bean peels, J. Hazard. Mater. 154 (1–3) (2008) 639–648.
[14]M.S. Tanyildizi, Modeling of adsorption isotherms and kinetics of reactive dye from aqueous solution by peanut hull, Chem. Eng J. 168 (2011) 1234–1240.
[15]M. Husseien, A.A. Amer, Azza EI- Maghraby, A. Nahla, J. Appl. Res. 3 (11) (2007) 1352–1358.
[16]Lina Shao, Zongming Ren, Gaosheng Zhang, Linlin Chen, Facile synthesis, char- acterization of $MnFe_2O_4$/activated carbon mag- netic composite and its effectiveness in tetracycline removal, Mat. Chem. Phys. 135 (2012) 16–24.
[17]V. Ranjithkumar1, A. Nizarul Hazeen, M. Thamilselvan, S. Vairam, Magnetic acti-vated Carbon-Fe_3O_4 Nanocomposites— synthesis and application in the removal of acid yellow dye 17 from water, J. Nanosci. Nanotechnol. 14 (2014) 1–11.
[18] Na Yang, Shenmin Zhu, Shi Xu Di Zhang, Synthesis and prop- erties of magnetic Fe_3O_4- activated carbon Nanocomposite par- ticles for dye removal, Mater. Lett. 62

(2008) 645–647.

[19] Lunhong Ai, Haiyan Huang, Zhonglan Chen, Xing Wei Jiang, Activated carbon/ $CoFe_2O_4$ composites: facile synthesis, magnet- ic performance and their potential application for the removal of malachite green from water, Chem. Eng. J. 156 (2010) 243–249.

[20]X.M. Liu, S.Y. Fu, L.P. Zhu, High-yield synthesis and character- ization of mono-disperse sub-microsized $CoFe_2O_4$ octahedra, J. Solid State Chem. 180 (2007) 461–466.

[21]T. Santhi, S. Manonmani, V.S. Vasantha, Y.T. Chang, A new alternative adsorbentfor the removal of cationic dyes from aqueous solution, Arab. J. Chem. 9 (2016) S466–S474.

[22]G. Zhang, J. Qu, H. Liu, A.T. Cooper, R. Wu, $CuFe_2O_4$/activated carbon composite: a magnetic adsorbent forthe removal of acid orange II and catalytic regeneration, Chemo- sphere 68 (2008) 643–647.

[23]R. Sivashankar, A.B. Sathya, K. Vasantharaj, V. Sivasubramani- an, Magnetic com-posite an environmental super adsorbent for dye sequestration –A review, Environmental Nanotechnology, Monit. Manage. 1–2 (2014) 36–49.

[24] Hu. Jing, M.C. Irene, Lo Guohua Chen, Comparative study of various magnetic nanoparticles for Cr(VI) removal, Sep. Purif. Technol. 56 (2007) 249–256.

[25] LiXia Wanga, Jianchen Li, Yingqi Wanga, Lijun Zhaoa, Qing Jiang, Adsorption capability for Congo red on nanocrystalline MFe_2O_4 (M = Mn, Fe, Co, Ni) spinel ferrites, Chem.Eng. J. 181–182 (2012) 72–79.

[26]J. Raffieabaseri, P.N. Palanisamy, P. Sivakumar, Adv. Appl.

[27]E. Guibal, C. Roulph, P. Le Cloirec, Infrared spectroscopic study of biomass and materials of biological origin, Envi. Sci. Technol. 29 (1995) 2496–2503.

[28]T.H. Liou, Development of mesoporous structure and high ad- sorption capacity of biomass-based activated carbon by phos- phoric acid and zinc chloride activation, Chem. Eng. J. 158 (2010) 129–142.

[29] Mambo Moyo, George Nyamhere, Edith Sebata, Upenyu Guyo, Kinetic and equili-brium modelling of lead sorption from aqueoussolution by activated carbon from goat dung, Desalin.Water Treat. (2014) 1–11.

[30] Atul Kumar, Pratibha Chaudhary, Poonam Verma, Adsorption of reactive red 194 dye from textile effluent by using class F fly ash, Sch. J. App. Med. Sci. 1 (2) (2013) 111–116.

[31] Mustafa T. Yagub, Tushar Kanti Sen, Sharmeen Afroze, H.M. Ang, Dye and its re-moval from aqueous solution by adsorption:a review, Adv. Colloid Interf. Sci. 209 (2014) 172–184.

[32]M. Otero, F. Rozada, L.F. Calvo, A.I. Garcı´ a, A. Moran, Elim-ination of organic water pollutants using adsorbents obtained from sewage sludge, Dyes Pigm. 57 (2003) 55–65.

[33]I.D. Mall, V.C. Srivastava, N.K. Agarwal, I.M. Mishra, Removal of congo red from solution by bagasse fly ash and activated car- bon: kinetic study and equilibrium isotherm analyses, Chemo- sphere 61 (2005) 492–501.

[34]M.J. Martin, A. Artola, M.D. Balaguer, M. Rigola, Activated carbons developed from surplus sewage sludge for the removalof dyes from dilute aqueous solutions, Chem. Eng. J. 94 (2003) 231–239.

Removal of Direct Dye using Activated Carbon Prepared from *Prosopisjuliflora* Bark: Isothermal, Thermodynamic and Kinetic Studies

N. Muthulakshmi Andal[1]* and K. S.Thangamani[2]

[1]Department of Chemistry, PSGR Krishnammal College for Women, Coimbatore-4
[2]Department of Chemistry, Dr. N.G.P Institute of Technology, Coimbatore-

Abstract

Activated carbon of *Prosopis juliflora* Bark (PJBAC) is prepared using muffle furnace and employed for the removal of Direct Brown MR (DBMR) dye from aqueous solutions. Surfacemorphological studies carried out for unloaded and loaded carbons; indicate the surface changes in the porous nature of the dye loaded PJBAC. Batch equilibration experiments pertaining to the study of operating factors viz., varying initial dyeconcentrations, preset time intervals and differentdoses of PJBAC, variable pH and temperature environments are verified to assess the sorptive nature of PJBAC. Optimized conditions have been setup as 400 mg/L initial concentration; 60 min contact time; 100 mg dosage, pH 2 and 30^0Ctemperature for a maximum of 90.2% dye re- moval with a sorption capacity of 180.4 mg/g. The isothermal plots for Langmuir, Freundlich and Tempkin equations are tested, wherein the best fit of linearity is observed for both Langmuir and Freundlich adsorption isotherms, suggesting monolayer/multilayer sorp- tion. Thermodynamic parameters are calculated to study the nature of sorption,observed as endothermic, irreversibleand spontaneous for DBMR-PJBAC sys-tem. Kinetic studies reveal the applica- bility of second order kinetic model.

Keywords: Direct Brown MR dye,*Prosopis juliflora* bark activated carbon, adsorption, Isotherms, Ki- netics.

Introduction

Effluents discharged from textile, paper, rubber, plastics, leather, cosmetics, pharmaceutical and food industries have been found to be the main sources of water pollution. Textile indus-tries discharge a lot of wastewaters containing different dyes like direct, reactive, mordant, acid and basic dyes. The direct dyes are the second largest dye class in the color index, which are used for dyeing rayon, linen, cellulosic fibers, wool and silk. Generally, dyes havea relatively low fixation degree on the textile substrate and hence 10-50% of the dye is un-exhausted and remains in the dye bath as a waste [1]. These dyes are recalcitrant molecules, which are difficult to degrade biologically [2]. They are carcinogenic, mutagenic and terato- genic affecting human beings, fish species and microorganisms. Hence, their removal from wastewaters has become environmentally important [3]. Thus, the removal of dyes from colored effluent of textile industries is of major concern. Various techniques like chemical coagulation [4], electrolysis, chemical oxidation [5], photochemical degradation [6], mem- brane filtration [7], biological treatment, adsorption by activated carbon [8] are known to be effective for the removal of dyes from polluted water. Amongst, the numerous techniques available for dye removal, adsorption is the procedure of choice giving excellent results as it can be utilized for decolouration of dye effluents [9, 10]. Most commercial systems currently use activated carbon as a sorbent to remove dyes from effluents because of its excellent adsorption ability. However, although activated car- bon is a preferred sorbent, its widespread use is restricted due to high cost. In order to de- crease the cost of treatment, various attempts have been made to find inexpensive adsor-bents. A sorbent can be considered low-cost if it requires little processing, being abundant in nature, a by-product or waste material generated from another industry [11]. Such alterna- tives include low cost activated carbons prepared from natural products like bagasse [12], nutshells [13], Jute fiber [14], rice husks

[15], plum kernels [16] and coconut shells [17]. Ac-tivated carbons prepared from various products are widely used as adsorbents due to their high adsorption capacity, large surface area, microporous nature, chemical and thermal stability. In this article, an attempt has been made by

employing phosphoric acid activated carbon prepared from *Prosopis juliflora* bark for the removal of Direct brown MR dye from aqueous solution, the first of its kind not reported elsewhere.

Materials and Methods

Preparation of Dye Solution

Direct brown MR was procured from Sigma Aldrich and was used without further purifi- cation. A stock solution of the dye (1000 mg/L) was prepared using doubly distilled water and the dilutions for the required concentrations were made. The characteristics and chemi- cal structure of the dye is listed and depicted in **Table 1** and **Figure 1** respectively.

Table 1 Characteristics of Direct Brown MR.

Molecular Formula	C29H19N5Na2O7S
Molecular Weight	627.54
C.I Name	Direct Brown 2
C.I Number	22311
CAS Number	CAS 2429-82-5
λmax	420nm

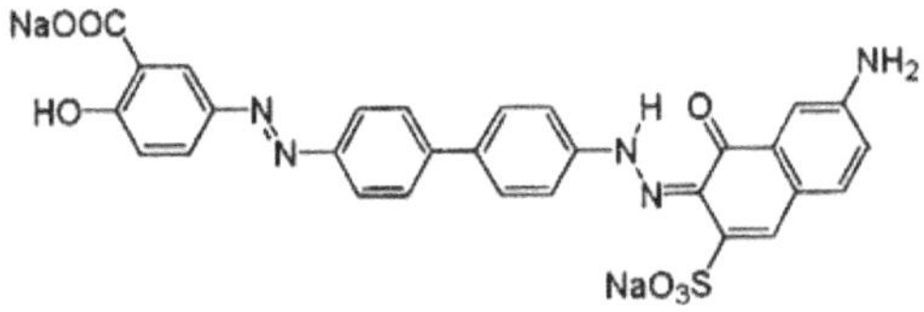

Figure 1 Structure of Direct brown MR.

Adsorbent Preparation/Characterization

The *prosopis juliflora* bark was collected from various localities in Coimbatore, washed wellwith distilled water and sun dried. The dried materials were soaked in boiling solution of 35% Phosphoric acid, subjected to charring and were separated, air dried and carbonized at 500^{0}C in muffle furnace for 1hr. Later, the material was powdered after cooling/activated at 800^{0}C for 10 min, washed well with distilled water. The pH was maintained at 7±0.2. The dried powder material was ground and stored in an airtight container and utilized for further experiments.

Results and Discussion *Morphological Studies*

Scanning Electron Microscopy

Scanning electron microscopy is an important tool employed for characterizing the surface morphology of the adsorbent before and after the adsorption process. **Figure 2a** shows large number of unoccupied pores present on the surface of PJBAC and layered structures were also visible, whereas in Figure 2b, almost all the pores are covered with dye molecules and the layered structure had disappeared, these differences obviously being visualised in the images supporting the sorptive affinity of PJBAC for DBMR dye molecule.

Batch Adsorption Experiments

Varying PJBAC doses (50 mg-200 mg: 50 mg intervals) were added to 50 ml of aqueous DBMR dye solutions(at pH-7 intervals) taken in 100 ml Erlenmeyer flasks with varying concentrations (50 mg/L- 400 mg/L: 50 mg/L intervals) and subjected to agitation in a me- chanical shaker at preset time interval(10 min-100 min:10 min interval). Residual concentra- tions of the filtered sample were estimated spectrophotometrically at a wavelength corre- sponding to maximum absorbance (λ_{max}=420 nm), using an UV/Vis spectrophotometer (La- bIndia UV 3000^{+} Model). The initial pH of the solution was adjusted by the addition of di- lute aqueous solutions of HCl or NaOH (0.1N). Experiments pertaining to different sorption conditions ie., pH (2 -12 : 2) and temperature (30^{0}C-60^{0}C: 10^{0}C intervals) were performed and the amount of dye adsorbed onto activated carbon was calculated as follows. Duplicate experiments were conducted to ensure the reproducibility ofvalues within ± 2%.

Figure 2 (a) SEM image of PJBAC Carbon (b) SEM images of dye loaded PJBAC Carbon$q_e = (C_0\text{-}C_e)\ V/W$ (1)

where, C0 and Ce are the initial and equilibrium concentrations (mg/L) of the dye respec- tively. V is the volume of thesolution and W is the weight of the adsorbent used.

Effect of Initial Dye Concentration and Contact Time

The adsorption process is highly dependent on initial dye concentration [18]. The increase indye concentrations from 50 mg/L to 400 mg/L decreased the percent removal from 91.8% to 83.6%, and the corresponding adsorption capacity calculated varying from 22.95 mg/g to

172.60 mg/g. It may be explained, at higher initial concentration, the ratio of initial number of dye molecules to the available surface area is high, hence fractional adsorption becomes dependent on initial concentration. For fixed adsorbent dose, the total available sites are lim-ited, adsorbing almost same amount of sorbate, resulting in a decrease in the percentage re- moval [18]. The curves depicted in **Figure 3** are single, smooth, continuous leading to satu- ration implied by plateau formation suggesting the possibility of monolayer coverage of dye molecules. Similar results have been reported in the employment of modified Mango Bark / Neem Bark powder for the adsorption of Malachite Green and Babul Seed Carbon Powder inthe Methylene Blue removal [19, 20].

The amounts of dye adsorbed in the initial stage of contact time, reached a maximum through steep rise upto a contact time of 60 minutes, later becoming almost constant denot- ing the attainment of equilibrium, which shall be due to the availability of large number of vacant surface active sites initially. After a lapse of time, when the adsorption of the dye molecules had occurred, the remaining unoccupied surface sites may be difficult to be filled in due to repulsive forces acting between the dye molecules and the carbon surface [21]. Thence, an initial concentration of 400 mg/L and 60 minutes contact time have been fixed as optimum conditions for further experiments.

Effect of Dosage

The influence of PJBAC dose furnished the information regarding sorption ability and an optimum requirement forthe maximum % removal as illustrated in **Figure 4**, which reveals that 100 mg itself exhibited a maximum dye removal of 86.3% at pH 7. Further increase in adsorbent dose, did not register significant increase, this may be attributed to saturation of adsorption sites due to particulate interaction such as aggregation. Such aggregation would lead to a decline in total surface area and increase in diffusion path length [22]. The adsor- bent dose was fixed as 100 mg.

Effect of pH

pH of the dye solution plays an important role in its degradation. Maximum adsorption of DBMR dye had occurred at pH 2, later a reduction in sorption was observed at increasing pH environments. This is observed obviously from the steep decline (**Figure 5**). Further re- duction at basic pH values can be explained by the fact that when the pH of thedye solu-tion is adjusted accordingly, the number of negatively charged sites will also increase, result- ing in least adsorption of anionic dyes [23].

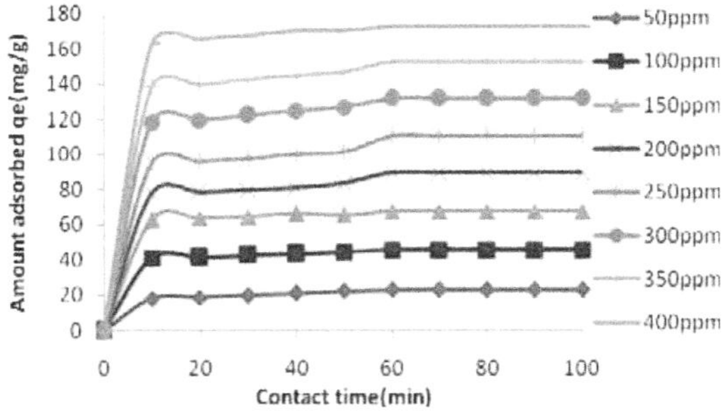

Figure 3 Effect of Initial Dye Concentration and Contact Time.

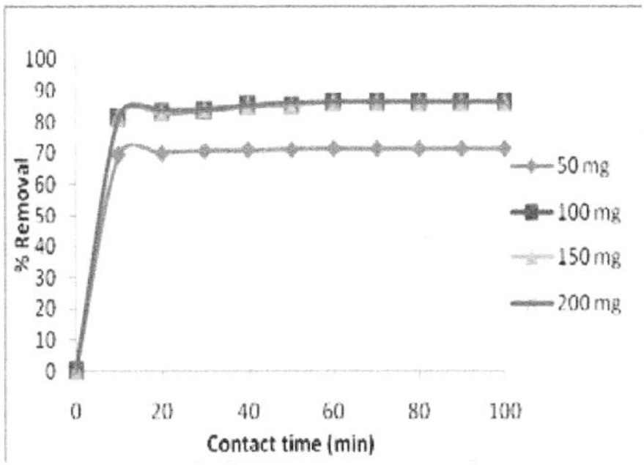

Figure 4 Effect of Dosage.

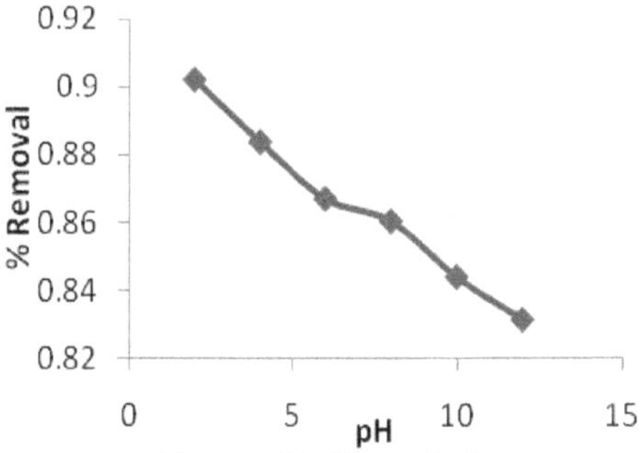

Figure 5 Effect of pH.

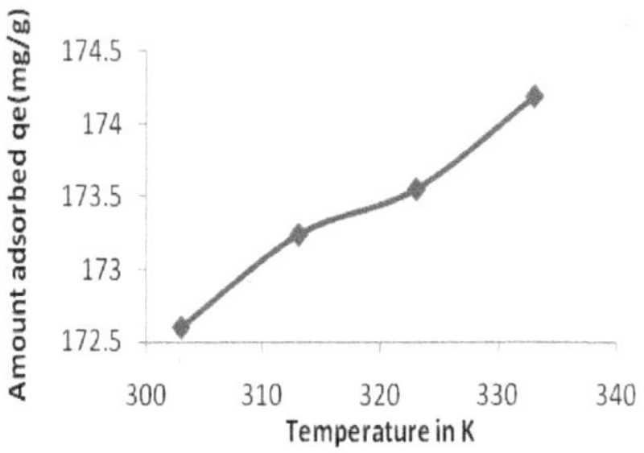

Figure 6 Effect of Temperature.

Effect of Temperature

Results pertaining to the influence of varying temperatures registered linear rise of q_e values (172.6 mg/g – 174.2 mg/g) with temperature (**Figure 6**). This trend reveals the increase inthe mobility of dye molecules and the decline ofretarding forces on the diffusing ions, there- by increasing the sorption capacity [18] of PJBAC.

Adsorption isotherm

The adsorption isotherm is fundamental in describing the interactive behaviour between ad- sorbent and adsorbate and is important in the design of adsorption system [24]. The most widely employed isotherms viz., Langmuir, Freundlichand Tempkin models were studied to predict the nature of the system at different initial dye concentrations with a dose of 100 mg.

Langmuir isotherm

In 1916, Langmuir developed a model that was based on the theory of adsorption of solute molecules on adsorbent at homogeneous sites with uniform energy levels [25]. The isothermdescribes that no further adsorption takes place once the active sites are covered, that is only monolayer sorption occurs on the solid surface. The linearised form of the Langmuir equa- tion is as follows

$$C_e/q_e = 1/(aQ_m) + C_e/Q_m \quad (2)$$

where, C_e (mg/L) is the concentration of dye solution at equilibrium, q_e (mg/g) is the amountof dye adsorbed at equilibrium, Q_m is the binding capacity and 'a' is binding constant. The Q_m and 'a' values are calculated from the slope ($1/Q_m$) and intercept ($1/a\ Q_m$) respectively of the linear plot(C_e/q_e vs C_e). The linearity of the plot observed for C_e/q_e vs C_e (**Figure 7**) re- veals the applicability of the Langmuir model.

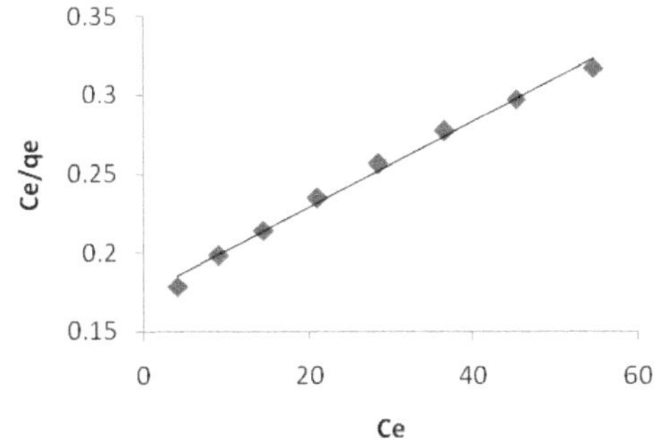

Figure 7 Langmuir isotherm.

Freundlich isotherm

Freundlich isotherm can be derived assuming a logarithmic decrease in the enthalpy of ad- sorption with increase in thefraction of occupied sites and the linearised form of equation:

$$\text{Log } q_e = \log K_F + 1/n \log C_e \qquad (3)$$

where, C_e, is the equilibrium dye concentration in solution (mg/L); A plot of log q_e vs log C_e is depicted in **Figure 8**, from which the constants K_F and 1/n are calculated. K_F and 1/n are isothermal constants indicating the sorption capacity and sorption intensity respectively. The studied system follows Freundlich isotherm, as is evident from the graph with good correla- tion. The favourability of the adsorption isothermal process can also be supported through dimensionless factor (R_L),

$$R_L = 1/1{+}bC_0 \qquad (4)$$

where, C_0 is the initial concentration (mg/L).

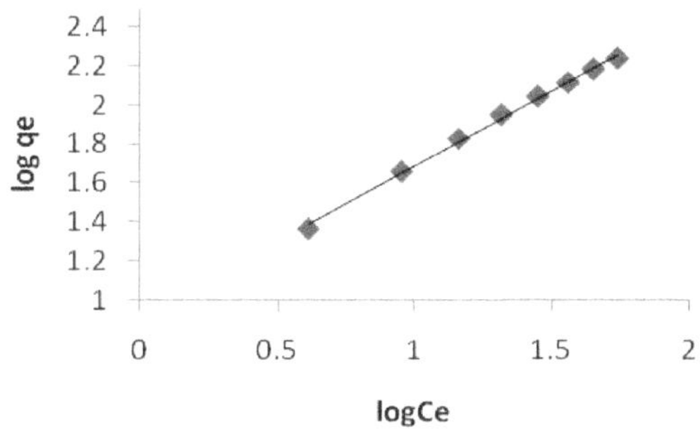

Figure 8 Freundlich isotherm.

The calculated R_L value for DBMR – PJBAC system was 0.17 indicating that the ad-sorption process to befavourable [26] as evidenced from **Table 2**.

Table 2 Equilibrium Parameter R_L.

R_L values	Nature of ad-sorption
$R_L > 1$	Unfavourable
$R_L = 0$	Irreversible
$R_L = 1$	Linear
$0 < R_L < 1$	Favourable

Tempkin isotherm

Tempkin and Pyzhey (1940) assume that the heat of adsorption would decrease linearly withincrease of coverage of adsorbent, which is due to some indirect adsorbate interactions on adsorption isotherm [27]. The linearised form of isotherm is

$$q_e = B_T \ln A_T + B_T \ln C_e \quad (5)$$

where, $B_T = RT/b_T$, T absolute temperature in Kelvin and R is the universal gas constant.The constant b_T is related to the heat of adsorption (kJ/mol) and A_T is the equilibrium bind- ing constant corresponding to the maximum binding energy. The Tempkin constants A_T and b_T were derived from the intercept and slope of the linear plots of q_e vs ln C_e shown in **Fig- ure 9**. The consolidated parametric values of Langmuir, Freundlich and Tempkin constants are listed in **Table 3**. The values for the constants were observed to be higher than those re-ported by other researchers as evident from **Table 4**. This proves the better sorption capacityof PJBAC in trapping the anionic DBMR dye molecules.

Thermodynamic Study

The feasibility and spontaneous nature of adsorption process is evaluated from the values of thermodynamic parameters (ΔG^0, ΔH^0& ΔS^0). These parameters were calculated using Van'tHoff equation [34].

$$\Delta G^\circ = -RT \ln K_d \quad (6)$$

$$\ln K_d = (\Delta S^0/R) - (\Delta H^0/RT) \quad (7)$$

where, ΔG^o is the free energy change (kJ/mol) and K_d is the distribution coefficient. ΔH^0&

ΔS^0 were calculated fromthe Van't Hoff Plot (**Figure 10**) and the corresponding values arelisted in **Table 5**. The negative value of Gibbs free

energy change (ΔG^0), positive values of enthalpy change (ΔH^0) and entropy change (ΔS^0)is indicative of thespontaneity, endothermicity and high disorderliness respectively.

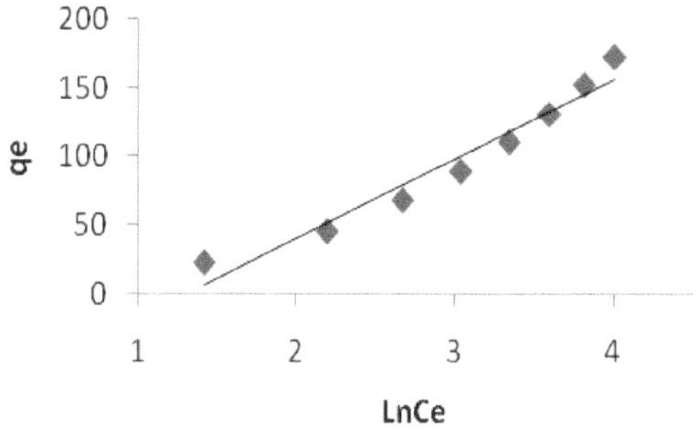

Figure 9 Tempkin isotherm.

Table 3 Isothermal Constants Data.

Isotherm models								
Lang-muir			**Freundlich**			**Tempkin**		
q_m (mg/g)	b(L/mg)	R^2	1/n	K_F(mg/g)	R^2	b_T	A_T	R^2
500	0.012	0.992	0.774	8.128	0.996	43.44	5.686	0.950

Table 4 Comparison of PJBAC against studied Activated Carbons.

Adsorbent	**Dye**	**Adsorption capacity (mg/g)**	**K_F (mg/g)**	**Reference**
Poly aniline coated Sawdust Composite	Direct Green 6	217.4	4.47	[28]
Pine Cone	Acid Black 26	62.89	19.47	[29]
	Acid Green 25	43.29	7.25	

	Acid Blue 7	37.45	7.95	
Prosopis juliflora Seed	Direct red 23	90.9	12.4	[30]
Rice Husk	Direct Red 31	129.87	12.34	[31]
	Direct Orange 36	66.67	4.581	
Pomegranate Peel	Direct Blue 106	58.14	2.301	[32]
Macore fruit Shell	Methylene Blue	10.6	1.728	[33]
	Methyl Orange	3.4	1.225	
Prosopis juliflora Bark	Direct brown MR	500	8.128	Present study

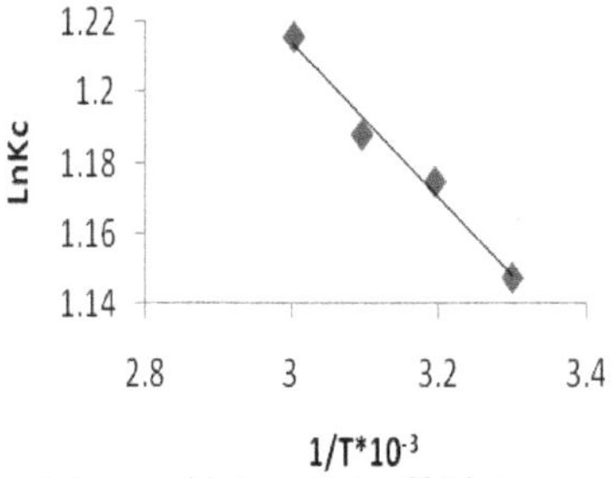

Figure 10 Van't Hoff Plot.

Table 5 Thermodynamic parameters.

Tem-p°C	**ΔG⁰ KJ/mol**	**ΔH⁰ KJ/mol**	**(ΔS⁰) KJ/mol**
30	-0.2890		
40	-0.3057	0.2221	1.878
50	-0.3190		
60	-0.3366		

Kinetic Studies

Pseudo-first-order and pseudo-second-order kinetic models have been investigated to assess the sorption dynamics of the current system. Pseudo-first-order equation employed to ex- plain the solid/liquid adsorption is as follows

$$\log(q_e - q_t) = \log q_e - k_{ad}\, t/2.303 \qquad (8)$$

where, q_e and q_t corresponds the amounts of DBMR adsorbed at equilibrium and time t re- spectively (mg/g), k_{ad} is the rate constant of pseudo-first-order adsorption (min^{-1}). A plot of $\ln(q_e - q_t)$ versus t is a straight line, where q_e and k_1 are calculated from intercept and gradientof the line, respectively [35].
Pseudo second order kinetic model is given by the following equation:

$$t/q_t = 1/k^2 q_e^2 + t/q \qquad (9)$$

where k_2 (g/mg /min) is second order rate constant. A plot of t/qt versus t (**Figure 11**) gives astraight line, and q_e and k_2 are calculated from gradient and intercept of the line, respec- tively [36]. Constants calculated from pseudo firstorder, pseudo second order kinetic mod- els are given in **Table 6**. When the calculated q_e values and experimentally verified q_e valuescalculated for pseudo first order are significantly different from each other, they are compa- rable in the case of pseudo second order kinetic model. On the other hand, when correlation coefficients of both models are compared, it is apparent that coefficients of pseudo second order model is higher and much closer to unity. Hence, itis obvious that DBMR adsorption on PJBAC is better described by pseudo second order model [37]. Similar trend was ob- served for the adsorption of direct dyes onto activated carbon prepared from saw dust [38], palm ash [39].

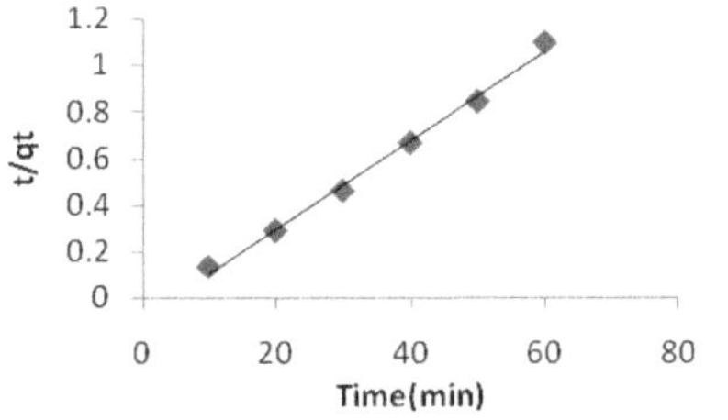

Figure 11 Pseudo second order plot.

Table 6 First-order and second-order kinetic constants.

		First order kinetic model			**Second order kinetic model**		
C_0 (mg/L)	$q_{e\ exp}$ (mg/g)	$q_{e\ cal}$ (mg/g)	k_1(min^{-1})	R^2	$q_{e\ cal}$ (mg/g)	k_2(min^{-1})	R^2
400	172.6	55.63	0.002	0.966	194.0	0.044	0.994

Conclusion

Phosphoric acid activated carbon was prepared from *Prosopis juliflora* bark, an ecofriendly tree waste. This carbon was characterized using SEM, which registered significant morpho- logical changes on the surface of both dye loaded and unloaded carbons. Its efficient role play as an adsorbent for removal of Direct Brown MR from aqueous solutions was experi- mentally verified by employing batch equilibration method. The maximum amount of dye adsorbed was optimized under different operating factors, initial dye concentration, contact time, PJBAC dosage, pH and temperature of the solutions. Verification of the isothermal studies viz., Langmuir, Freundlich and Tempkin revealed that both Langmuir and Freundlich isotherms had better linear fit compared to Tempkin model. Variations in free energy, enthalpy and en-tropy registered favourable nature of adsorption. The calculated rate of adsorption for the sorptive system follows second order kinetic model. Comparison of the efficiency of PJBACwith other prepared activated carbons exhibited better chelating property in the removal of anionic DBMR dye, justified by the isothermal constants. Thus, it is concluded that PJBAC possesses good sorption ability in the removal of dyes.

References

[1] Lazearidis N K, Darapantsios T D & Geiogantas D,Water Res 37(2003) 3023

[2] Ould Brahim et al., Discolouration of Azo dye solutions by Adsorption on Activated Carbon prepared from Cryogenic Grinding of used tires ,journal of "the Italian associ- ation of chemical engineering 38 (2014)

[3] Konicki.W et al., Equilibrium and kinetic studies on acid dye acid red 88 adsorption by magnetic ZnFe2O4 spinel ferrite nanoparticles. J. Colloid Interf. Sci 398 (2013)152-160,

[4] Eren Z and Acar FN; Adsorption of Reactive Black 5 from an aqueous solution: equilibrium and kineticstudies. Desalination, 194 (2006)1-10.

[5] Lin S H and Lo CC; Treatment of textile wastewater by foam flotation.Environ, Tech- nol., 17(1996), 841-9.

[6] Arsalin IA and Baicioglu Bahnemann DW; Advanced chemical oxidation of reac-tive dyes in Simulated dyehouse by ferrioxalate –fenton / UV-A and TiO2/UV-A process. Dyes & pigments, 47(3) (2000) 207-18

[7] Baoub M H V, Gauthier R, Gauthier H, Chabert B & Rammah M E, J Appl Polym Sci., 77 (2000) 171.

[8] Walkar G M & Weatherley L P, Environ Pollution, 108 (2008) 219.

[9] Jain, A.K., Gupta, V.K., Bhatnagar, A., Suhas,. Utilization of industrial waste prod- ucts as adsorbents for the removal of dyes, Journal of Hazardous Materials. B 101 (2003) 31–42.

[10] Derbyshire, F., Jagtoyen, M., Andrews, R., Rao, A., Martin-Gullon,I., Grulke, E.,. Carbon materials in environmental applications. In: Radovic, L.R. (Ed.), Chemistry and Physics of Carbon, Marcel Dekker, New York, 27 (2001) 1–66.

[11] Bailey, S.E., Olin, T.J., Bricka, M., Adrian, D.D., A review of potentially low-cost sorbents for heavy metals. Water Res. 33 (1999) 2469–2479.

[12] Tsai W.T. Chang C.Y. Lin M.C. Chien S.F. Sun H.F. Hsieh M.F., Chemosphere, 45(2001) 51.

[13] Aygun A. Yenisoy-Karakas S. Duman I.,Microporous and Mesoporous Materials66(2003) 189.

[14] Senthilkumar S. Varadarajan P.R. Porkodi K. Subbhuraam C.V., Journal of Colloidand Interface Science.,284 (2005) 78.

[15] Yalcin N. and Sevinc V., Carbon, 38 (2000) 1943.

[16] Wu F.C. Tseng R.L. Juang R.S., Journal of Hazardous Materials B, 69 (1999) 287.

[17] Su W. Zhou L. Zhou Y.P., Letters to the editor, Carbon, 41 (2003) 861.

[18] Theivarasu C, Chandra S, Performance of Elephant (Eliphas maximus) Dung Activat- ed Carbon for the Removal of Crystal Violet from Aqueous Solution, International Journal of Chem Tech Research. 6(2) (2014)1362-1376

[19] Ruchi Srivastava, D.C Rupainwar, A comparative evaluation for adsorption of dye on Neem bark and Mango bark powder, Indian journal of Chemical Technology 18(2011)67- 75

[20] M. Sujatha, A. Geetha, P. Sivakumar and P.N. Palanisamy "Orthophosphoric acid ac- tivated Babul seed carbon as an adsorbent for

the removal of Methylene Blue" E.journal of Chemistry 5(4) (2008)742-753.

[21] Malarvizhi T.S, Santhi T and Manonmani S. "A Comparative Study of Modified Lig- nite Fly Ash for the Adsorption of Nickel from Aqueous Solution by Column and

Batch Mode Study" Research Journal of Chemical Sciences 3(2) (2013) 44-53.

[22] Crini, G., Peindy, H. N., Gimbert, F., & Robert, C. "Removal of C.I. Basic Green 4 (Malachite Green) from aqueous solutions by adsorption using cyclodextrin based ad- sorbent: Kinetic and equilibrium studies".,Separation and Purification Technology, 53(2007) 97–110.

[23] Raffica Baseri J, Palanisamy P.N and Sivakumar P, Adsorption of reactive dye by a novel activated carbon prepared from Thevetia peruviana, International Journal of Chemistry Research, 3 (2012) 36-41.

[24] Sh. Wang, Z. H. Zhu, Characterisation and environmental application of an Australiannatural Zeolite for basic dye removal from aqueous solution, Journal of Hazardous Materials 136(2006) 946- 952.

[25] I. Langmuir, The constitution and fundamental properties of solids and liquids. Part I. Solids, Journal of American Chemical Society. 38 (1916) 2221–2295.

[26] Xiong, L., Yang, Y., Mai, J., Sun, W., Zhang, C., Wei, D., et al. Adsorption behaviourof methylene blue onto titanate nanotubes. Chemical Engineering Journal 156 (2010) 313–320.

[27] M.J. Temkin, V. Pyzhev, Recent modification to Langmiur isotherms, Acta Physi- ochim., USSR 12 (1940) 217–222.

[28] A. Geetha and P.N. Palanisamy Kinetics and Equilibrium Studies on The Removal of Anionic Dyes using Polyaniline Coated Sawdust Composite , International Journal of Chem Tech Research 7(5) (2014-2015) 2439-2447

[29] Niyaz Mohammad Mahmoodi , Bagher Hayati , Mokhtar Arami , Christopher Lan, Adsorption of textile dyes on Pine Cone from colored wastewater: Kinetic, equilibri- um and thermodynamic studies, Desalination 268 (2011) 117–125

[30] N. Gopal , M. Asaithambi , P. Sivakumar , V. Sivakumar, Adsorption studies of a di- rect dye using polyaniline coated activated carbon prepared from Prosopis juliflora , Journal of Water Process Engineering 2 (2014) 87– 95

[31] Yusra Safa, Haq Nawaz Bhatti, Kinetic and thermodynamic modeling for the removal of Direct Red-31 and Direct Orange-26 dyes from aqueous solutions by rice husk, De-salination 272 (2011) 313–322

[32] Nevine Kamal Amin, Removal of direct blue-106 dye from aqueous solution using new activated carbons developed from pomegranate peel: Adsorption equilibrium and kinetics, Journal of Hazardous Materials 165 (2009) 52–62.

[33] Xiong, L., Yang, Y., Mai, J., Sun, W., Zhang, C., Wei, D., et al. Adsorption behaviourof methylene blue onto titanate nanotubes. Chemical Engineering Journal 156 (2010) 313–320.

[34] Albadarina, Ahmad B., Mangwandi, Chirangano, Al-Muhtaseb, Ala'a H., Walker, Gavin M., Allen, Stephen J., Ahmad, Mohammad N.M, Kinetic and thermodynamics of chromium ions adsorption onto low-cost dolomite adsorbent. Chem. Eng. J.179 (2012)193–202.

[35] Lagergren S, Sven K & Sogenannten Z T D, Vetenskapsakad Handl, 24 (1898)1.

[36] Y.S. Ho, G. McKay, Kinetic models for the sorption of dye from aqueous solutionby wood, Journal of Environmental. Science. Health Part B: Process Saf. Environ. Prot. 76 (1998)183–191.

[37] H.B. Senturk, D. Ozdes, A. Gundogdu, C. Duran, M. Soylak, Removal of phenolfrom aqueous solutions by adsorption onto organo modified

Tirebolu bentonite: equilibri- um, kinetic and thermodynamic study, Journal of Hazardous Materials. 172 (2009)353–362.

[38] P.K. Malik, Dye removal from wastewater using activated carbon developed from sawdust: adsorption and kinetics, Journal of Hazardous Materials, B 113 (2004) 81– 88.

[39] A.A. Ahmed, B.H. Hameed, N. Aziz, Adsorption of direct dye on palm ash: kinetic and equilibrium modeling, Journal of Hazardous Materials. 141 (2007) 70–76.

Publication History

Received	15th Apr 2016
Revised	13th Nov 2016
Accepted	20th Nov 2016
Online	30th Dec 2016

A Comparative Study on the Sorption of Divalent ions by Bivalves Shells: Equilibrium and Statistical Studies

N. Muthulakshmi Andal*, S. Charulatha, Gayathri N.S and J. Anuradha Department of Chemistry, PSGR Krishnammal College for Women, Coimbatore-4

Abstract

The present study aims at examining the potenti- ality of the identified no- cost sorbent i.e. Bi- valves shells to remove Cu(II) and Zn(II) ions from the aqueous solutions. These shells are col- lected from seashores, washed with double dis- tilled water, dried, crushed and treated using 0.1N HCl. The characteristics and functional groups present in the treated shells are supported by SEM, EDAX and Physio-chemical parameter analyses. Various operating factors influencing the adsorption of divalent ions onto the treated shells are experimentally verified by Batch Equi-libration method. The experimental results de- rived the following optimized conditions : 0.18mm size, 1g dosage, 10 min agitation time, pH 7 environment for the trapping of Cu(II) ions (98.1% removal) and 0.18mm size, 1g dosage, 60min agitation time, pH 5.5 environment for that of Zn(II) ions (97% removal) at an initial concentration of 1000 mg/L in both the cases

Statistical analysis using SPSS software is carried out to assess the correlation of experimental values for both the systems with the theoretical approach to justifythe trapping ability of TMSP.

Keywords: bivalve shell, characteriza- tion, divalent ions, biosorption, statisti- cal studies.

***Correspondence**

Author: Muthulakshmi Andal
Email: muthulakshmiandal@psgrkc.com

Introduction

Utilization of heavy metals in industries viz., paints, pigments, batteries, ceramic glazes, electroplating and textiles [1], over the past few decades has led to serious environment problems on account of improper untreated disposal [2]. Copper is widely employed in min- ing/ metallurgical, paints, pigments and electroplating industries [3]. According to WHO standards, the permissible concentration of divalent copper in effluents and drinking water are 0.05 and 1.0 mg/L [4] respectively, beyond which it is carcinogenic and leads to several ailments viz., gastrointestinal problems, headache, fatigue, depression, skin rashes, learning disorders through the accumulation in kidneys, brain, skin, pancreas and heart [5].

Environmental Protection Agency (EPA) has declared Zn^{2+} to be one of the priority pollu- tant due to its nature, leading to dehydration, electrolyte imbalance, stomach ache, nausea and dizziness.

Various methods have been reported for the sorption of metal ions from aqueous solutions, the common being solvent extraction, filtration, coagulation, chemical precipitation, ion ex- change, electrolysis, membrane process and adsorption. Adsorption is an effective process for the removal of heavy metal ions. Low cost materials prepared froma variety of raw ma- terials as wood, peat and animal origin viz., chitin and chitosan have been employed in trap- ping toxic metal ions by many researchers [6].

The objective of the present study is to assess the adsorption capacity of Mussel shell for Cu(II) and Zn(II) ions. Mussel shell is belonging to the family bivalve molluscs from saltwa-ter and freshwater habitats, is a waste product generated during shellfish processing.

Experimental Section

Preparation/Treatment of Adsorbent Materials

Mussel shells were collected from Sea shores, washed thoroughly with double distilled wa- ter and crushed partially using mortar and pestle. The crushed MSP were then soaked in 0.1 N HCl for 4 hours, washed with double distilled water and sun dried, thereafter referred toas TMSP. TMSP was ground well in a laboratory blender and sorted into varied particle sizes (85 BSS, 72 BSS, 52 BSS, 36BSS and 22 BSS) using standard test molecular sieves (JAYANT Scientific Instruments Co., Mumbai) and stored in air tight containers. The picto- rial representations of the raw and treated MSP (85 BSS mesh size) are shown in **Figure 1.**

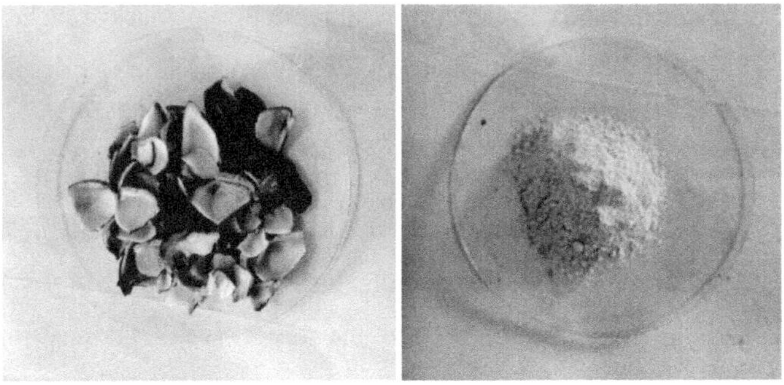

Figure 1

Particle Size Determination

The sieved MSP of varying mesh sizes were subjected to image microscopic analysis in or- der to determine the particle sizes using Binocular Microscope (OLYMPUS make, Model- CX21I). Fifteen different granular particles of each mesh size were chosen for the measure- ment of lengths and breadths to arrive at an average, as no two single particles are alike. By applying multiplication factors, the particle sizes were calculated. The calculated particle sizes corresponding to the mesh sizes of 85 BSS, 72 BSS, 52 BSS, 36 BSS and 22

BSS are 0.18 mm, 0.21 mm, 0.30 mm,
0.42 mm and 0.71 mm respectively. The microscopic structure of treated MSP (0.18mm par- ticle size) is depicted in
Figure 2.

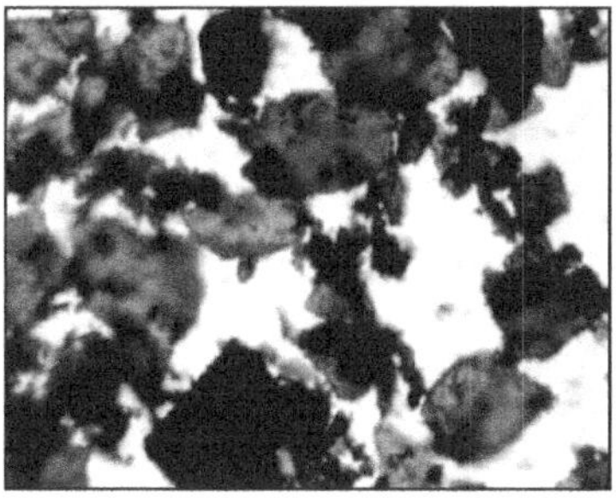

Figure 2

Physio- Chemical and Surface Characterizations (TMSP)

The physio chemical characteristic studies viz., pH (*DELUXE pH Meter (LABTRONICS*), Conductivity (*ELICO (CM 180) Digital Conductivity Meter*), Moisture content, (Xylene- extraction test method (ASTM D 2867-95), Bulkdensity, Specific gravity, Porosity and ash content were determined. The surface area and volume of the pores were obtained by N_2 physisorption at 77 K. These analyses were performed with a Micromeritics *BEL, Japan, Inc*Surface Area Analyzer using the BET method. The surface morphology was analyzed using Scanning Electron Microscopy (SEM) coupled with energy-dispersive X-ray spectroscopy (EDAX) (*JEOL JFM- 6390*).

Choice of Metal ions

The heavy metals of interest investigated for the present study include: Copper [Cu(II)] and Zinc [Zn(II)]. The choice of heavy metals was decided based on the literature collection conducted where the prevalence of these metals were observed in the effluents discharged from various electroplating industries in and around Coimbatore.

Batch Mode Adsorption Studies

The agitation of Cu(II) and Zn(II) ions with TMSP in a mechanical shaker (KEMI- KR3-1) was experimentally verified to define the role of variable parameters viz., particle sizes (0.71 mm, 0.42 mm, 0.30 mm, 0.21 mm and 0.18 mm), dosages of TMSP (200 mg, 300 mg, 400 mg and 500 mg), initial concentrationsof the aqueous Cu(II) and Zn(II) solutions (100-1000 ppm: 100 ppm interval), agitation time intervals between the sorbate and sorbent species (30 min, 60 min, 120 min), pH environ- ments (3, 5, 7, 9 and 11) and temperatures of the system (293- 333 K: 10 K interval) to op- timize the conditions for the systems ensuring maximum sorption.

Duplicate experiments were conducted to ensure the reproducibility of values within ± 2%. The initial and residual concentrations of corresponding metal ions were analyzed using Atomic Absorption Spectrophotometer: *Shimadzu (AA 6200) Model*

Data Evaluation

The percentage of adsorption of metal ions from aqueous solutions was estimated by using thefollowing equation [7].

$$\%\ \text{adsorption} = \frac{(C_i - C_e)}{C_i} \times 100 \qquad (1)$$

The amounts of metal ions removed by adsorbent (q) were calculated using the mass balance

$$q = \frac{V(C_i - C_e)}{W} \qquad (2)$$

where, V is the volume of the solution (L), W is the mass of the adsorbent (g), C_i and C_e arethe initial and equilibrium metal concentrations (mg/L) respectively.

Statistical Analysis

The relationship between adsorbed Cu(II) and Zn(II) ions and variable parameters was cor- related using Pearson Moment Coefficient Method. The extent of statistical fit was verified using SPSS 20 software, the output variable being ANOVA, Pearson Correlation and de- scriptive analysis with a significance based on 95% confident level.

Results and Discussions

Surface Characterization

The sorption activity of any sorbent mainly depends on its chemical nature and pore struc- ture. **Table 1** shows the various physiochemical properties of TMSP. The ash content of TMSP exhibits very less value, indicative of the least presence of inorganic matter and high quantity of carbon content (≈ 45%). The specific gravity of TMSP (1.86) is the ratio of the masses of equal volumes of biomass and distilled water. The lower bulk density (1.67) against thestandard value of 2.20 shows that TMSP has a large number of pores, which are more suitable for adsorption. Since, the high pore size has a greater potential to store/ ad- sorb particles. The presence of water vapours considerably inhibits the sorption activity, wherein the calculated Moisture content for TMSP being 7.24%. The surface area of TMSP is 1.52 m^2/g, indicates that the extent of surface occupancy by adsorbate has a decisive influ-ence upon the removal efficiency of TMSP. The porosity and mean pore diameter values implies that TMSP is mesoporous in nature [8]

Table 1 Physicochemical Characteristics of the Adsorbent.

Factors	Values
pH of 1 % solution	10.87
Conductivity	39.23
Moisture (%)	7.24
Bulk density (g/L)	1.67
Specific gravity	1.86
Porosity	10.22
Ash content (%)	4.3
Surface area (m^2/g)	1.52
Mean Pore diameter (A^0)	240
Carbon (%)	43.69
Nitrogen (%)	2.17
Hydrogen (%)	5.50
Sulphur (%)	-

SEM Analysis

The surface morphologies of the unloaded, Cu(II) / Zn(II) loaded TMSP are depicted in **Fig- ures 3 (a-c)**. Figure 3a represents a highly porous nature of the unloaded TMSP surface withcoarse textured pores of different shapes. The smooth covered pores on the surfaces after Cu(II) and Zn(II) sorptions as observed in Figure 3b and 3c respectively indicate the bindingof the respective ions onto the surface active sites of TMSP, where the decline of ruggednessis visible. This can be evidenced from the presence of surface pores only in Figure 3a, not distinct in Figure 3b and 3c.

Figure 3 (a) Unloaded TMSP (b) Cu(II) loaded TMSP (c) Zn(II) loaded TMSP.

EDAX Analysis

The EDAX spectra were recorded to analyze the elemental constitution of TMSP qualita- tively. The spectra of TMSP before and after adsorption are depicted in **Figures 4(a –c)** in- dicative of the presence of O, C, Ca and Cl, being reported as the principle elements of any adsorbent [9]. The appearance of new peaks at an energy range of 8-10 KeV (Figure 4b and 4a), confirms the Cu(II) and Zn(II) adsorption onto TMSP.

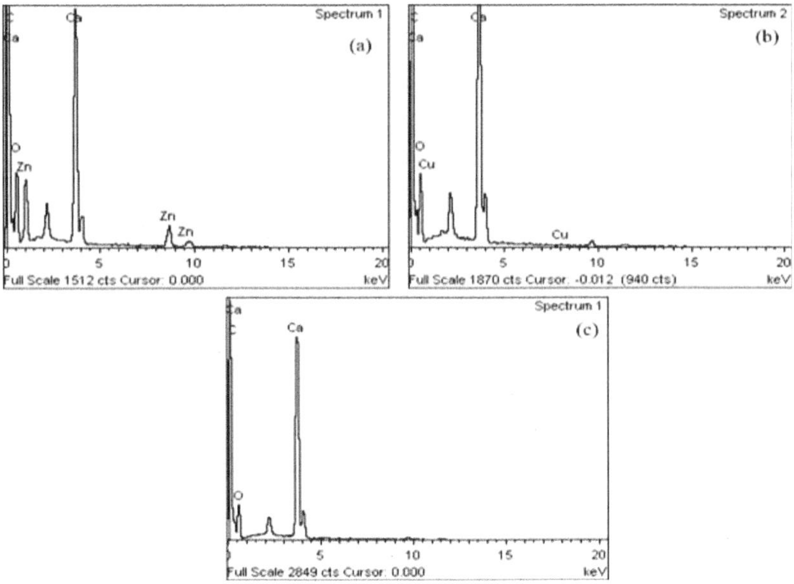

Figure 4 (a) Zn(II) loaded TMSP, (b) Cu(II) loaded TMSP, (c) Unloaded TMSP

Effect of Particle Size

Figure 5 illustrates the adsorption of Cu(II) and Zn(II) onto TMSP under studied particle sizes. The smooth and steep declining of the curves indicate a systematic trend followed in the adsorption pattern, supported by the corresponding values given in **Table 2**. It is under- stood from the table that as the particle size increased from 0.18 mm to 0.71 mm, the amountof divalent ions adsorbed declined in both the cases. It shall be justified that the increase in sorption may depend on the large external surface area for smaller particle sizes [10]. Since 0.18 mm exhibited maximum amounts adsorbed of 86.09 mg/g and 83.76 mg/g for Cu(II) and Zn(II) ions respectively, this size (0.18 mm) is fixed as the optimized size for further experiments. The P-values of Cu(II)- 0.0137 and Zn(II)- 0.0260 at 95% confident level ($P<0.05$) emphasizes statistically that there is significant correlation in adsorption pattern. The negative correlation values of Cu(II)- r = -0.8624 and Zn(II)- r = -0.7395 observed between the particle sizes and amounts adsorbedconfirms the decreased adsorption capaci- ty as the sizes of the particles were increased.

Effect of Dosage

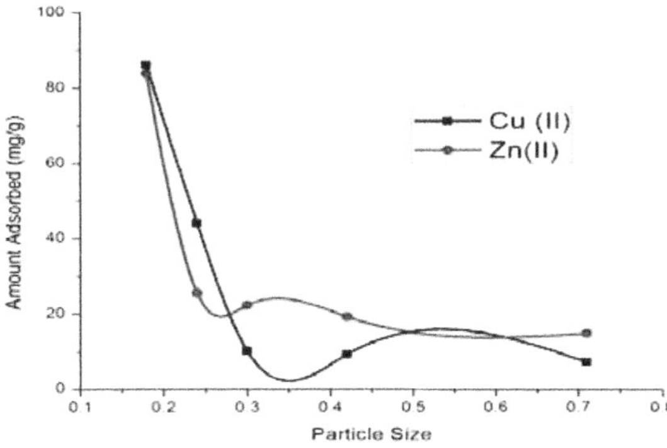

Figure 5 Effect of particle size.

Table 2 Effect of particle size.

S.No	Particle Size	Amount adsorbed mg/g	
		Cu(II)	Zn(II)
1	0.18mm	86.09	83.76
2	0.24mm	44.09	25.52
3	0.30mm	10.29	22.29
4	0.42mm	9.32	19.32
5	0.71mm	7.26	14.87

.The results for the influence of varied doses on the divalent ions – TMSP systems are pre- sented in **Table 3** and depicted in **Figure 6**. The data revealed that a steady increase in the amounts of Cu(II) and Zn(II) adsorption upto 1000 mg dosage, beyond which a slight de- cline in the removal was registered [11]. This is obvious from the curve patterns henceforth, 1000 mg is fixed for further experiments. The P-values of Cu(II) = 0.0455 and Zn(II) =

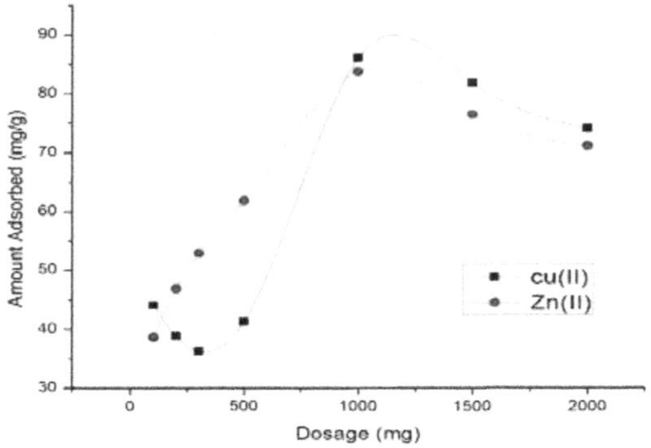

Figure 6 Effect of Dosage.

0.02605at 95% confident level ($P<0.05$) indicates statistically that there is extremely signifi- cant correlation in adsorption pattern. The positive correlation coefficients (for Cu(II)- r = 0.8858 and Zn(II)- r = 0.9942) indicating that the dosages and the amounts adsorbed are di- rectly proportional to each other.

Table 3 Effect of Dosage.

S.No	**Dosag- es**	**Amount adsorbed mg/g**	
		Cu(II)- TMSP	**Zn(II)- TMSP**
1	100 mg	44.09	38.65
2	200 mg	38.83	46.87

3	300 mg	36.23	52.87
4	500 mg	41. 23	61.87
5	1000 mg	86.09	83.76
6	1500 mg	81.87	76.54
7	2000 mg	74.08	71.17

Effect of Initial Concentration and Contact Time

The data pertaining to the influence of the chosen material upon different initial concentra- tions agitated at predetermined time intervals is listed in **Table 4**. In the case of Cu(II) sys- tem, a contact time of 10 mins itself was

found to be sufficient for adsorbing a maximum of 86.09 mg/g, but a contact time of 60 minswas observed for a maximum 83.76 mg/g amount of Zn(II) being adsorbed after the con- ductance of series of pilot studies. This can be due to Zn(II) ions produce net positive chargein the solution, therefore would make it difficult for negative electrons to dissociate, while comparing with Cu(II) ions [12]. A further increase in the contact time had resulted in a dip for all the concentrations of [Cu(II) and Zn(II)], as it is obvious from the curve pattern (**Fig- ure 7**). This decrease can probably be due to the slow pore diffusion of the solute ion intothe bulk of the adsorbent [13].

Table 4 Effect of Initial Concentration and Contact Time

	Time (min)	Amount Adsorbed (mg/g)				
		100ppm	250ppm	500ppm	750ppm	1000ppm
	10	20.19	35.05	43.06	54.09	86.09
Cu(II)-TMSP	20	9.82	24.75	32.87	42.99	78.28
	30	9.41	14.79	22.64	32.99	43.43
	30	45.87	17.98	41.98	60.98	64.87
Zn(II)-TMSP	60	56.81	25.87	52.98	70.87	83.76
	90	46.87	15.87	41.87	61.87	46.87

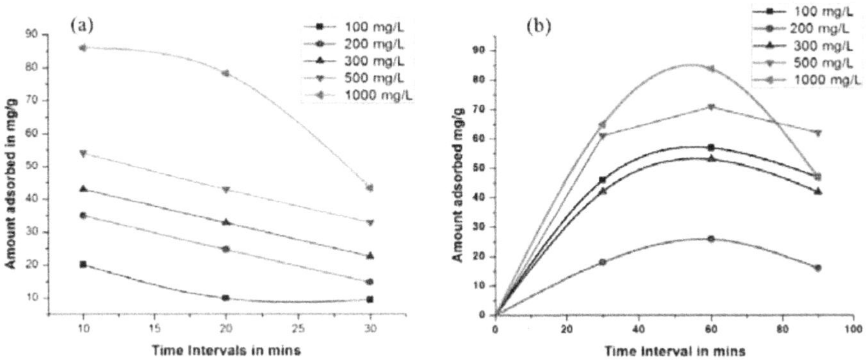

Figure 7 Effect of Initial Concentration and Contact Time (a) [Cu(II)], (b) [Zn(II)] .

Effect of pH

The pH dependence of Cu(II) and Zn(II) biosorption under so far optimized conditions is represented as bar chart (**Figure 8**). The reduction in the amounts of the adsorbed divalent ions can be attributed to the fact that theprecipitation of ions as hydroxide ions might occurin alkaline solutions. Also, the lower q_e values observed in acidic conditions indicate the competence between the proton (H^+) and metal ions (Cu(II) and Zn(II)) to get sorbed on the surface [12] of TMSP. The calculated P-value (Cu(II) = 0.0387 and Zn(II) = 0.0455) at 95% confident level ($P<0.05$) statistically prove that there is significant correlation in adsorption pattern. From the observation, it is obvious that pH 7 and 5.5 recorded higher q_e values, thereby maintained in all further experiments.

Descriptive statistics are a set of brief descriptive coefficients that summarizes a given data set, which can either be a representation of the entire experimental work (**Tables 5** and **6**). The observed significant relationship between the parameters and the amounts of Cu(II) and Zn(II) adsorbed were substantiated by performing Pearson Correlation Analysis, indicated in **Tables 7** and **8**. The prediction of its relationship for both the ions with TMSP is verified through statistical analysis which can be a strong tool for monitoring heavy metal pollution.

Comparison of Preferential order of adsorption

The order of preferential adsorption of metal ions onto TMSP was found to be Cu(II)>Zn(II). The relative abilities of the solute- ion species to compete for surface sites of the sorbents are governed by intrinsic factors such as valence, ionic radius, pH, and the solu-tion activities [14]. Both the employed ions being divalent in nature, the selectivity depends entirely on their hydrated ionic radii. Smaller the size of the ion, greater is the degree of hy- dration. Thus, for ions of similar charge, hydrated ionic radii determine the order of prefer- ence of adsorption. Cu(II) being smaller in size than Zn(II), its hydrated ionic radius is smaller (0.419 nm) than Zn(II) (0.430 nm) [15] leading to greater access of Cu(II) adsorp- tion in preference to Zn(II).

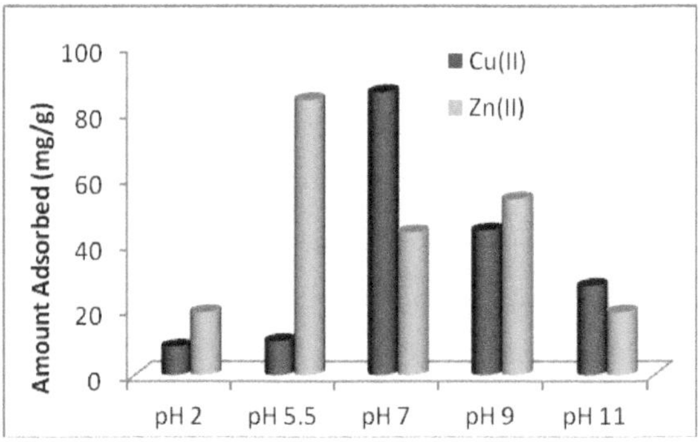

Figure 8 Effect of pH.

Table 5 Statistical Analyses [Cu(II)].

Descriptive	**Parameters (Amount adsorbed in mg/g)**			
	Particle Size	**Dos- age**	**Initial Con.**	**pH**
Mean	37.448	55.4	42.43	71.24
SD	36.235	16. 92	31.69	20.99
Maximum	86.09	73.56	73.64	88.1
Minimum	9.325	28.15	13.45	57.3
Degrees of free- dom	3	4	4	3
Sum of Squares	1006.5	12.788	558.61	812.1
Mean Square	503.25	4.156	292.87	706.07
Variance	948.38	269	350.26	110.84
Range	64.44	41.16	30.19	20.8

Table 6 Statistical Analyses [Zn(II)].

Descriptive	Parameters (Amount adsorbed in mg/g)			
	Particle Size	Dosage	Initial Con.	pH
Mean	37.72	56.8	52.43	81.24
SD	30.79	17.29	21.69	10.99
Maximum	83.76	83.76	83.64	98.1
Minimum	19.32	38.65	33.45	67.3
Degrees of freedom	3	4	4	3
Sum of Squares	1289.3	13.788	758.61	1812.1
Mean Square	644.66	4.596	252.87	906.07
Variance	948.38	299	470.26	120.84
Range	64.44	45.11	50.19	30.8

Table 7 Pearson correlation among selected parameters and amount of Cu(II) ion ad-sorbed.

	Particle size	**Dos-age**	**pH**	**Tempera-ture**
Pearson corre-lation	-0.739	0.994**	0.094	0.146
Sig. (2 tailed)	0.261	0.001	0.906	0.815
N	4	5	4	5

Table 8 Pearson correlation among selected parameters and amount of Zn(II) ion adsorbed.

	Particle size	**Dos-age**	**pH**	**Tempera-ture**
Pearson corre-lation	-0.786	0.984**	0.091	0.134
Sig. (2 tailed)	0.224	0.001	0.916	0.892
N	4	5	4	5

**Correlation is significant at the 0.01 level (2-tailed).

Conclusion

The selected animal waste (Mussel shells) was powdered and modified using 0.1 N HCl (TMSP) for trapping Cu(II) and Zn(II) ions. The physio chemical characteristic studies viz., pH, conductivity, moisture content, bulk density, specific gravity, porosity, ash content and elements (C,H,N,S) were determined for the prepared TMSP. The surface morphological changes, elemental constitutions and surface area of TMSP as observed from using SEM, EDAX and BET analysis respectively for unloaded / loaded material, registered a significantchanges in porosity structure (SEM) and the appearance of Cu(II) and Zn(II) peaks in the EDAX images supporting the adsorption positively. The Batch equilibration conditions for maximum percentage removal (98.1%) of Cu(II)- TMSP system was optimized at: 0.18mm particle size, 1000 mg dosage, 10 minutes agitation time, 1000 mg/L initial concentration of Cu(II) ions, pH 7 of the solution medium at 30^0C and for Zn(II)- TMSP system (97%) was optimized at: 0.18mm particle size, 1000 mg dosage, 60 minutes agitation time, 1000 mg/L initial concentration of Zn(II) ions, pH 5.5 of the solution medium at 30^0C. The statistical output data for viz., particle sizes, initial concentrations of the metal ions (Cu(II) and Zn(II)), dosages and pH of the solutions had a significant effect on Cu(II) and Zn(II) adsorption. A very good correlation between the input and output variables is observed by applying SPSS 20 software.

Acknowledgment

The authors extend their acknowledgment to Defence Research and Development Organisa-tion (DRDO), New Delhiin execution of the research work.

References

[1] O.S. Amudaa,. A.A. Giwa, I.A. Bello, Biochem. Eng. J. 2007, 36, 174–181.

[2] W. Mohammad, I. Amer, Fawwaz Khalili, M. Akl Awwad. J. Env. Chem. & Ecotoxi.2010, 2(1), 001-008.

[3] C. Karthika, M. Sekar, Res. J. Chem. Sci. 2013, 3(3), 65-69.

[4] WHO, Guidelines for Drinking- Water Quality 1984.

[5] Sureshkumar; Halnor Maqdoom Farooqui; Milind Ubale, Inter.J. Appl. Inno. Engin.&

Manag. 2013, 2(3), 125-135.

[6] M. Ivona, J. Astvan, Int.J.Curr.Microbiol.App.Sci, 2014, 20(2), 283–293.

[7] HakanDemiral, IlknurDemiral, FatmaTumsek, BelginKarabacakoglu, Chem. Eng. J.,2008, 144, 188-196.

[8] www.mike.blaber.org

[9] Efficient adsorption of Cr (VI) from aqueous solution on low cost adsorbent developedfrom Limoniaacidissima (Wood apple) shell. Ads. Sci.& Tech., 2010, 28 (6), 547-560.

[10] M. Makeswari and T. Santhi, Arabian Journal of Chemistry, 2014.

[11] S.F. Montanher, E.A. Oliveira, and M.C. Rollemberg, J. Hazard. Mater., 2005, 117, 207-211.

[12] M. Nameni, M.R. AlaviMoghadam, Arami. M., Inter.J. Env. Sci.& Tech., 5, 2008, 161-168.

[13] T. Karthikeyan, S. Rajagopal and Lima Rose Miranda, J. Hazard. Mater. 2005, 124, 192-199.

[14] MD Bethedsa, Registry of toxic effects of chemical Substances.

Adsorption Dynamics of Ecofriendly Litter WastesFor Zn(II)/Cr(VI) From Electroplating Effluents - Continuous Column Method

N Muthulakshmi Andal[1], S Charulatha[2], J Anuradha[2] and Gayathri NS[2]

[1] *Assistant Professor, Department of Chemistry, PSGR Krishnammal College for Women, Coimbatore.* [2] *Research Scholar, Department of Chemistry, PSGR Krishnammal College for Women, Coimbatore.*

Received: September 18, 2016; Accepted: October 05, 2016; Published: October 20, 2016

***Corresponding author:** *N Muthulakshmi Andal, Assistant Professor, Department of Chemistry, PSGRKrishnammal College for Women, Coimbatore-4, E- mail: muthulakshmiandal@psgrkc.com*

Abstract

Electroplating industries, play a momentous role in the development and growth of numerous metal manufacturing, but also equally pollute the environment through vari- ous means. Most of the electroplating industries operated in Coimbatore are reported un- der Red category list, their main source of pollution being heavy metal leaching into aquatic streams. The present work deals with the sorption of Zn(II) and Cr(VI) being highly prevalent electroplating effluents. The process of biosorption has many attractive features compared to the conventional effluent treatment methods adopted so for. Batch studies are conducted at varying operating factors to assess the best sorption efficiency amongst the chosen cost effective and eco-friendly materials viz., Mussel Shell Powder (MSP), Prosopis juliflora Bark (PJB), Terminallia cattapa Seed Shell (TCSS) and Aegle marmelos correa (AMC). Based on the experimental data the Treated Mussel Shell Pow- der (TMSP) and Treated Prosopis juliflora Bark (TPJB) are found to be the best sorbent for Zn(II) and Cr(VI) removal respectively. The scaling up process through column ex- periments are performed to quantify the efficiency of the fixed sorbents under optimized conditions. The long term analysis at the laboratory levels, reveal

100% and 92% metal removal for Zn(II)-TMSP and Cr(VI)- TPJB systems respectively.Introduction

Industrial sectors expansion involves increased usage of non-biodegradable heavy met- als and leading of excess toxic pollutants through liquid effluents. This is an issue of growing concern for humanity and its ecosystem. The wastewaters emanating from such industries viz., plating, tanning, metallurgy require treatment before discharging, but least cared, wherein heavy metal ions like Cr(VI) and Zn(II) are present.

The tolerance limit for Cr(VI) and Zn(II) ions from the effluent discharges into surface waters are 0.1 mg/L and 5.0 mg/L respectively. In potable water, it is 0.05 mg/L and 10.0 mg/L correspondingly [1]. The health risks of Cr(VI) and Zn(II) ions while exceeding the permissible limits results in skin allergy,liver, stomach problems and also found to be carcinogenic [2]. It is essential to sequestrate these ions from polluted areas, suggesting suitable methodology.

Various methodologies such as electrochemical, ion exchange, membrane filtration, evaporation, solvent extraction, emulsion per traction technology, reverse osmosis and chemical coagulation, are available for trapping of heavy metal ions. But these methods possess their own disadvantages, often involving high capital/operational costs and gen- eration of secondary wastes [3, 4].

Biosorption process is one of the efficient methods in trapping toxic metal ions, due to its simplicity, sludge-free operation, easiness in handling, availability of various adsor- bents and also efficient removal of heavy metals at lower-concentration levels itself [5]. Biosorbents are prepared from natural waste biomasses that are available in large quanti- ties or certain waste collected from agricultural operations. These low/ no cost materials are reported to have excellent sorption potential that are affordable and eco-friendly.

The current study evaluates the biosorption capacity of litter wastes in trapping Cr(VI) and Zn(II) ions via batch / column method. MSP, PJB, TCSS and AMC were identified, prepared, treated before experimental verification. Removal of heavy metal ions employing the above said eco-friendly materials have not been reported elsewhere in literature.

Materials and Methods

Collection/Treatment of adsorbent materials

The identified litter waste materials Mussell Shell (MSP), Prosopis juliflora Bark (PJB), Terminallia cattapa Seed Shell (TCSS) and Aeglemarmeloscorrea (AMC) (Figure 1) were collected from localities of Coimbatore, Kangeyam, Salem and coastal areas of Tamil Nadu. Collected PJB, MSP and AMC were washed, sun dried, crushed down into small pieces and then sieved. The sieved particles of different mesh sizes for the four materials were treated with 0.1N HCl for 3 hours, washed several times to neutralize and then air dried. The several washingsduring

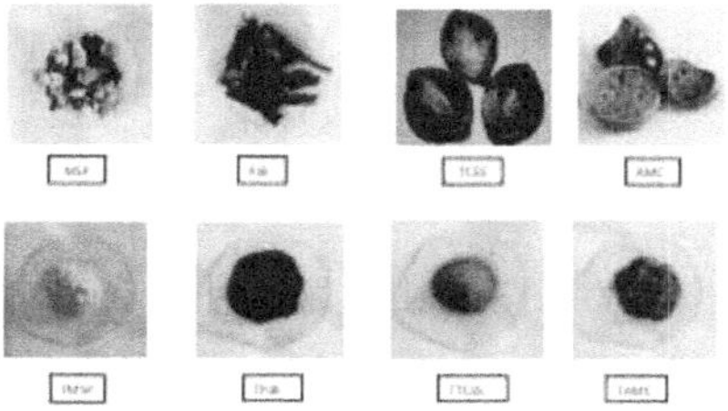

Figure 1: Raw and treated adsorbent materials.

neutralization procedureensure the increase in surface area while soaking the materials. The acid treated Mussell Shell powder (TMSP), Prosopisjuliflora Bark (TPJB), Terminallia cattapa Seed Shell (TTCSS) and Aegle marmeloscorrea (TAMC)Removal percentage (%) = $(C_1 - C_2) / C_1$ X100 Amount adsorbed (mg/g) = $(C_1 - C_2)$ X V/M where,

C = initial metal concentration (mg/L), C = final metal ere stored in air tight containers

Chemicals

All the chemicals employed were of Analytical Reagent grade. Doubly Distilled (DD) water was used for preparation and dilutions of solutions. A stock of 1000 ppm metal solu-tions was prepared by dissolving required amount of K Cr O and Zn(NO)concentration (mg/L), V= volume (ml) and M= mass of adsorbent (mg).

Results and DiscussionEffect of Particle Size

Tables 1(a)-1(b) show the influence of variable particle ^{3} ^{2}sizes of the sorbent materials on the varying systems indicating in 1000 ml of standard flask. The pH of the solutions were adjusted using 0.1N HCl and 0.1N NaOH.

Batch equilibration studies

The equilibration of Cr(VI) and Zn(II) aqueous solutions (50ml volume) with the mate- rials were experimentally verified in a mechanical shaker (KEMI) to define the role of variable parameters viz., particle sizes (0.18mm, 0.24mm, 0.30mm, 0.42mm and 0.52mm) and dosages (200 mg, 400mg, 500mg, 1g, 2g) of the treated adsorbents, initial concentra- tions of the aqueous Cr(VI) and Zn(II) solutions (100-1000mg/L : 100 mg/L interval), preset time intervals between the sorbent and sorbate species (10-120 minutes : 30 minutes interval), pH of the medium (2,4,7,9 and 11) and temperature (293K-333K : 10 K inter- val) in order to assess the sorption efficiencies under laboratory conditions. The agitated samples were filtered and the residual metal ion (Cr(VI) and Zn(II)) concentrations were analyzed using Atomic Absorption Spectrophotometer: Shimadzu (AA 6200) (Figure 2) asper standard methods.

All experiments were carried out in duplicate and mean values are presented with the maximum deviation of 2.0%. The percentage removal and amount adsorbed were deter- mined as follows:0.18mm size to be the favorable size for maximum removal. This is
owed to the surface phenomenon, smaller adsorbent size offers larger surface area for metal binding.

Effect of Contact Time/Initial Concentration

The effect of contact time at varying time intervals and that of initial concentrationsfor the systems are listed in tables 2(a) and 2(b), wherein the registered data indicate the enhancement in amount adsorbed only up to certain metal ion concentrations. This may be of the saturation in the sorption sites on the sorbent as the concentration of the metal ions increased. The varying time intervals for the systems indicate the attainment of equi- librium up to specific contact time, further increase in time interval registered a decline.

Effect of Dosage

The amounts of metal ion adsorbed at different dosages are listed in tables 3(a) & 3(b).The maximum sorptive ability of the materials under varying conditions depict their po- tentiality as able adsorbents in sequestering heavy metal ions.

Effect of pH

Maximum uptake was observed at pH 2.5 for Cr(VI). At increasing pH environments,

a sharp decline in uptakewas observed, which is in good agreement with previous reports. The speciation studies of Cr(VI) in aqueous solution shows that H_2CrO_4 predominates

at pH less than 1.0, $HCrO_4^-$ for pH between 1.0 and 6.0 and CrO_4^{2-} at pH above 6.

This is due to the electrostatic attraction between the positively charged surfaces of the

adsorbent with $HCrO_4^-$ ions. But in highly acidic medium (pH=1.0), H_2CrO_4 (neutral form) is the predominant species of Cr(VI). Hence, percentage removal decreased due to the involvement of less number of $HCrO_4^-$ anions to the positive surface. At higher pH value, the reduction in adsorption may be due to the dual competition of both OH^- and CrO_4^{2-} ions to get adsorbed on the surface of the adsorbent among which OH^- predomi- nates. These results are in agreement with several previous investigations on metal re- moval by a variety of materials.

The pH dependence for Zn(II) systems registered maximum removal between 5-7; the other ranges exhibit irregular curves. The reason being that at low acidic medium, the H^+ ions compete with the metal ions to get sorbed on the materials and under higher basic conditions metal ions combine with OH^- ions to formrespective precipitates.

The optimized conditions resulted from the batch studies for the chosen materials are, Cr(VI) system: 0.18 mm, 200 mg, 1000ppm, 60 min, pH 2.5 , 30^0C ; Zn(II) system: 0.18 mm, 1.5g,

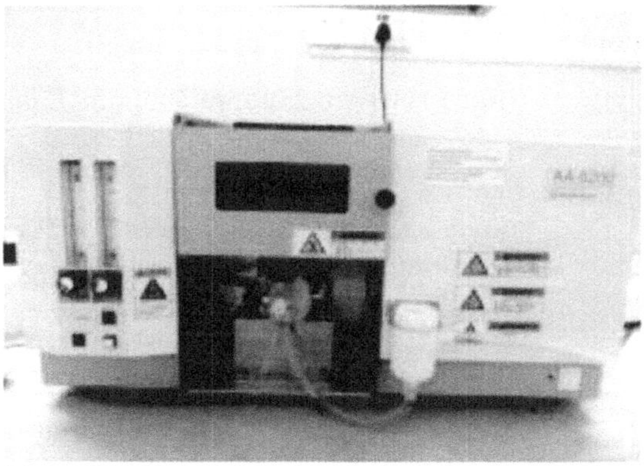

Figure 2: Atomic Absorption Spectrophotometer Shimadzu (AA 6200).

Table 1(a): Cr (VI)-Particle Size.

Adsorbents	Amount Adsorbed (mg/g)				
	0.18 mm	**0.24 mm**	**0.30 mm**	**0.42 mm**	**0.71 mm**
TTCSS	16.8854	4.335	4.3340	4.542	4.452
TPJB	22.6303	6.876	5.93834	5.826	5.8337
TMSP	13.1202	7.85	6.9756	6.7262	4.3826
TAMC	11.3931	9.384	6.837	6.383	5.3836

Table 1(b): Zn(II)-Particle Size.

Adsor-bents	Amount Adsorbed (mg/g)			
	0.18 mm	**0.24 mm**	**0.30 mm**	**0.42 mm**
TTCSS	38.23	28.21	19.74	22.23
TPJB	55.43	44.43	33.17	27.72
TMSP	83.76	25.52	22.29	19.32
TAMC	41.87	34.32	27.28	28.38

Table 2(a): Cr (VI)-Contact Time/Initial Concentration.						
Adsor-	**Time**	**Amount Adsorbed (mg/g)**				
bents	**(min)**	**100 ppm**	**250ppm**	**500ppm**	**750ppm**	**1000 ppm**
TTCSS	30	2.7165	3.827	5.726	6.837	5.342
	60	3.9482	5.9147	9.2536	10.5213	16.8854
	120	2.9726	3.98216	6.726	8.8726	10.765
TPJB	30	3.726	4.7236	7.837	10.872	6.865
	60	4.1617	9.9872	15.009	17.921	22.6303
	120	3.5262	5.837	14.763	12.2726	7.976
TMSP	30	0.564	1.987	4.676	4.765	6.876
	60	1.2867	3.4775	8.5776	9.6274	13.1202
	120	0.9875	2.234	6.765	6.9765	10.976
TAMC	30	0.9876	1.234	3.654	4.876	7.9876
	60	2.4485	2.9872	7.1107	7.5983	11.3931
	120	1.2343	1.7654	4.7656	5.764	8.654

Table 3(a): Cr (VI)-Dosage.					
	Amount Adsorbed (mg/g)				
Adsorbents	**200 mg**	**300 mg**	**400 mg**	**500 mg**	**1000 mg**
TTCSS	16.8854	11.625	10.298	9.937	8.282
TPJB	22.6303	17.2726	15.765	11.7598	10.282
TMSP	7.282	8.2822	9.765	11.272	13.1202
TAMC	11.3931	9.2826	8.875	5.3837	5.3837

Table 3(b): Zn (II)-Dosage.

Adsorbent	Amount adsorbed (mg/g)				
	100mg	200mg	300mg	500mg	1000mg
TTCS	24.85	26.87	29.76	31.87	38.23
TPJB	11.23	29.87	31.98	37.45	55.43
TMSP	38.65	46.87	52.87	61.87	83.76
TAMC	26.85	33.76	34.98	39.87	41.87

750ppm, 60 min, pH 5.6, 30^0C. Further increase in agitation time, beyond the attainment of equilibrium, registered least changes in the residual concentrations of the metal ions. The most suitable adsorbent for maximum removal of Cr(VI) is observed to be TPJB, since it is a natural bio accumulator of Cr(VI) by the transaction of metal to the aerialpart of the plant [5]. It is an active transport mechanism and the preferential order being TPJB>TTCSS>TMSP>TAMC. TMSP is found to be best for Zn(II) system, for the rea- son that the TMSP is more porous in nature when compared to other adsorbents and the preferential order being TMSP>TPJB> TTCSS> TAMC.

Column Studies

The Batch mode studies are the basic pilot studies performed to screen the biosorbents before adopting the adsorption method to the field levels. The results of the Batch equili- bration method insist the feasibility and compatibility of the chosen systems for their promising application. Based on batch studies results, column studies were carried out to quantify the adsorbents' efficiencies in the continuous columns' running with the aqueous Cr(VI) and Zn(II) solutions with TPJB and TMSP respectively [6]. Later, the column schemes are extended to the effluent discharges.

Column packing

The column is made up of cylindrical glass tube, the inner diameter is 2.5 cm and the height is 30 cm. It was packed with sorbent materials between two supporting layers of glass wool, spread with the glass beads at the top of the already packed glass wool layer placed at the bottom. The step by step packing is as: Glass wool layer (3cm thickness), glass beads (2cm thickness), TPJB /TMSP (50/300 g: 6cm height), glass wool (1cm thickness). These materials were loaded from the top of the column and allowed to settle by gravity force. The bottom of the column was fitted with rubber with flow adjustable knob and thus the outflowrate was controlled by adjusting the knob (Figure 5).

1000 mg/L aqueous solutions of Cr(VI) and Zn(II) were prepared and poured from the top of columns slowly and the flow rate through the columns were fixed as 100 ml/5 mins for Cr(VI)-TPJB system and 100ml/20 seconds for Zn(II)-TMSP system after sev- eral trials. Maximum removal of 98% and 100% were registered for Cr(VI)-TPJB & Zn(II)-TMSP respectively, after analysis in AAS. The percentage removal was observed to decline upto 90% after passing 20 litres of inlet solutions. Further decline was envis- aged upto 30 litres and then the column went exhausted.

Collection of Effluent Samples

Wastewaters containing Cr(VI) and Zn(II) ions above the tolerance levels were found at various electroplating industries located in Kurumbapalayam region of Coimbatore. 10litres of these effluent samples were collected in pre cleaned PET bottles of 1 litre capac-ity and analyzed for Cr(VI) and Zn(II) ions of initial concentrations (AAS) after a se- ries of dilutions, Cr(VI): 720 mg/L, Zn(II): 40 mg/L. The pH and conductivity valuesof the samples were also recorded using LABTRONICS pH meter and conductivity me- ter respectively (Table 4).

Zinc plating bath, the processing, uncoated/ Zn plated materi- als and the effluents generated are represented in figure 4.

Analysis of Industrial Effluents- Batch/Column Methods

Effluent concentrations of 50 mg/L were fixed for the batch studies and experimentally verified (TPJB and TMSP) under the optimized conditions. The results revealed an appre- ciable 70% removal of Cr(VI) and Zn(II) ions under batch mode. Sorptive nature of the sorbent materials was quantitatively estimated by the performance of the column studies, on the basis of the long term analysis at the laboratory conditions. The values recorded are 100% and 92% for Zn(II) and Cr(VI) ions respectively.

Table 4: Effluent Samples - Initial Concentrations

S. No	Name of Industries	Cr(VI) (mg/L)	Zn(II) (mg/L)	Conductivity (mv)	pH
1	Unit I	71.52	25.87	21.76	2.3
2	Unit II	695.75	3.398	32.98	1.98
3	Unit III	211	2.817	29.82	2.98
4	Unit IV	6.4647	23.298	87.87	5.35
5	Unit V	5.875	5.876	90.79	0.57
6	Unit VI	449.6	4.383	24.4	1.82
7	Unit VII	720.05	40.08	-0.693	3.45

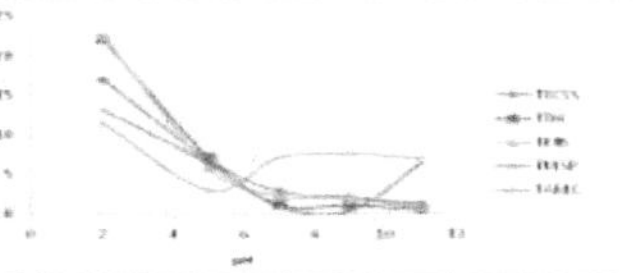

Figure 3a: Effect of pH - Cr(VI.

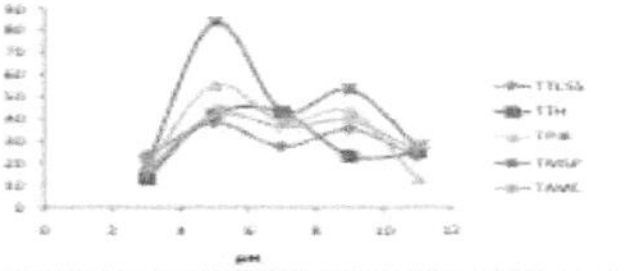

Figure 3b: Effect of pH - Zn(II).

Figure 4a: Zinc Plating bath.

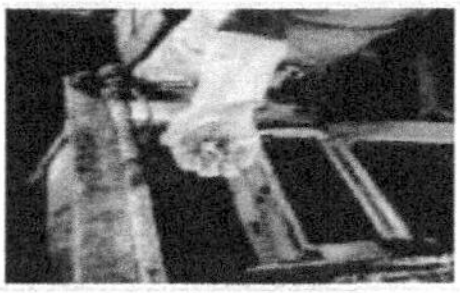

Figure 4b: Work at the bath plating.

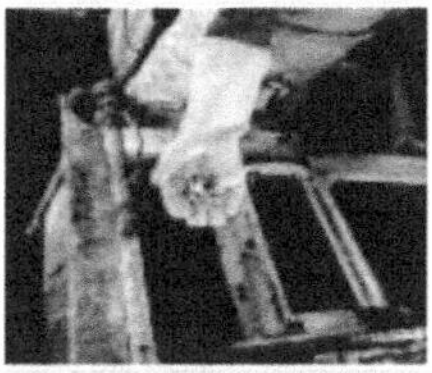

Figure 4b: Work at the bath plating.

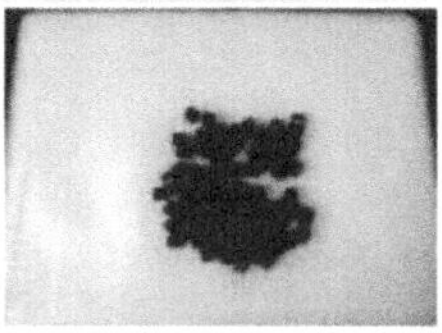

Figure 4c: Material before platting.

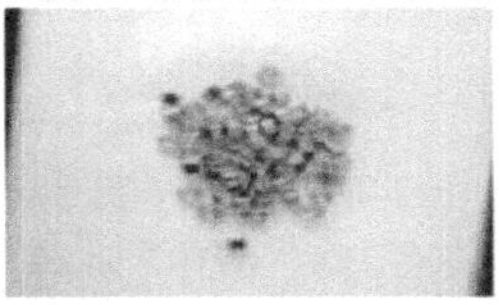

Figure 4d: Material after platting

Figure 4e: Effluent generated from the bath.

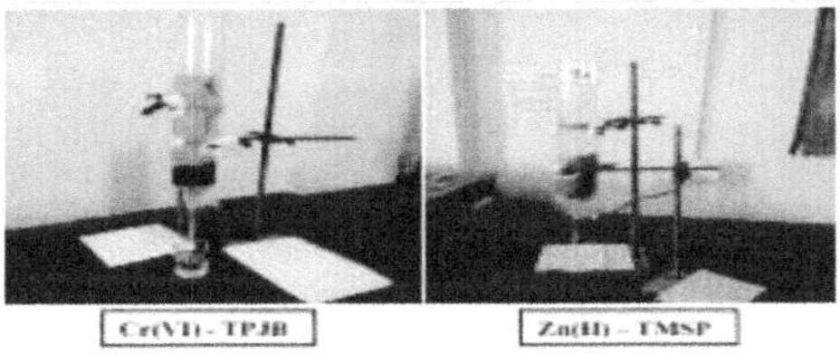

Figure 5: Column packing

Conclusion

Batch equilibration method was adopted to analysis the effect of the variables viz., particle sizes and dosages of the adsorbents (TTCSS, TPJB, TMSP, TAMC), initial concentrations of the adsorbate solutions (Zn(II), Cr(VI)), predetermined time intervals between the sorbent - sorbates materials, pH of the solution media and temperatures. TPJB and TMSP were derived to be the best sorbents amongst all the identified materialsbased on the optimized conditions through batch equilibrium studies. Quantification of these results was justified by column method for the aqueous metal solutions with the fixed sorbent materials. Various electroplating industries located in and around Coimba- tore were identified and surveyed for the extent of metal pollution. The effluent samples containing 40mg/L of Zn(II) and 720mg/L of Cr(VI) were collected from Electroplating Industries at

Kurumbapalayam area and the performance of the column for these effluentdischarges revealed 100% and 92% removal for Zn(II) and Cr(VI) ions respectively. The concluding remarks holds good for TPJB and TMSP to be promising litter waste materialsin the sequestrating Cr(VI) and Zn(II) ions at field levels.

Financial Support

The authors submit their due acknowledgment to Defence Research & Development Organization (DRDO), New Delhi for the timely Financial Support in execution of the research work

References

1. Owlad M, Aroua MK, Daud WAW and Baroutian S. Removal of hexavalent chromi- um-contaminated water and wastewater: a review. Water, Air, and Soil Pollution. 2009;200(1):59–77.

2. Volesky B. Detoxification of metal-bearing effluents: biosorption for the next century. Hydrometallurgy. 2001;59(2-3):203–216. Doi: doi. org/10.1016/S0304- 386X(00)00160-2

3. Wilde EW, Benemann JR, Wilde EW and Benemann JR. Bio removal of heavy met- als by the use of microalgae. Biotechnol Adv. 1993;11(4):781-812.

4. Monser L and Adhoum N. Modified activated carbon for the removal of copper, zinc, chromium and cyanide from waste water. Separation and Purification Technology. 2002;26(2-3):137–146.

5. Sugashini S and Meera Sheriffa Begum KM. Column Adsorption Studies for the Re- moval of Cr(VI) Ions by Ethylamine Modified Chitosan Carbonized Rice Husk Com- posite Beads with Modeling and Optimization. J of Chem. 2013;2013:pp.11. Doi: doi.

Assessment of Physicochemical Characteristics and the Level of NutrientCon-tents as Well as Heavy Metal Ions in Waters of Three Lakes at Coimbatore, Tamil Nadu, India

Ezhili N.[1], Muthulakshmiandal N.[2]

[1]Department of Zoology, [2]Department of Chemistry, PSGR Krishnammal College forWomen, Peelamedu, Coimbatore-04,Tamil Nadu, India.

Abstract

Lake water quality degradation and heavy metals pollution have been serious environ- mental threats for last decades, especially in those fresh water lakes situated in the vicini- ty of the cities and human habitations. The present study aimed to investigate the physi- cochemical parameters, level of nutrient contents and heavy metals to determine the water quality of Ukkadam, Kuruchi and Singanallur Lakes. All the physicochemical parameters except dissolved oxygen, sulphate, silicate and nitrate values were found above the pre- scribed limit as recommended by WHO (2011) and BIS (2003) drinking water quality. The concentrations of Ca, Cr, Cu, Fe, Ni and Pb in waters of the selected three lakes ranged from 0.014 – 0.038, 0.145 – 0.063, 0.025 – 0.552, 0.100 – 3.460, 0.028 – 0.117 and 0.463 – 2.172 mg/L respectively. The concentrations of heavy metals except Cu werefound to be higher than the permeable dose. This investigation revealed that the water in these lakes is partially contaminated with high level of heavy metals. It is recommended that proper implementation of new wetland waste management system and constant mon- itoring is urgently required to maintain good water quality in these lakes for future gener- ation.

Keywords: Physicochemical parameters; Nutrient; Heavy metals; Coimbatore

1. Introduction

Lakes are important natural resources for freshwater, replenishing of ground water and al-so habitats for wide variety of flora and fauna. Worldwide lakes act as a key factor for climatic changes and are considered as one of the most versatile ecosystems [1]. They are facing variety of pressures such as eutrophication and heavy metal pollutions due to an- thropogenic activities. In recent years, rapid growth of human population, intensive de- velopment of industrial, agricultural based sectors have resulted in surplus discharge of these toxic wastes alongwith the municipal waste materials into the wetland lakes. This inturn has increased the level of nutrient inputs against the safe level of their natural occur- rence, resulting in accelerated eutrophication [2].

In worldwide more than one billion people do not have suitable drinking water, and two to three billions lack access to basic sanitation services. About three to five mil- lions peoples are dying annually from water related diseases [3, 4]. Surface waters (fresh water lakes, rivers, streams) are the principal natural water resources [5]. Nowadays heavy metals have become common pollutants throughout the world due to their toxic nature towards all living organisms. Heavy metals are being discharged into the aquatic ecosystems through natural and anthropogenic sources [6]. These pollutants are playing a major role in the degradation of wetlands and directly affect biotic inhabitants [5].

Coimbatore is referred popularly to as Southern Manchester of India, an important industrial city in Tamilnadu. There are around 28 wetlands in and around the city limits which are mostly fed by Noyyal river. The lake water has been used for agricultural land irrigation purposes and fish culturing by local farmers. These wetlands serve as a com- mon reservoir of untreated industrial effluents, dumping of agricultural as well as munic- ipal wastes bring more concern for the aquatic environment. Ukkadam lake has the larg- est water spread areawith urbanized level. Hence water irrigation from this wetland is lowbut, the lake is being used for regular fishing. A number of migratory birds are encoun- tered in the lake during winter. The Kuruchi wetland has a water spread area of 343.96 acres and holds the least water storage capacity because of its shallowness. Singanallur Lake is highly affected by eutrophication and is almost filled with *Eichhornia crassipes,* weeds. Although *Eichhornia crassipes*is praised for its ability to sequester nutrients and other chemicals from water, it is considered undesirable for wetlands. However, no detailed study has been reported so far in the selected three wetlands viz., Ukkadam, Ku- ruchi and Singanallur Lakes of Coimbatore city regarding the ecological risk assessment factor due to heavy metal contamination of their water quality. Therefore, the present investigation was aimed to assess the physicochemical parameters, level of nutrient con- tents and heavy metals present in the water samples from the selected three lakes.

2. Materials and Methods

2.1 Description of study area

Figure 1 represents the three Lakes chosen for study. Ukkadam Lake is an urban lake sit- uated close to a municipal market and bus stand which has the largest water area spread amongst the Coimbatore wetlands. This Lake lies between 10°59' 40" and 76°46' 18" longitude. Kuruchi Lake, a shallow basin lake, is situated in Kuruchi at latitude of 10°58'38'' and longtitude of 76° 55' 44''. This lake is a point for the inflow of received munic- ipal sewage and a famous site for dumping garbage inhabitations around its vicinity [7]. Singanallur Lake is an urban lake situated towards, the Southwest of Coimbatore, Tamil Nadu, India. It is located within the geographical coordinates of 10° 56'46'' latitude and 77°01' 11'' longtitude. The lake is divided into two equal halves, but connected by a railway line running through it. The southern part of the lake receives freshwater only during the rainy season from the Noyyal river and the northern part of the lake receives industrial wastewater, domestic sewage and urban solid wastes [8]. The present study fo- cuses only on northern part of lake due to the enhanced pollution by various industrial, urban wastes and domestic sewage.

2.2 Sample collection, physicochemical and nutrient content analysis

Surface water samples were collected during the periods from September 2012 to August

4

2013 from the three sampling stations at a distance of 500 m for each lake. Samples were

collected in pre cleaned plastic bottles and labeled. The water samples were taken to the laboratory and filtered through Whatmann No.1 filter paper and kept in a refrigerator at 4°C with addition of HNO_3 in order to preserve samples until analysis [9]. The pH valueswere measured using a digital pH meter. Various physicochemical parameters viz., Total Dissolved Solids (TDS),Total Solids (TS), Biochemical Oxygen Demand (BOD), Chemical Oxygen Demand (COD), Dissolved Oxygen (DO), Chloride $(Cl)^-$, Sulphate (SO^{2-}), Nitrates (NO^-), Silicates (Si^{2+}) were analyzed using the standard methods [9, 10]. Calcium (Ca^+) and Magnesium (Mg^{2+}) concentrations were determined by using the method of Katz and Navone [11]. Phosphate (PO^{3-}) was determined by molybdate method [12]. Sodium (Na^+) and Potassium (K^+) content was analyzed using Flame Pho- tometer [13].

2.3 Heavy metal analysis

Surface water samples were digested with 10 ml of diacid mixture (HNO_3: HCl, 7:3) on ahot plate for 10 min, thedigested samples were filtered using Whatman No.1 filter paper and made up to 25ml using distilled water. These samples were used for analysis to de- tect the presence of six heavy metals namely Cd, Cr, Cu, Fe, Ni and Pb using Atomic Ab-sorption Flame Emission Spectrophotometer (AAFES – 6200 (Shimadzu).

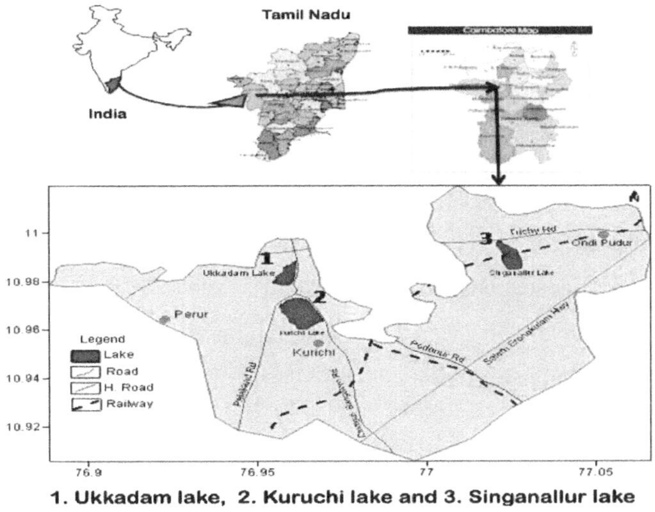

Figure 1: Study area map.

2.4 Statistical analysis

Mean metal concentrations were evaluated according to site and season by analysis of variance (ANOVA) and Duncan's multiple range tests [14]. The statistical analyses were performed according to the methods described by Gomez and Gomez [15], using SPSS- 11 computer software.

3. Results and Discussion

The variations in the physicochemical parameters of the surface water samples collected from the three lakes during the monsoon, pre and post monsoons and summer seasons aresummarized in Table 1. These parameters were compared with International [16] and Na-tional [17] standards. The temperatures obviously recorded higher during summer season (30.0 to 31.3°C) and lower in the monsoon season (27°C). The pH values were observed tobe in the range of 8.03 – 8.60, 7.20 – 7.94 and 7.80 – 8.10 for Ukkadam Lake, Kuruchi Lake and Singanallur Lake respectively, during the

study periods. Among the three lakes,the pH values were found to be high in Ukkadam Lake during the summer season, which exceeds the WHO [16] permissible limits (6.5 – 8.5). This can be possibly related to an- thropogenic activities as well as natural processes. Similar results have been reported by Mohanraj et al. [18] and Chandra et al. [7]. In Kuruchi and Singanallur Lakes the pH lev- els recorded were found to be within the WHO [16] recommendations. Sankand and Patil
[19] have documented that the higher levels of pH has enhanced the photosynthetic activ-ity during summer and pre monsoon seasons in aquatic ecosystems.

Total suspended solids (TSS) and Total Solids (TS) in surface waters of three lakes ranged from 1.23 -7.82 mg/L and from 603 – 1953 mg/L during the study periods, respec- tively. These parameters registered higher in Ukkadam Lake during the summer season, compared to the other two lakes. This may be attributed to the fact of dumping lot of sus- pended solid wastes containing effluents received from textile and dyeing industries [20]. The TDS values in Ukkadam Lake, Kuruchi Lake and in Singanallur Lake varied be-tween 1219 – 2171 mg/L, 599 – 5033 mg/L and 1228 – 4669 mg/L respectively. The ob- servations reveal that the TDS values in Kuruchi Lake are much higher than that of Ukka- dam and Singanallur Lakes. However, all the TDS values exceed the WHO [16] permissi- ble limit (1000 mg/L). The present values of TDS in all the water samples from the three lakes were found to be higher than the previous results reported by Mohanraj et al.
[18] and Chandra et al. [7]. Rahman et al. [20] have discussed that, the dissolution of low molecular weight organic bases originating from dye industries may result in the higher TDS value of the samples.

The Total Alkalinity (TA) and Total Hardness (TH) values in all the water samples ranged from 251 – 562 mg/L and from 202 – 450 mg/L respectively. Among the three lakes, Singanallur Lake registered greater TA and TH values.

The highest TA value (562 mg/L) wasnoticed in Singanallur Lake at the summer season. This is considerably higher than the WHO

[15] guidelines limits and the previous report by Chandra et al. [7], who have documented that higher level of TA in wetlands is likely due to the accumulation of domestic wastes and open defecation. The maximum level of TH (450 mg/L) was recorded in Singanallur Lake during the summer season. Against the WHO [16] recommitted limits of 300 mg/L. Almost, similar kind of results was reported in Ukkadam and Singanallur Lake by Mohanraj et al. [18]. High levels of TH in lake water may be attributed to the mechanical engineering wastes, hospital and sewage wastes [21].

Dissolved Oxygen (DO) is mainly involved in maintaining the oxygen balance of aquatic ecosystems based on the tropical and pollution levels. DO acts as a good indicator of healthy status of freshwater systems and this value above 5 mg/L is a support for all the livingorganisms. However, this value if observed to be less than 3 mg/L shall cause health risk to human [22]. As evident from Table 1, the DO levels varied from 0.48 – 5.54mg/L, 0.68 –

6.19 mg/L and 0.19 – 1.15 mg/L in Ukkadam Lake, Kuruchi Lake and Singanallur Lake re- spectively. However, the highest DO level was noticed in pre monsoon while lower level was recorded in monsoon season in all the three Lakes. Reports of Iqbal et al. [23] indicate that the DO contents were found to be high during winter season (5.20 mg/L) than summer (4.09 mg/L) in Rawal Lake, Pakistan. The present DO results may be attributed due to the tempera- ture changes as well as increased anthropogenic activities in and around theselakes during the study periods.

BOD values ranged between 20.6 – 60.0 mg/L, 19.0 – 48.7 mg/L and 20.3 – 42.0 mg/Lin Ukkadam, Kuruchi and SinganallurlLakes respectively. The BOD values recorded were above the WHO [16] tolerance level in all the three lakes. Highest BOD value (60.0 mg/L) was observed in Ukkadam Lake. The highest value of BOD has been reported by Mohanrajetal. [18] and Chandra et al. [7]. The

COD levels varied from 103 – 220 mg/, 82.0 – 197 mg/L and 104 – 187 mg/L L in Ukkadam, Kuruchi, and Singanallur Lakes, respectively with Ukka- dam Lake, being the maximum. The highest levels of BOD and COD are associated with the increased concentrations of organic and inorganic wastes being released into these lakes.

3.1 Nutrient properties of water

Seasonal variations of some nutrients in surface waters of three lakes are depicted as bar charts in Fig. 2. From the bar diagram it is obvious that Na^+ and K^+ values varied between 14.08–55.26 mg/L and 0.27–12.17 mg/L respectively, during the monitoring period (Fig.2 a, b). The maximum concentrations of Na^+ (55.26 mg/L) and K^+ (12.17 mg/L) ions were ob- served in Ukkadam Lake during the pre monsoon and minimum level was noticed in Kuruchi Lake during the post monsoon season. The levels of Ca^{2+} and Mg^{2+} ions ranged from 14.0 mg/L to 60.0 mg/L and 7.7 mg/L to 17.8 mg/L, respectively (Fig. 2 c & d). The highest Ca^{2+} and Mg^{2+} concentrations were observed during the summer season. In this study, the mean values of Cl^- ion range between 291–1038 mg/L (Fig. 2g). Seasonal variation data imply that during the monitoring period, the chloride ion concentrations were found to be high in the monsoon season at Ukkadam Lake which may be probably due to the accumulation of more municipal wastes. Similar results were reported by Chaturvedi and Pandey [24]. The chloride levels in all the lakes were observed to be higher than the permissible limits recommended byWHO[16]. Sulphate, Silicate and Nitrate values did not vary significantly (Fig. 2 e, f & h).but within the permissible limits (Sulphates 200 mg/L and nitrates 45 mg/L) prescribed by WHO [16].

3.2 Heavy metals concentration in water of three lakes

Seasonal variations of heavy metal concentrations (Cd, Cr, Cu, Fe, Ni and Pb) in the surface waters of the threelakes in comparison with National/ International guidelines are presented in Table 2. The concentrations of Cd ranged between 0.015-0.038 mg/L, 0.014 – 0.038 mg/L and

0.014 – 0.037 mg/L for Ukkadam, Kuruchi and Singanallur lakes respectively. The maximum concentrations of Cd recorded was 0.038 mg/L, 0.038 mg/L and 0.037 mg/L) for Ukkadam, Kuruchi and Singanallur lakes respectively during the post mon- soon season exceeding WHO [16] and BIS [17] recommended standards (0.003 mg/L) for drinking water.

Table 1. Physicochemical parameters of the surface waters in three lakes and comparison with National and International Guidelines

Sampling Sites	Seasons	Temp.(°C)	pH	SS (mg/L)	DS (mg/L)	TS (mg/L)	TA (mg/L)	TH (mg/L)	BOD (mg/L)	COD (mg/L)	DO (mg/L)
Ukkadam Lake	Monsoon	27.0±0.0	8.35±0.3	6.52±1.1	1789±33	1790±33	390±73	342±52	60.0±8.1	220±35	0.48±0.1
	Postmonsoon	27.0±0.0	8.05±0.1	5.43±0.4	1604±35	1609±34	425±75	314±46	31.0±9.0	140±40	0.72±0.2
	Summer	31.3±0.6	8.60±0.3	7.82±0.2	2171±15	1953±22	456±46	365±19	44.3±6.8	218±20	1.15±0.2
	Pre monsoon	30.6±1.3	8.03±0.2	2.86±1.2	1219±54	1223±54	329±10	299±11	20.6±19	103±79	5.54±1.8
Kuruchi Lake	Monsoon	27.0±0.0	7.85±0.2	1.77±0.5	5033±21	7244±19	249±34	202±10	48.7±10	197±37	0.68±0.2
	Postmonsoon	27.0±0.0	7.2±0.2	1.59±0.7	4557±34	4559±34	337±62	248±10	19.0±1.0	100±20	1.00±0.1
	Summer	30.6±1.3	7.94±0.0	6.05±0.4	1352±15	1646±30	391±4.9	374±4.2	42.6±9.3	194±2.9	1.23±0.1
	Pre monsoon	30.6±1.3	7.50±0.0	1.23±1.2	599.1±27	603.1±27	251±73	292±11	19.3±19	82.0±82	6.19±2.2
Singanallur Lake	Monsoon	27.0±0.0	8.01±0.1	2.55±1.0	4669±17	1196±26	376±46	367±14	42.0±8.9	147±25	1.35±0.1
	Postmonsoon	27.0±0.0	7.80±0.2	1.26±0.7	4542±34	934±13	562±12	288±16	20.3±6.0	104±16	0.92±0.3
	Summer	30.6±1.3	7.78±0.1	5.29±0.2	1228±86	1598±43	465±55	450±58	38.6±9.8	187±15	1.11±0.0
	Premonsoon	30.6±1.3	8.10±0.1	5.06±0.5	1294±81	1428±13	503±61	376±57	25.3±13	171±28	1.15±0.1
WHO Permissible limit (mg/L) (2011)		—	6.5 - 8.5	—	500	—	—	300			
BIS Permissible limit (mg/L) (2003)		—	6.5 - 8.5	—	500	—	—	300	2.0		6.0

Data are mean ± SE (n=3), Water temperature (Temp.), Suspended Solid (SS), Dissolved Solids (DS), Total Solids (TS), Total Alkalinity (TA), Total Hardness (TH), Biological Oxygen Demand (BOD), Chemical Oxygen Demand (COD) and Dissolved Oxygen (DO).

Table 2: Heavy metal concentrations of lake waters and comparison with National and International Guidelines

Sampling Sites	Seasons	Heavy metals content (mg/l) in lake water					
		Cd	Cr	Cu	Fe	Ni	Pb
Ukkadam Lake	Monsoon	0.0157±0.000	0.0310±0.001	0.3501±0.006	3.4600±3.060	ND	ND
	Postmonsoon	0.0389±0.000	0.0595±0.000	0.1810±0.005	0.3001±0.100	ND	0.6143±0.000
	Summer	0.0223±8.819	ND	0.5523±0.012	0.4333±0.145	0.0283±0.000	0.4635±0.005
	Premonsoon	0.0277±5.773	ND	ND	1.1000±0.458	0.1177±0.005	2.0921±5.775
Kurinchi Lake	Monsoon	0.0145±0.001	0.0145±5.773	0.3179±0.000	0.1000±0.000	ND	ND
	Postmonsoon	0.0380±3.333	0.0305±0.000	0.3390±0.003	0.1666±0.066	ND	0.5690±0.000
	Summer	0.0200±5.770	ND	0.0509±0.010	0.9001±0.400	0.0400±3.333	0.4810±0.000
	Premonsoon	0.0240±0.000	ND	ND	0.8333±0.523	0.1011±0.000	2.1728±0.005
Singanallur Lake	Monsoon	0.0140±0.001	0.0475±0.090	0.3240±0.005	0.2666±0.176	ND	ND
	Postmonsoon	0.0374±0.001	0.0633±0.000	0.1020±0.000	0.1000±0.057	ND	0.5310±0.000
	Summer	0.0220±0.001	ND	ND	1.4000±0.305	0.0429±3.333	0.4926±0.001
	Premonsoon	0.0263±0.002	ND	0.0254±0.000	1.5666±0.375	0.0944±0.000	2.1046±0.002
WHO Permissible limit (mg/L) (2011)		0.003	0.050	2.000	0.300	0.070	0.010
BIS Permissible limit (mg/L) (2003)		0.003	0.050	1.500	--	0.200	0.050

Data are mean ± SE (n=3), ND – Not Detected, Cadmium (Cd), Chromium (Cr), Copper (Cu), Iron (Fe), Nickel (Ni), Lead (Pb), World Health Organization (WHO) and Bureau of Indian - Standard (BIS).

Similar observations were also reported for the study of Bangshi River, Bangladesh by Rahman et al. [20]. The higher levels of Cd in lake water may be due to the discharging of wastewaters from electroplating and textile printing industries. The concentrations of Cr in Ukkadam and Singanallur lakes during the post monsoon seasonwere 0.059 mg/L and 0.063 mg/L respectively exceeding the WHO [16] and BIS [17] standards as men- tioned in the Table 2. Similar result was reported by Singh et al. [25] in Naukuchiatal Lake, India. The concentration ofNi in water samples ranged from 0.094 to 0.117 mg/L for pre monsoon and 0.023 to 0.042 mg/L for summer season. The maximum Ni con- centration found in these lakes during the pre monsoon season is higher than the rec- ommended WHO [16] value (0.070 mg/L). Virha et al. [21] has observed higher level of Ni in water samplesof Upper Lake, Bhopal, India.

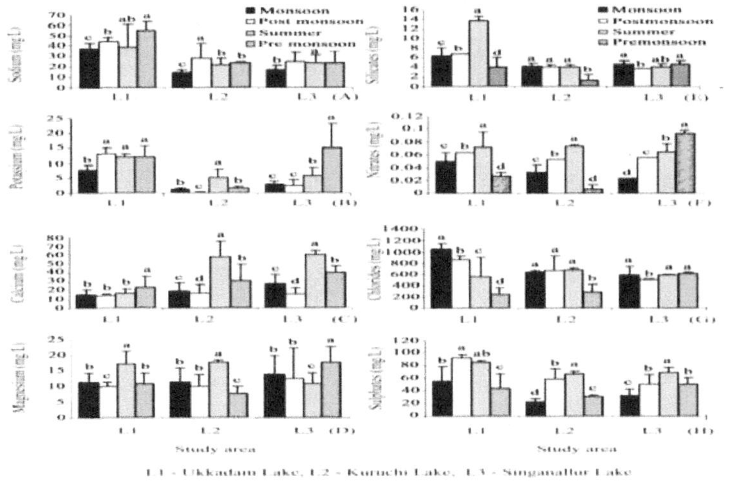

Figure 2. Nutrient properties of studied three lake waters,Data are means SE (n=4), bars with different letters indicate significant differences at *P* ≤0.05. Sodium (A), Potas-sium (B), Calcium (C), Magnesium (D), Silicates (E), Nitrates (F), Chlorides (G) and Sulphates (H).

From this investigation, the occurrence of Cr and Ni ions is confirmed in all the three lakes. It was noticed that, some electroplating and textile industries have been estab- lished and are discharging their effluents into these lakes. However, the Cu concentra- tion was found to be below the WHO [16] permissible dose (2 mg/L) in all the collectedsamples.

The concentrations of Fe ranged from 0.266 - 3.460 mg/L. The maximum Fe concentra-tion was recorded in Ukkadam Lake (3.460 mg/L) during the monsoon season. Occur- rence of ions in almost all the lakes at any one season or others seemed to be higher than the WHO's [16] recommendations (0.300 mg/L).

This may be attributed due to the direct discharge of mechanical engineering wastes, hospital and domestic wastes in and around these lake areas. The mean concentrations of Pb was ranged from 0.463 mg/L to 2.172 mg/L. Relatively maximum level of Pb wasobserved in Ukkadam lake (2.092 mg/L), Kuruchi lake (2.172 mg/L) and Singanallur lake (2.104 mg/L) during the pre monsoon season, against the WHO prescribed limit of

0.010 mg/L. This might be due to receiving of wastewaters from automobile and metal industries. Rahman et al. [20] have documented higher concentrations of Pb in freshwater ecosystems due to the long – term accumulation of Pb from motor vehicles emission be- ing the major source for enhanced Pb contamination in the aquatic environments. From the heavy metal analysis, it is evident that the heavy metal concentrations (Cd, Cr, Cu,Fe, Ni and Pb) in Ukkadam and Singanallur lake waters much higher than that of previ- ous report made byMohanraj et al in 2000.

Conclusion

The present study of the selected lakes revealed that the observed values for most of the parameters seemed to be much higher than the recommended level. This may be due to the accumulation of heavy metals, climate change in the recent years, discharge of wastes in folds due to the population growth in and around the Coimbatore city. The modernization of industries leads to the accumulation of heavy metal contaminants in water stream by industrial effluents due to lack of proper waste water treatment proce- dures. Results clearly showed that the heavy metal level was found to be permissible limit and therefore, an urgent wetland phytoremdiation system could be adopted to cleanthe pollutants. The present study together suggested that there will be proper implemen- tation of new waste management systems to provide better water quality in the urban ar-eas and to ensure the availability of safety drinking water from natural water bodies to consumers in the years to come.

Acknowledgement-Financial assistance received from University Grants Commission [UGC Major Project No. 41-13/2012 (SR)] Government of India, New Delhi is grateful-ly acknowledged.

References

1. Hou D., He J., Lub C., Renb L., Fand Q., Wang J., Xie Z. *Ecotoxi. Enviro. Safety*.(2013) 93:135–144
2. Zan F., Huo S., Xi B., Li Q., Liao H., Zhang. *J. Environ. Earth. Sci.*(2011), 62: 1643-1653.
3. Zamani A. A., Yaftian M. R., Parizanganeh, A. *Iran J Environ Health SciEng*.(2012),9:29
4. Ozturk M., Ozozen G., Minareci O., Minareci, E. *Iran J Environ Health Sci Eng*.(2009),6:73–80
5. Bem H., Gallorini M., Rizzio E., Krzemin, S. M.*Environ. Int.* (2003), 29 (4): 423–4286. Gupta K., Sharma A. J. *Environ. Bio.*2005,261(2): 205–211
7. Chandra R., Nishadh K. A., Azeez, P. A. *Environ. Monit. Assess.* (2010), 169: 671-676
8. Shanthi K., Ramasamy K., Lakshmana, P. *Bull. Environ. Contam.Toxicol.*(2003), 70:372-378
9. APHA. *(21st ed.). Washington, D.C.*(2005)
10. Tandon H. L. S.*5rd edn, (New Delhi,),*(2001), 143
11. Katz H.,Navone, R.*J. Amer. Water.Works. Assoc. (*1964), 56: 121
12. Edwards G. P., Molof A. H., Schneeman, R. W. *J. Amer. Water. Works.Ass.*(1965), 57:917-925
13. Thompson K. C., Reynolds, R. J.*2nd edn. John Wiley and Sons, New York,*(1978)
14. Duncan D. B.*Biometrics*, 1955, 11: 1–42
15. Gomez A. K., Gomez, A. A. *II edn. John Wiley and Sons, New York,*(1984)
16. WHO, *World Health Organisation. Geneva*, (2011)

17. B.I.S. (Bureau of Indian Standards drinking water specification). *2nd revision, IS10500,*(2003)

18. Mohanraj R., Sathishkumar M., Azeez P. A., Sivakumar, R. *Bull. Environ. Contam.Toxicol.* 64 (2000) 638

19. Sunkand B. N., Patil H. S.*J. Environ. Bio.* (2004), 25 (1): 99-102

20. Rahman M. S., Saha N., Molla, A. H.*Environ. Earth. Sci.*(2013), 12665: 013-2631-5

21. Virha R.,Biswas A. K., Kakaria V. K., Qureshi T. A., Borana K., Malik, N.

Bull. Environ.Contam.Toxicol.(2011), 86: 168-174

22. Radhakrishanan R., Dharmaraj K., Renjithakumari, B. D.*J. Environ. Biol.*(2007), 28:105-108.

23. Iqbal J., Shah H. M., Akhter, G. *J. Geochemi. Explor.*(2013),125: 94-101

24. Chaturvedi J., Pandey, N. K. *Curr. World. Environ.* (2006), 2: 177-179

25. Singh A. P., Srivastava P. C., Srivastava, P. *Water Air and Soil Pollu.*(2008), 188:181–193.

Experimental Optimization of an Agricultural Waste in the Uptake of Cu(II) Ions

N. Muthulakshmi Andal* and S.Charulatha[1]

Department of Chemistry, PSGR Krishnammal College for Women, Coimbatore – 4,

***Correspondence** N. Muthulakshimi Andal Email: andaldr@gmail.com

Abstract

The potential of tamarind hull, a low costagricultural waste, to remove copper ions from aqueous solutions is evaluated in a batch process. The tamarind hulls are crushed, pow- dered and modified using 0.5 N sulfuric acid and formaldehyde. The experiments were performed to assess sorption capacity of the sorbent. The influences of variable parame- ters viz., particle sizes and doses of the sorbent material, agitation time and the pH are studied for Cu(II)-MTHPsystem. The optimum conditions form the maximum removal of Cu(II) ions by modified tamarind hull powder is established, from the experimental results and is presented in this paper.

Keywords: Adsorption, copper ions, agricultural waste, modification, parameters

Introduction

Water bodies are extremely polluted by heavy metal ions released from diverse indus- tries such as metal plating, mining, painting, wire drawing, batteries and printed circuit board manufacturing, as well as agricultural sources. Heavy metals, such as Cu, Zn, Pb and Cd, are prior toxic pollutants leached out by industrial wastewater [1]-[2]. Heavy metals are indispensable for many organisms in small amounts, but their enhanced dos-es induce acute or chronic poisoning. Dosage of heavy metal ions exceeding the toler- ance limits, may disrupt the normal function of the central nervous system, cause changes in the blood content and adversely affect the functions of lungs, kidneys, liver and other organs. The long-term action of heavy metals may cause the development of cancer, allergy, dystrophy, physical, neurological degenerative processes, Alzheimer's and Parkinson's diseases. [3]

Copper ions (Cu^{2+}) are water-soluble. They are employed in industries like min-
ing, metallurgical, electrical, electroplating, paints, pigments and electronics. The max- imum permissible limit of Cu(II) in effluent discharges into water bodies is 0.25 mg/ L^2 [4]. The excess Cu(II) in aqueous waste discharges has become one of the prior pollu- tants. These excess Cu(II) accumulates in liver and brain. Copper toxicity is afunda- mental cause of Wilson's disease [5]. Many conventional techniques such as chemical precipitation, membrane filtration, electrolysis, ion exchange, co- precipitation and ad- sorption have been employed for the removal of heavy metals in wastewater treatment [6]. Among all this, the adsorption technique has gained momentum and popularity in recent times. Agricultural wastes are natural abundant in the environment, economic and readily available for the sorption of heavy metal ions from aqueous solutions. Ag- ricultural wastes may therefore serve as an excellent alternative to activated carbon, which is quite unaffordable especially in developing countries [7].Adsorbent materials derived from Biomass can be used for the effective removal and recovery of heavy metal ions from wastewater: from algae, bacteria, sea-weeds, some higher plants, and agricultural wastes [8]. In view of abundant availability of these naturally occurring materials, a systematicstudy to assess the potential use of modified tamarind hull pow- der for the removal of Cu(II) from aqueous solutions is carried out in this paper.

In India, tamarind (*Tamarindus indica L.*) is an economically important tree which grows abundantly in the dry tracts of Central and South Indian states. Indian production of tamarind is about 3 lakh (0.3 million) tonnes per year. The hard pod shellis removed (deshelled) when the fruit is ripe and the fruit is the chief acidulate used in the preparation of foods. The shells (hulls) are discarded as waste and pose greater dis- posable problem [9]. These hills are available at no cost and so far, no studies have been reported for the removal of Cu(II) ions using TH powder.

Hence, the present study aims at the exploitation and utilization of Tamarind Hull Powder (THP) for the sorption of Cu(II) ions.

Materials and Methods Chemicals

All the chemicals used were of analytical reagent grade. Doubly Distilled (DD) water wasused to carry out the experimental studies. Stock solution of 1000ppm copper solutions were prepared by dissolving 2.6826g of $CuCl_2 \cdot 2H_2O$ in 1000 ml water. Aliquots of the adsorbate solutions of varying Cu(II) ions concentrations (2-14 ppm: 2ppm) were pre- pared by progressive dilutions of stock copper solution. 0.1 N HCl and NaOH solutions were employed to adjust the pH of the aliquots.

Procedure of adsorbent preparation and modification

Tamarind hulls were collected from various localities in Coimbatore, washed well with distilled water and sun dried. Then the dried hulls (Fig. 1) were

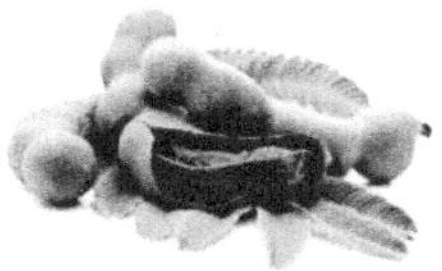

ground in a laboratory
blender and sorted into different particle sizes using standard test molecular sieves. The varied sizes of the THP were treated with formaldehyde followed by H_2SO_4 to modify thenature of the sorbent materials. After modification, the material was washed well, dried and kept it in air tight container, there after called as Modified Tamarind Hull Powder (MTHP).

Batch Equilibration Studies

Batch Equilibration experiments were carried out at room temperature, in order to inves- tigate the nature of metal – adsorbent interaction. The influence of various parameters viz., particle sizes and dosages of MTHP, agitation time between MTHP and sorbate solu-tion, pH of the medium were investigated. An optimum initial concentration of Cu(II) ionswas fixed for all the studies. The flasks with the contents (50ml of sorbate solution with varying particle size, dosage and pH values) were agitated at preset time intervals with140 rpm using a Kemi Orbital mechanical shaker.

The agitated samples were filtered and the residual concentrations of the Cu(II) ions were analyzed using AAS (AA- 6200, Shimadzu make). The percentage removal was calculat- ed from the recorded values. The quantity of Cu(II) ions adsorbed onto the adsorbent at the time of equilibrium (q_e) was calculated using the following expression:

$$q_e = (C_o - C_e) V/m \quad (1)$$

where C_o and C_e are the initial and equilibrium concentrations (mg/L) of Co(II) ions in solution. V is volume (L) of solution and m is weight of adsorbent (g) taken for experi- ment.

The percentage of adsorption of metal from aqueous solution was estimated by using the following equation [10].

$$\% \text{ adsorption} = \frac{(C_i - C_e)}{C_i} \times 100$$

Results and DiscussionsEffect of Particle Sizes

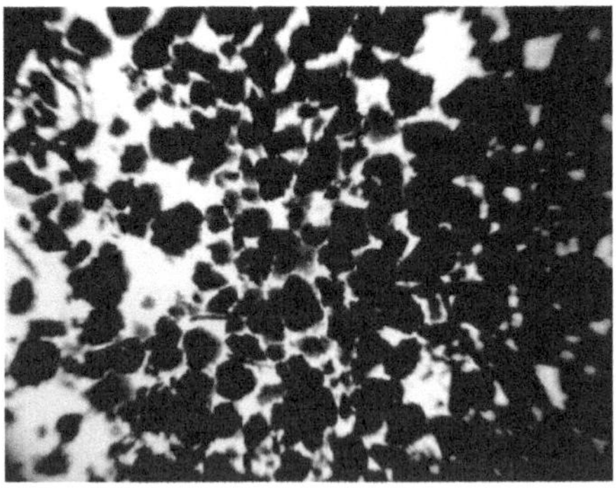

Figure 2: Microscopic Structure of MTHP

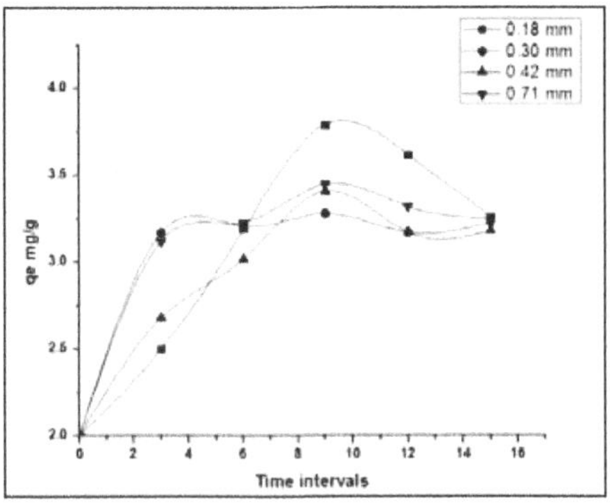

Figure 3: Effect of particle size

Adsorbent particle size has a significant influence on the kinetics of sorption due to change in number of adsorption sites. The sorption experiments of Cu(II) carried out for five different particle sizes of MTHP viz., 0.18, 0.21, 0.30, 0.42, 0.71 mm are shown in Fig. 3.The steep rise in the curve indicate that the smaller 0.18 mm particle size of MTHP exhibited maximum percentage removal of 93% for Cu(II) ions. The number of binding sites for adsorption increases with surface area. This is due to higher surface area of unit weight for lesser particle size and thence greater percentage removal [11,12]. In view of these experimental results, it was decided to limit the discussion of further experiments using 0.18mm particle size for Cu(II)-MTHP system. The MTHP microscopic structure of0.18mm particle size is depicted in Fig. 2

Effect of Agitation Time

The amount of adsorption of Cu(II) increased with the contact time. However, the adsorp-tion tends towards attaining equilibrium at 9 min. The gradation in the amount adsorbed isrepresented by the continuous smooth rise in the curve (Fig.4) plotted against the preset time intervals (3-15 minutes: 3 minutes interval). After 9 minutes, the semi parabolic curve indicates that the system tend towards saturation and had attained equilibrium at 9 minutes. The rapid initial adsorption is probably due to the abundant availability of active sites on the sorbent surface for the adsorption of these metal ions. As the surface sorption sites become exhausted, the rate of uptake is controlled by the rate of transport from the exterior to the interior sites of adsorbent particles which may be responsible for the slower stage [13]. Further, increase in contact time did not register an increase in adsorp- tion. Thus, 9 minutes agitation time was considered to be the optimum contact time for the present system.

Effect of Dosage

Adsorbent dose seems to have a great influence on sorption process. Dose of adsorbent add- ed into the solution determine the number of binding sites available for adsorption. The ad- sorption density decreased from 3.79 to 1.8 mg/g when the adsorbent dose was increased from 50mg to 200mg at 50mg intervals as evident from Fig. 5. It is obvious from the figure that 50mg of MTHP had maximum adsorption of Cu(II), the corresponding curve being atop in the graph. The decrease in adsorbent density is basically due to the adsorption sites remaining unsaturated during the adsorption process. This may be due to the overlapping and aggregation of adsorption sites which occurs as a result of overcrowding of adsorbent particles due to increase in the dose [14].

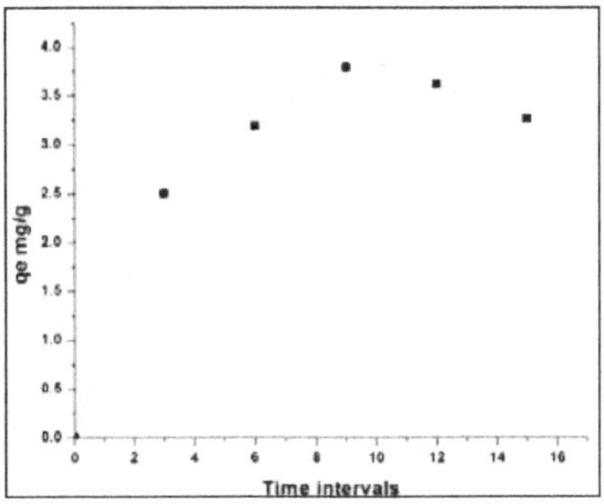

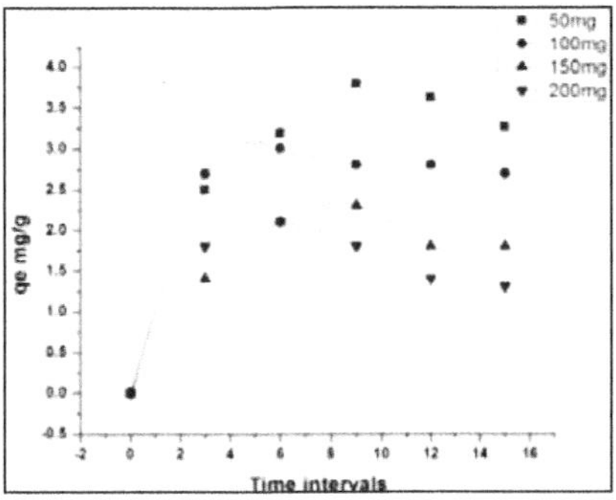

Figure 4 Effect of agitation time Figure 5 Effect of adsorbent dosage

Effect of pH

The pH dependence for maximum removal of Cu(II) ions onto MTHP was studied at varying pH environments viz., 3-

11. It was observed that the ability of trapping Cu(II) ions by MTHP is pH dependent as evi- dent from the Fig. 6 wherein the maxima had occurred at pH 5.6 in the inverted parabolic curve. Binding sites are generally protonated

orpositively charged (by the hydronium ions) at low pH values. This leads to repulsion between the metal cation and the adsorbent at a higherpH value [15]. The precipitation of insoluble copper hydroxide might have occurred at pH values greater than 6.

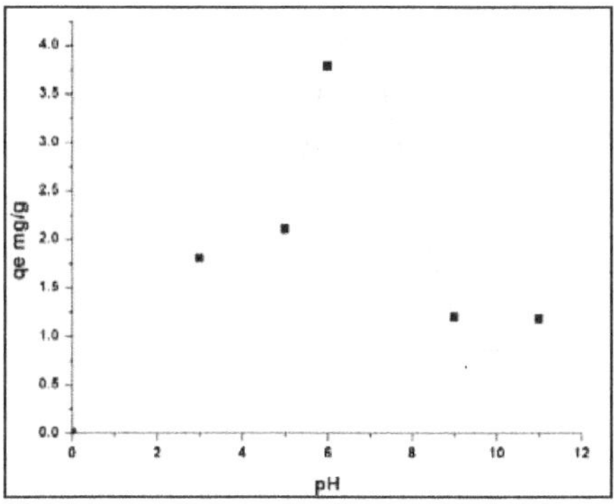

Figure 6 Effect of pH

Conclusion

The experimental results have led to the conclusion that the modified tamarind hull powder, has 93% efficiency to adsorb Cu(II) ions from the aqueous solutions. The modification has enhanced the sorption capacity of the chosen material. 0.18mm particle size, 50 mg dosage,9 minutes agitation time, 4ppm initial concentration of Cu(II) ions, pH
5.6 of the solution medium, were optimized for the significant removal of divalent ion by MTHP.

References

[1] J. Peric, M. Trgo, N. V. Medvidovic, —Removal of zinc, copper andlead by naturalZeolite—a comparison ofadsorption isotherms,WaterRes., 38, 1893–1899, 2004.

[2] M. A. Stylianou, M.P. Hadjiconstantinou, V.J. Inglezakis, K.G.Moustakas, M.D. Loiz- idou, Use of natural clinoptilolite for theremoval of lead, copper

and zinc in fixed bed column, J. Hazard.Mater., 143, 575–581, 2007.

[3] Kvesitadze G, Khatisashvili G., ,Sadunishvili T., Ramsden JJ., Biochemical Mecha- nisms of Detoxification in Higher Plants Basis of Phytoremediation, Springer-Verlag Berlin Heidelberg, 2006.

[4] V. Tharanitharan, A. Gayathri and K.B. Nagashanmugam, Removal of Cu(II) from Wa- ter and Wastewater using Thermally Activated Gingelly Oil Cake Carbon, Coromandal Journal of Science, Vol. 1, No. 1, pp.40-47, 2012.

[5] B. S. Khangarot and P. K. Ray, Environmental Copper and Human Health Science, Re- porter, 352, 1988.

[6] EduInam, UbongEtim, UbongEduok and Joseph Essien, Heavy Metals Sorption Poten- tial of Calcareous shells ofAnimal Origin.Int. J. ofChe.Envir.&Phar. Res., 3(3): 184-194,2012.

[7] Abdelkrim Cheriti, Mohamed Fouzi Talhi, Nasser Belboukhari1 and Safia Taleb, Copper Ions Biosorption Properties of Biomass Derived from Algerian Sahara Plants, Expanding Issues in Desalination, 2011.

[8] N. Ahalya, , R.D. Kanamadi and T.V. Ramachandra, Biosorption of chromium (VI) by Tamarindusindica podshells. International Journal of Environmental Science Research, 1(2):77-81.

[9] Xue Song Wang, ZhiZhong Li, Sheng Rong Tao, Removal of chromium(VI) from aque- ous solution using walnuthull, J. Environ. Manage. 90, 721-729, 2009.

[10] AhmetOrnek, MahmutOzacar, I. AyhanSengil, Biochem. Eng. J., Vol. 37, 192-200, 2007.

[11] Muhammad Riaz, RaziyaNadeem, Muhammad Asif, Tariq Mehmood Ansari, Khalil-ur- Rehman, J.Hazard. Mater., Vol. 161, 88-94, 2008.

[12] ZainulAknarZakaria, MarliniSuratman, Nurfadilah Mohammed, Wan Azlina Ahmad,Chromium(VI)removal from aqueous solution by untreated rubber wood sawdust, De- salination.244, 109-121, 2009.

[13] J. Anandkumar, B. Mandal, Removal of Cr(VI) from aqueous solution using Bael fruit (Aeglemarmeloscorrea) shell as an adsorbent, J. Hazard.Mater.

168, 633-640, 2009.

[14] T. Shanthil and V. M. Selvarajan, Removal of Cu (II) Ions from Aqueous Solution by
Carbon Prepared fromHenna Leaves, Pungam Bark and CAC, Journal of Chemical andPharmaceutical Research, 4(9):4296- 4306, 2012.

Publication History
Received 21st Sep 2013
Revised 22nd Dec 2013
Accepted 15th Feb 2014
Online 29th Apr 2014

Sorption kinetics and equilibrium studies on the removal of toxic Cr(VI)ions employing modified Indian almond nut shells

N.Muthulakshmi Andal*[1]& G.Gohulavani[2&3]

1. Assistant Professor, Dept. of Chemistry, PSGR Krishnammal College forWomen, Coimbatore, Tamilnadu,India
2. Research Scholar, Dept. of Chemistry, PSGR Krishnammal College for Women, Coimbatore, Tamilnadu,India
3. Assistant Professor, Dept. of Chemistry, Hindusthan Institute of Technology, Coimbatore, Tamilnadu, India

Abstract

Exploring the use of novel agricultural based adsorbents in controlling Cr(VI) pollution in waste water is one of the main areas of research on pollution control due to its high toxicity, posing threat to the human and environment. Adsorption has attracted attention as a cost-effective tool for the removal of such heavy metal ions. This paper presents the experimental results carried out for the sorp- tion of Cr(VI) ions from aqueous solutions employing modified *Terminalia catappa* nut shell (TTCNS). The characterization of TTCNS is investigated us- ing BET, BJH, FTIR, SEM and EDAX analysis. The effect of variable parame- ters viz., particle size, contact time, initial concentration, sorbent dose and pH were studied. The system has been optimized to achieve suitable parameters for maximum removal of Cr(VI) which are: 0.18mm particle size and 200 mg ad- sorbent dose of selected sorbent material,11 mg/L initial concentration of ad- sorbate species, 30 minutes contact time and pH 2 for the solution medium. Ad- sorption equilibrium and kinetics were verified by Langmuir, Freundlich iso- therms and Lagergren first order, Pseudo second order models. The applicabilityof the methodology developed is also tested with real effluent sample collected from chrome plating industry.

Keywords: adsorption, isotherm, kinetics, equilibrium, nut shells

1. Introduction

Chromium is one of the heavy metals of concern due to its toxic and non-biodegradable nature. Industrial activities such as mining, electroplating, tanning, metallurgical operations and manufacturing have led to the release of this pollutant into the environment. It exists in two stable forms as Cr(III) and Cr(VI), the trivalent ion is known to be an essential trace element in plants and animal metabolism, the hexavalent chromium is carcinogenic and mutagenic. This is due to its easy penetration into biological membranes resulting in health hazards such as skin disorders, respiratory tract problems, lung carcinoma, acutetubular necrosis of kidney and even death in extreme cases(R.Mehra 2003). As per USEPA, the permissible potable limit of Cr(VI) in drinking water is 0.1ppm and irrigation water is 100 ppm (US Department of Health and Human Services,1991). It is therefore imperative to determine this metal ion concentration in wa-ter bodies.

Environmental concerns have motivated many physico-chemical techniques for heavy metal removal from aqueous solutions like reduction, coprecipitation, ion-exchange, reverse osmosis, dialysis and resin chelation(J.N.Egila et al., 2011). These techniques however, posses their own disadvantage like incomplete metal removal, high reagent cost and energy requirements. Therefore, the need for costeffective alternative technologies is essential for this task. Adsorption using nat- ural adsorbents, an innovative and economical method, has evolved the front- line of defense especially for metals that could not be effectively removed by othertechniques.

Several natural sorbents from agricultural, aquatic and forest wastes have been utilized as such and in modified forms by various researchers in the sequestration of chromium ion from aqueous solutions and industrial effluents. Husk of Toor dal (*Cajanuscajan*) was investigated for the removal of Cr(VI) and

Fe(III) ions from aqueous solution by N.Ahalya et al(2007).F.N.Acar and E.Malkoc(2004) employed beech saw dust *Fagusorientalis L*, to elaborately ex- amine the removal of Cr(VI).Uptake of Cr(VI) and Ni(II) from aqueous solu- tions had been investigated by O.O.Kehinde et al.,(2009),usinglow cost agricul- tural by-products like coconut husk and teak tree bark. Recently low cost adsor- bents are activated/modified by various methods for metal removal. Zinc chlo- ride activated *Terminaliaarjuna* nuts(KaustubhaMohantyet al.,2005) and phos- phoric acid activated tamarind hulls(A.Verma et al.,2006) have been utilized to trap Cr(VI). Dried biomass of *Oscillatorialaete-virens* was found to posses the potential to remove heavy metals like Cr(VI) and Ni(II) from polluted water (Suman Das, 2012).

The objective of the present study is to investigate the feasibility of employing modified *Terminalia catappa* nut shells for the adsorption of Cr(VI) from aqueous solutions. Until this present work, little or no information is available on the suitability of using this selected agricultural waste in the removal of toxic Cr(VI) and seldom work has been reported in literature so far.

1. MATERIALS

1.1 Sorbent and Sorbate

The tropical Almond *Terminalia catappa* (Indian almond) belongs to the family Combrataceae. It is a large spreading tree distributed throughout the tropics and coastal environment (Species profiles, 2006). The fruit is a sessile, laterally compressed, ovoid to ovate and smooth skinned drupe. The oil containing seeds are en- cased in a tough fibrous husk with a fleshy pericarp. Thiscorky fibrous endocarp (nut) of the fruit was collected from various localities in Coimbatore, Tamilnadu, India

The nut shells were thoroughly washed with double distilled water,dried in air, then powdered and sieved to various sizes viz., 0.18mm, 0.24mm, 0.30mm, 0.42mm and 0.71mm. The powdered sieved adsorbent was first treated by boiling in 0.1 N HCl for three hours. After decanting the solution, the residue was boiled again with 0.1 N NaOH for three hours. The treated sorbent of varying sieve sizes were washed well several times with double distilled water. Later, they were soaked in water for sufficient time interval,to ensure swelling, as it would make more sorption sites available and finally, the sorbent materials weredried in sunlight. The chemically treated *Terminalia catappa* nut shell powder was used for further experiments and henceforth shall be denoted as TTCNSin the forthcoming discussions.

HCl, NaOH and $K_2Cr_2O_7$of AR gradewere purchased from RanbaxyChemi- cals.A stockCr(VI) solution of 1000 mg/L was prepared by dissolving 2.828 g of$K_2Cr_2O_7$in 1000 ml of double distilled water for the adsorption experiments.

1.2 Characterization of Sorbent

Table-1 depicts the various physicochemical parameters of TTCNS. Surface ar- ea, micropore area, micropore volume and average pore diameter were deter- minedby *Micrometrics Surface Analyzer*, (facilitated with Belsorp adsorp- tion/desorption software; using nitrogen adsorption at 77 K).The FTIR spectrum of the native and Cr(VI) loaded adsorbent were recorded with a *Thermo Nicolet Infrared Spectrophotometer.* The surface morphology of the treated adsor- bent,both loaded and unloaded was examined using the *Scanning Electron Mi- croscopeJEOL JFM- 6390* (Japan). After drying, the samples were covered with a thin layer of platinum (10 nm) using a sputter coater (SCD 0050 –Baltec, Lie- chenstein) and observed using the microscope (20 kV) under thevacuum of 1.33 x10^{-6} m Bar.

Table 1 Physiochemical Characteristics of TTCNS

Parameters	Values
Surface area (m^2/g)	83.14
Mean Pore diameter (nm)	38.69
Porosity %	94.95
Specific gravity	4.45
Bulk Density	0.25
Ash Content %	9.09
Ion exchange capacity(meq/g)	0.40
pHzpc 6.20	
Moisture %	8.51
Pore volume cm^3/g6.80	

2. Methods

3.1. BET surface analysis

BET and BJH methods are used for the evaluation of the particle and mesopore size distribution. Pore sizes are classified in accordance with the classification adopted by the International Union of Pure and Applied Chemistry (IUPAC manual., 1982)i.e [micropores diameter(d<20A°),mesopores (20A°< d< 500A°) and macropores (d > 500A°].The adsorbents for liquid phase adsorbates should have predominantly mesopores in their structure, due to the larger sizes of liq- uid molecules(V.C.Srivastava et al., 2006). Surface characteristic values de- picted in table 1, proves that, TTCNS lie in the mesopore category and has wide pore size distribution (Fig.1) providing wide distribution of surface area.

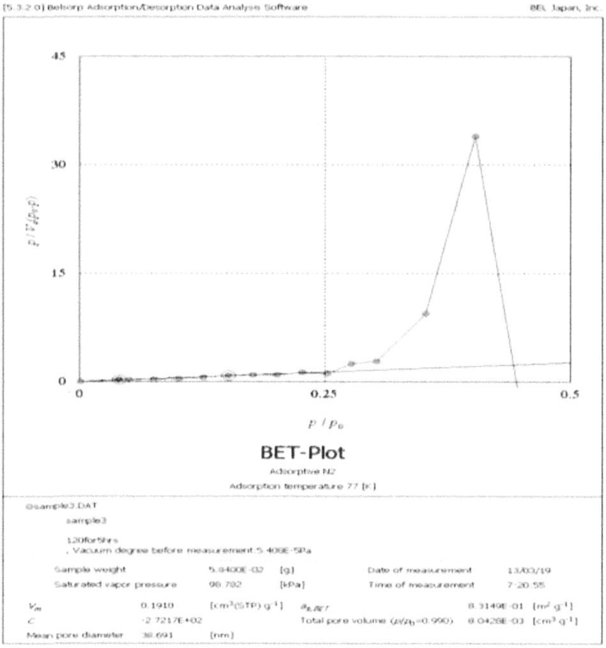

Figure 1:BET plot for TTCNS

3.2 Equilibration Experiments

Batch equilibration studies were carried out at $25+1^0$ C. At varying initial concentrations, TTCNS dosage varying in the range of 0.18mm-0.71mm particle sizes, 50-250 mg dosages were added to samples separately and the iodine flasks were agitated in an orbital mechanical shaker at predetermined time intervals.pH of the solutions were adjusted to different values accordingly using 0.1 M HCl and 0.1 M NaOH. The contents of the flasks were then filtered and the re- sidual metal ion concentrations were determined using *Shimadzu (AA 6200) Atomic Absorption Spectrophotometer.*The solution volume V ml was kept con- stant (50ml). The metal ion sorption percentage at any instant of time was de- termined by using the following equation (1)

$$\% \text{ adsorption} = \frac{(C_i - C_e)}{C_i} \times 100 \quad (1)$$

where C_o and C_e are the initial and final concentrations of Cr(VI) respectively.

3.3 Specifications of Isothermal and Kinetic Studies

The equilibrium study on adsorption has provided sufficient information on the capacity of the adsorbent. An adsorption isotherm is characterized by certain constant values that express the surface properties and affinity of the sorbent (Ahmed and Mustafa, 2008). It can also be used to compare the adsorptive capacities of the adsorbent for different pollutants. Theadsorption isotherm modelswere applied to the experimental data obtained from parameter study conducted for various initial concentrations of the Cr(VI) ion. The linear equation (2) was used to plot the Langmuir model

Ce Ceqe qm1 bqm(2)

where q_e is the equilibrium metal ion concentration on the sorbent (mg/g), C_e is theequilibrium metal ion concentration in solution (mg/L), q_m is the maximum mono- layer adsorption capacity of the sorbent (mg/g) and b is the Langmuir sorption con- stant (L/mg) related to the free energy of sorption. The Langmuir constants q_m and bcan be determined from the linearplot of C_e/q_e versus C_e, which has a slope of $1/q_m$and intercept of $1/bq_m$.

Freundlich isotherm was plotted by applying its standard straight-line equation (3) $\log q_e = \log K_F + 1/n \log C_e$

(3)

where, K_F and 1/n are the Freundlich's constants related to the adsorption capacity (mg/g) and adsorption intensity of the adsorbent respectively, which vary with the heterogeneity of the adsorbent.

The adsorption kinetics were investigated with an aim of obtaining a deep insight into how the amount of adsorbed metal changes with time and the process time required to achieve equilibrium between the aqueous and the solid phase. In order to clarify the adsorption kinetics of Cr(VI) onto TTCNS, two ki- netic models viz., Lagergren's Pseudo-first –order and Pseudo-second-ordermodels were applied to the experimental datalog(q_e - q_t) = log q_e -k_1t

2.303

where, q_e and q_t are the adsorption capacity (mg/g) at equilibrium and at time t, respectively and k_1 is the rateconstant for pseudo-first-order adsorption (min^{-1}).

The linear form of the Pseudo-second-order rate equation can be ex- pressed as

$$\frac{t}{q_t} = \frac{1}{k_2 q_e^2} + \frac{t}{q_e}$$

where, k_2 is the rate constant for pseudo-second-order adsorption (g/mg min). The slopes and intercepts of plotsof t/q_t versus t were used to calculate q_e and k_2.

3.4. FT-IR Characterization

Examination of the spectra of TTCNS after Cr(VI) adsorption (figure 2b) show a shift in absorption of the–OH stretching frequency from 3386.48$cm^{-1}$to 3369.33cm^{-1} corresponding to unloaded spectra(figure 2a). This can beinterpret-ed as that O-H stretching is associated with metal adsorption. The –CH stretch- ing at around 2923.09 cm^{-1} is also slightly shifted in the Cr(VI) loaded spectra. The peak at 1423.37 cm^{-1} characteristic of–COO stretching is also found to shiftto around 1425.10 cm^{-1} in the corresponding metal loaded spectra, which is in- dicative of the participation of these groups in metal binding.

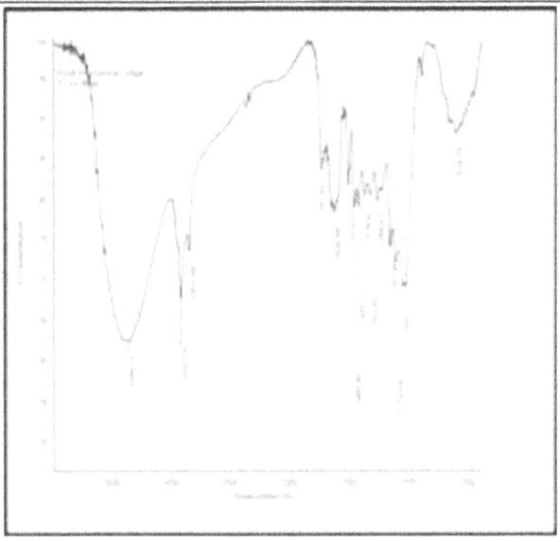

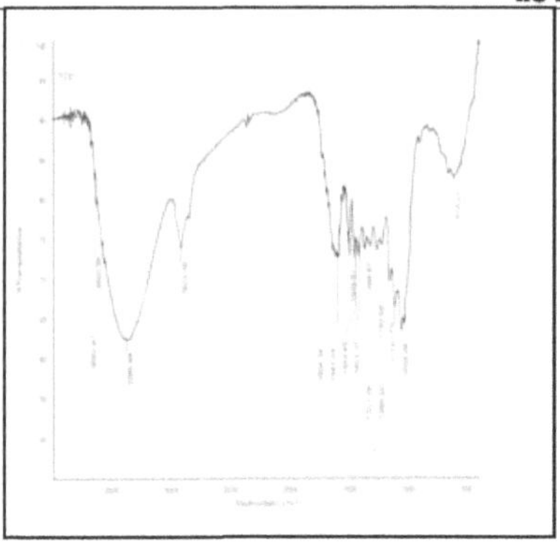

Figure 2a: FT-IR of unloaded TTCNS Figure2b: FTIR of Cr(VI) loaded TTCNS

SEM and EDAX Analysis

It can be observed from the SEM figures (3a&3b) that the TTCNS is found to posses' surface rugosity in such a way to adhere the solute species onto the surface of the adsorbent. Therefore, the adsorptive characteristics of TTCNS is expected to be effective on Cr(VI) adsorption. EDAX (figure 4b) shows the presence of new peak representing chromium so that it substantiates that the adsorption had occurred.

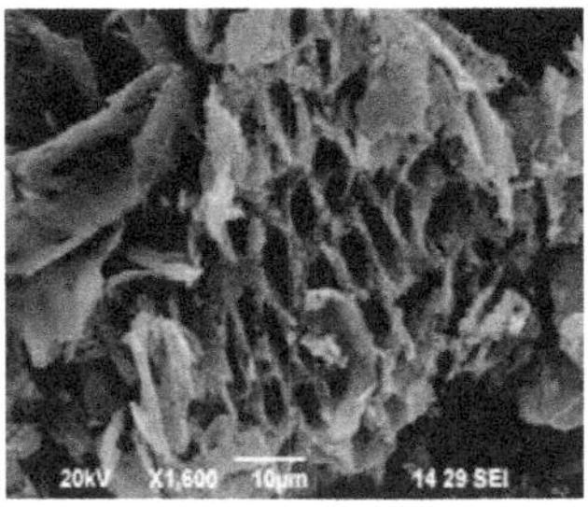

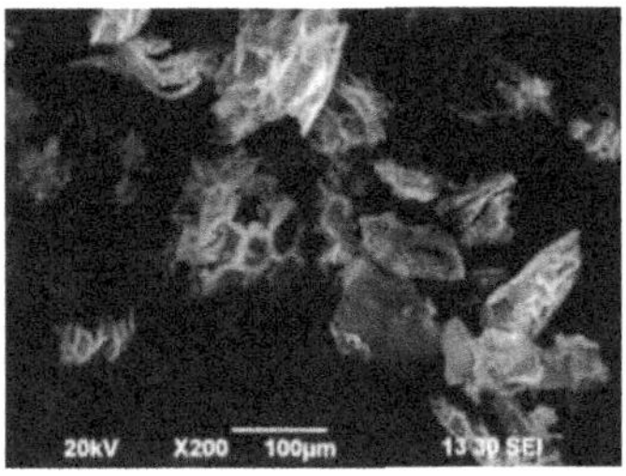

Figure 3a: SEM of unloaded TTCNS Figure 3b: SEM of loaded TTCNS

Figure 4a: EDAX spectrum of unloaded TTCNS

Figure 4b: EDAX spectrum of loaded TTCNS

3. Results and Discussion

3.1 Effect of Particle Size

Five different particle sizes of the sorbent, ranging from 0.18 mm to 0.71mm were utilized to investigate the influence of the adsorbent particle size for the metal removal. It is observed from the smooth and continuous curves (figure 5) that 0.18mm sizes of TTCNS had registered enhanced sorption capacity, show- ing a maximum uptake of 26.8 mg/g to be the optimum particle size.The rela- tively higher adsorption with smaller adsorbent particle size may be attributed tothe fact that smaller particles yield large surface areas and indicate that chromi- um ion adsorption occurs through a surface mechanism(IAGC, 2010). So, the studies pertaining to other parameters were carried out employing 0.18mm TTCNS.

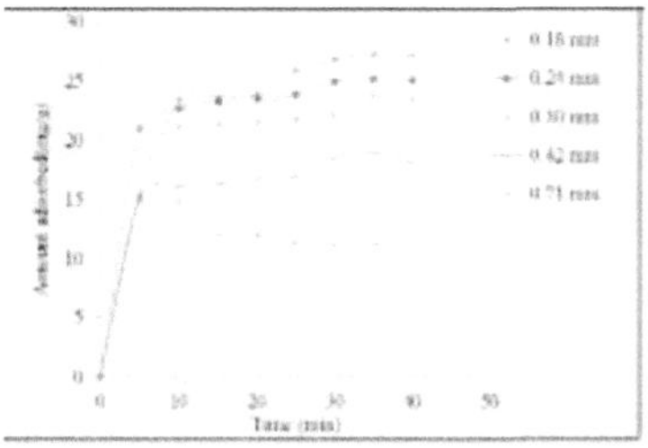

Figure 5: Effect of particle Size (TTCNS)

3.2 Effect of Contact Time and Initial Concentration

The agitation period for the adsorption of Cr(VI) by the sorbentwas investigated to identify the rate of metal removal. The amount adsorbed per gram of the adsorbent for Cr(VI) at different initial concentrations ranging from 3 to11 ppm are listed in table 2. At 30 minutes, higher amounts of Cr(VI) was adsorbed ontoTTCNS and the values were 11.85, 22.8, 23.8, 27.2 and 37.4 mg/g for initial concentrations of 3,5,7,9 and 11 ppm respectively(Fig.6).The amount of Cr(VI) getting adsorbed increased with contact time but at a particular point of time, the system attained equilibrium (i.e) the % removal of chromium was observed upto 30

minutes of agitation time later no further increase was registered(Carlos Green-Ruiz,2006). In this view point, optimum initial concentration of 11 ppm of Cr(VI) and an optimum contact time of 30 minutes has been fixed for the forthcoming experiments.

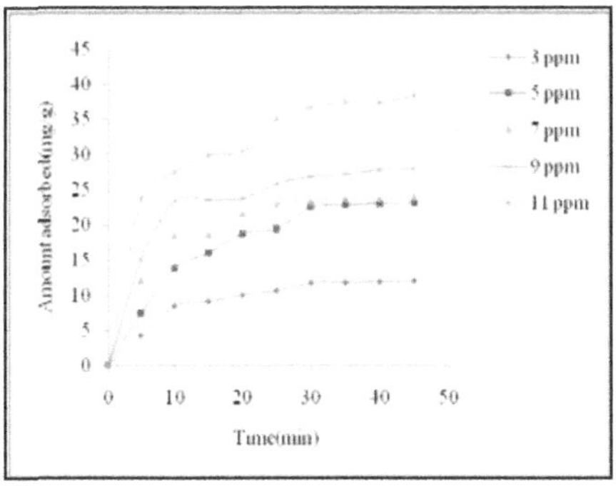

Figure 6: Effect of contact time and initial concentration [Cr(VI)onto TTCNS]

3.3 Effect of Dosage

The variation in doses (50-250mg; 50mg interval) was experimentally veri-fied. The decrease in the amount adsorbed beyond 200 mg was observed to de- crease from 54.4 to 52.3 mg/g, as evident from the steep rise in the curve upto 200mg and later a decline for 250 mg (fig.7). This may be basically due to electro-static interactions between theadsorbent particles and interference between the bind- ing sites(Hameed and Khaiary,2008).

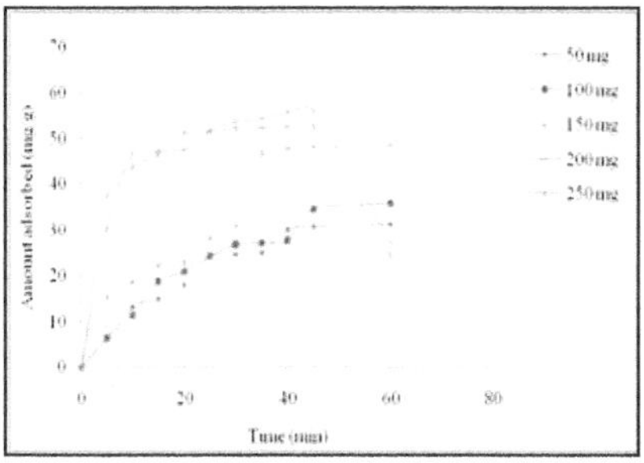

Figure 7: Effect of Dosage(TTCNS)

3.4 Effect of pH

Cr(VI) removal by TTCNS was investigated as a function of solution pH overthe range 1-8, for two different initial concentrations, as depicted in figure 8.Maximum uptake was observed at pH 1.78. At increasing pH environments, a sharp decline in uptake was observed, which is in good agreement with previousreports (Emine Malkoc,2007 &HuiGao, 2008). The speciation studies of Cr(VI) in aqueous solution shows that H_2CrO_4 predominates at pH less than 1.0,$HCrO^{4-}$ for pH between 1.0 and 6.0 and CrO $^{2-}$ at pH above 6.2 (Lyn Patrick,2006).

Adsorption of Cr(VI) at pH 1.78 onto TTCNS is due to the electrostatic attraction between the positively charged surface of the adsorbent with $HCrO^{4-}$ ions. But, in highly acidic medium (pH=1.0), H_2CrO_4 (neutral form) is the predominant species of Cr(VI). Hence, percentage removal decreased due to the involvement of less number of $HCrO^{4-}$ anions to the positive surface. At higher pH value, the reduction in adsorption may be due to the dual competition of bothOH^- and CrO_4^{2-} ions to get adsorbed on the surface of the adsorbent among which OH^- predominates (Manjeet Bansal et al., 2006)

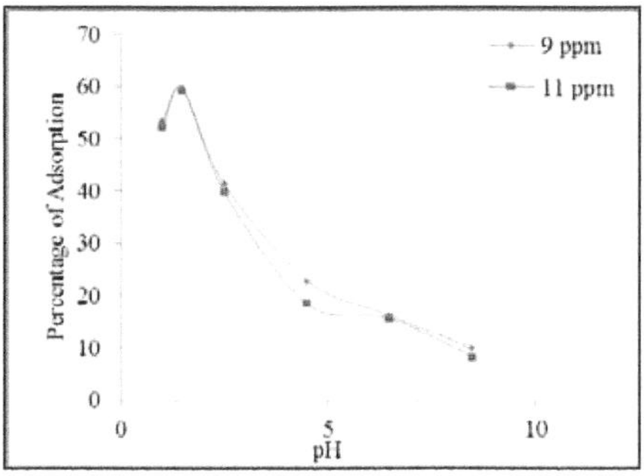

Figure 8: Effect of pH[Cr(VI)onto TTCNS]

3.5 Langmuir adsorption isotherm

The Langmuir constant q_m which is a measure of the monolayer adsorption capacity of TTCNS,Langmuir constant b, (free energy of sorption) and the correla-tion coefficient R^2 for Cr(VI) ion with the adsorbent,presented in table 3, indi- cates a good agreement between the experimental values and the isothermal pa- rameters. This also confirms the monolayer adsorption of the Cr(VI) by adsor- bent surface. The R_L (table 2) values calculated with initial concentration range 3-11 mg/L were between (0<RL<1) which is consistent with the requirement for a favourable adsorption process.

Table 2 Equilibrium parameter (R_L)

Conc. of metal ion (mg/L)	(R_L) Cr(VI)-TTCNS
3	0.47
5	0.34
7	0.27
9	0.23
11	0.19

3.6 Freundlich isotherm model

The plot of log q_e versus log $C_e$for the adsorption of Cr(VI) onto TTCNS (fig.9) was employed to generate the intercept value of K_F and the slope of 1/n, which illustrates that adsorption of Cr(VI) onto TTCNS obeyed Freundlich isotherm

very well. K_F derived from the equation (3) is an indicator of adsorption capacity of a given adsorbent. The results obtained indicate that TTCNS have higher capacity to adsorb Cr(VI).The value of 'n' greater than 1(Table 3) implies fa- vourable nature of adsorption.

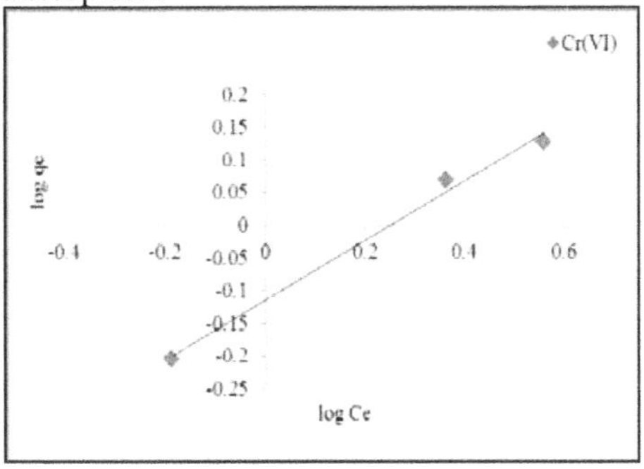

Figure 9: Freundlich isotherm plot

3.7 Lagergren First order kinetic model

The pseudo-first-order rate expression of equation (4) was used to test the exper-imental data for Cr(VI) onto TTCNS. The values of k_1, pseudo-first-order rate constant and q_e calculated obtained by the plot of log (q_e-q_t) versus t (Fig.10) arepresented in table 4

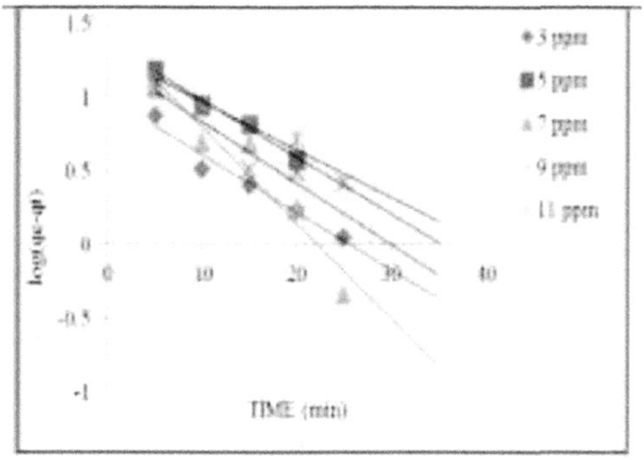

Figure 10: Pseudo-first order Kinetic model

3.8 Pseudo second order Kinetic model

The values of pseudo-second-order rate constants obtained by the plot of t/q_t versus t (Fig.11). The R^2 values observed for pseudo-second-order model were greater than the other employed systems. The decrease in K_2 and increase in q_e values with an increase in metal concentration was observed. The increase in q_e with concentration may be due to the more efficient utilization of the sorptive capacities of the adsorbent due to greater driving force (Amudha et al.,2009). As the increase in the metal concentration reduced the diffusion of metal ions at the boundary layers, K_2 decreased with concentration (Anirudhan,2008)

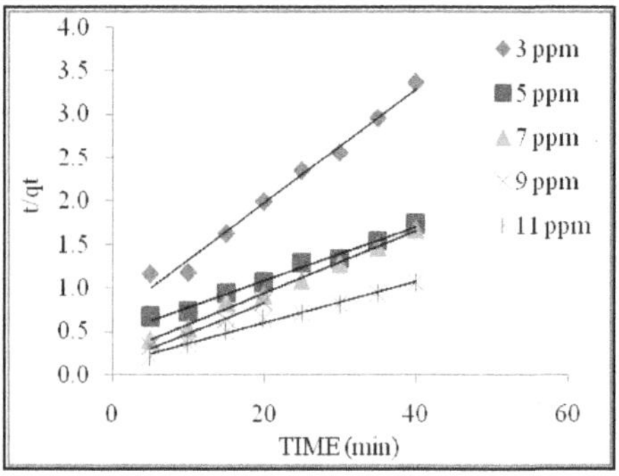

Figure 11: Pseudo-second order Kinetic model

Table 4 First-order and second-order kinetic constants for adsorption of Cr(VI)ion

First Pseudo-order			Second Pseudo order		
q_e	$K_1 X 10^{-3}$	R^2	q_e	$K_2 X 10^{-3}$	R^2
19.70	7.53	0.9947	42.37	6.3	0.9922

(Initial metal ion concentration:3-11 ppm, pH 1.78, Temp:303K)

3.9 *Effect of TTCNS on Industrial Effluent containing chromium*

The industrial effluent sample collected from Chromium plating industry was di-luted thrice and a volume of 50 ml was employed for the batch study. The effi- ciency of the selected adsorbent materials on industrial effluent was analyzed forthe optimised particle size of 0.18mm at a contact time of 30 minutes by batch study. Batch adsorption studies were carried out with different adsorbent dosag- es (50-500 mg; 50 mg interval). The effluent pH was adjusted to nearly 2.0 for Cr(VI),as low pH conditions favoured Cr(VI) removal. A dosage of 450 mg of TTCNS was sufficient for approximately 50% removal of Cr(VI) as depicted in Fig.12

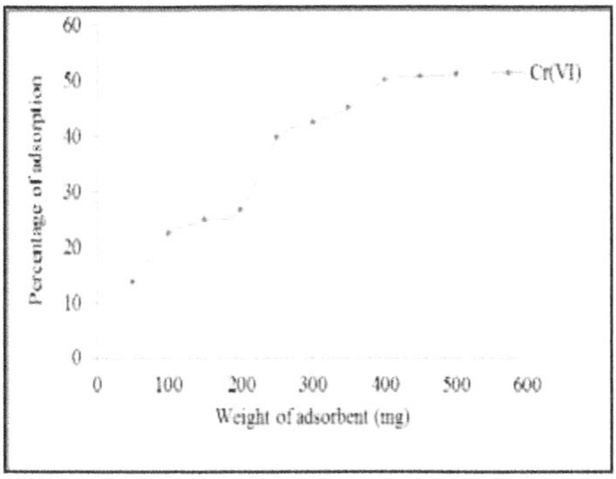

Figure 12. Effect of TTCNS on industrial effluent containing chromiumConclusion

This study identified Indian almond (*Terminalia catappa)* nut shell as a suitable biosorbent forCr(VI) removal employing batch equilibration method. The adsorption process was found to be pH dependent and optimized as 1.78 for Cr(VI)-TTCNS system. The probable mechanism of adsorption was surface complexation, as observed from Langmuir isothermal data and Lagergren first order kinetic model. Present research also confirmed the potentiality of Indian almond nut shell for the removal of toxic chromium from electroplating industry effluent.

References

US Department of Health and Human Services,1991, Profile for Chromium, Public Health Service Agency for Toxic substances and Diseases, Washington,DC.1991 R.Mehra, M.Juneja, Indian Journal of Biochemistry and Biophysics, 2003, 40, 131-135

N.Ahalya, R.D.Kanamadi and T.V.Ramachandra, Cr(VI) and Fe(III) removal using *Cajanus cajan* husk, J.Environ.Bio 28(4) (2007) 765-769.

F.N.Acar, E.Malkoc, The removal of chromium (VI) from aqueous solutions by *Fagus Orientalis*L Bioresour.Technol. 94(2004)13-15.

O.O.Kehinde, T M Oluwatoyin, O.OAderonke, Comparitive analysis of the effi- ciencies of two low-cost adsorbents in the removal of Cr(VI) and Ni (II) from aqueous solution, Afr.J.Environ.Sci .Tech.3(2009) Kaustubha Mohanty, Mousam Jha, B.C. Meikap, M.N.Biswas, Removal of chromium (VI) from dilute aqueous solutions by activated carbon developed from *Terminalia arjuna* nuts activated with zinc chloride, Chem. Eng.Sci. 60 (2005) 3049-3059.

A.Verma, S.Chakraborty, J.K. Basu, Adsorption study of hexavalent chromium using tamarind hull-basedadsorbents, Sep. Purif. Technol. 50 (2006) 336-341.

Suman Das, Biosorption of chromium and nickel by dried biomass of *Cyanobacterium Oscillatorialaete-virens,*

Int J.Env.Sci 3(1) (2012)341-352.

Species Profiles for Pacific Island Agroforestry.(2006)

sAhmet Sari, Mustafa Tuzen, Biosorption of Pb(II) and Cd(II) from aqueous so- lution using green alga (*Ulva lactuca*) biomass, J. Hazard. Mater. 152 (2008) 302-308.

IUPAC Manual of Symbols and Terminology of Colloid Surface, Butterworths, London, 1982.p. 1.

V.C. Srivastava, B. Prasad, I.D. Mall, M. Mahadevswamy, I.M. Mishra, Adsorp- tive removal of phenol by bagasse fly ash and activated carbon:Equilibrium, ki- netics and thermodynamics, Colloid Surface A: Physicochem. Eng. Aspects 272 (2006) 89–104

The 1st International Applied Geological Congress, Department of Geology, Is lamic Azad University - Mashad Branch, Iran, 26-28 April 2010 Use of sawdust of Aspen tree for the removal of Chromium(VI) from aqueous solution

Carlos Green- Ruiz, Mercury(II) removal from aqueous solutions by nonviable *Bacillus sp.* from a tropical estuary, Bioresour. Technol. 97 (2006) 1907-1911.

B.H. Hameed, M.I. El-Khaiary, Malachite green adsorption by rattan sawdust: Isotherm, kinetic and mechanism modeling, J. Hazard. Mater.159 (2008) 574-579.

Emine Malkoc, Yasar Nuhoglu, Potential of tea factory waste for Chromi-um(VI) removal from aqueous solutions: Thermodynamic and kinetic studies, Sep.Purif. Technol. 54 (2007) 291-298.

HuiGao, Yunguo Liu, Guangming Zeng, Weihua Xu, Ting Li, Wenbin Xia, Characterization of Cr(VI) removal from aqueous solutions by a surplus agricul-tural waste-Rice straw, J. Hazard. Mater. 150 (2008) 446-452.

N.D. Lyn Patrick, Lead Toxicity, A Review of the Literature - Part I: Exposure, Evaluation and Treatment, Alt. Med. Rev. 11 (2006) 1-22.

Manjeet Bansal, UmeshGarg, Diwan Singh, V.K.Garg, Removal of Cr(VI) from aqueous solutions using pre- consumer processing agricultural waste: A case study of rice husk, J. Hazard. Mater. 162 (2009) 312-320 V.Vinodhini and

N. Das, Relevant approach to assess the performance of sawdust as adsorbent of chromium(VI) ions from aqueous solutions, Int.J.Environ.Sci.Tech.7(1) (2010) 85-92
O.S.Amuda, F.E.Adelowo, M.O.Ologunde, Kinetics and equilibrium studies of adsorption of chromium(VI) ion from industrial wastewater using *Chrysophyllum albidum* (Sapotaceae) seed shells, Colloids and Surfaces B: Biointerfaces. 68 (2009) 184-192.
T.S. Anirudhan, L. Divya, M. Ramachandran, Mercury(II) removal from aqueous solutions and wastewaters using a novel cation exchanger derived from coconut coir pith and its recovery, J. Hazard. Mater.157 (2008) 620-627s

Sorption Ability of Modified Fruit Nut Shells in the Removal of Co (II) ions from Aqueous Solutions

N. Muthulakshmi Andal[1*] and S. Charulatha[2]

[1*]Department of Chemistry, PSGR Krishnammal College for Women, Affiliated to Bharathiar University, Coimbatore- 641 004, Tamil Nadu, India.

[2]Department of Chemistry, PSGR Krishnammal College for Women, Affiliated to Bharathiar University, Coimbatore- 641 004, Tamil Nadu, India

*** Corresponding author: andaldr@gmail.com**

Abstract:

The present work deals with the efficiency of modified *Terminalia catappa seed shell (TCSS)*, an agricultural waste, for the removal of Co(II) from aqueous solutions. Three methods of modifications of the sorbent material have been done and the experiments were carried out by employing batch equilibration method. The influences of variable parameters viz., particle sizes and adsorbent dosages of the sorbent material, agitation time, initial concentrations of the sorbate solutions, pH, cations, anions, co-ions and temperatures have been studied. The surface characteristic study of the modified TCSS has been supported by FT-IR, SEM and EDAX techniques. The linearity of the isothermal plots suggests that the Co(II) – TCCS systemobeys Langmuir and Freundlich adsorption models.

Keywords: adsorption, *Terminalia catappa*, modification, characterization, parameters

1.0 Introduction:

Heavy metal contamination in the aqueous waste streams of many industries, such as metal plating, mining operations, tanneries etc are gaining much concern in recent times towards remedial approach. Some of the metals which are associated with these activities are cadmium, chromium, cobalt, lead, copper and mercury. Heavy metals are non biodegradable and tend to accumulate in living organisms, causing various diseases and disorders (Susan *etal.*, 1999, Netzer and Hughes, 1984, Gomez-Lahoz *etal.*, 1993).Cobalt, a natural element present in certain ores of the Earth's crust, is essential for life in trace amounts. Important natural sourcesof cobalt in the environment are soil, dust and sea water. It exists in the form of various salts. Pure cobalt is an odourless, steel-grey, shiny, hard metal. Every one of us is exposed to low levels of cobalt in air, water and food. Cobalt has both beneficial and harmful effects on health. The metaland its salts are used in nuclearmedicine, enamels and semiconductors, grinding wheels, painting on glass and porcelain, hygrometers andelectroplating, as a foam stabilizer in beer, in vitamin B12 manufacture, as a drier for lacquers, varnishes and paints, and as a catalyst fororganic chemical reactions (Rengaraj and Seung- Hyeon Moon, 2002).

Cobalt is one of the priority pollutants in aqueous waste discharges. Cobalt containing compounds are widely used in many industries such as mining, metallurgical, paints, pigments and electronics. Exposure to higher concentrations of the metal may cause paralysis, diarrhea, low blood pressure, lung irritation and bone defects. The tolerance limit in potable water has been fixed as 0.05mg/L (Neha Gupta *etal.,*2011). Many conventional techniques suchas chemical precipitation, membrane filtration, electrolysis, ion exchange, co- precipitation and adsorption have been employed for the removal of heavy metalsin wastewater treatment (Edulnam *etal.*, 2012). Amongst all the removal techniques, adsorption is one of the most popular and effective process for the removal of heavy metals from wastewater.

The adsorption process offers flexibility in design and operation and in many cases produces treated effluent suitable for reuse, free of colour and odour. In addition, as adsorption is a reversibleprocess, regeneration of the adsorbent addsto the economy of the operation. Adsorption is the process of accumulating substances that are in solution on a suitable interface. It is a surface phenomenon,which involves a mass transferoperation, in that a constituent in the liquid phaseis transferred to the solid phase. Adsorbate is the substance that is removed from the liquid phase at the interface and adsorbent is the solid phase ontowhich the adsorbate accumulates. The adsorption process takes place in different steps (1) bulk solution transfer (2) film diffusion transfer (3) pore transport and (4) adsorption. Bulk solution transfer involves the movement of the material to be adsorbed through the bulk liquid to the boundary layer of fixed film of liquid surrounding the adsorbent. Film diffusion transport involves the transport by diffusion of the material to be adsorbed through the stagnant liquid film to the entrance of the pores of the adsorbent. Pore transport involves the transport of material to be adsorbed through the pores by a combination of molecular diffusionthrough the pore liquid and/or by diffusion along the surface of the adsorbent. Adsorption involves the attachment of thematerial to be adsorbed onto adsorbentat available adsorption sites (Ira N. Levine etal 1995).

Naturally occurring waste materials have been effectively employed in the sorption method due to their abundance, cheap, biodegradability and reusability characteristics. Therefore, they serve a better choice to commercially available activated carbons, which is highly expensive and pose greater disposable /regeneration problems after use. Some waste materials which have been reported are peat (Coupal and Spiff, 1999, Brown *etal* ., 2000, Ho *etal* ., 1996), living and death biomass (An *etal* ., 2001, Volesky and Holan, 1995, Rae and Gibb, 2003), clay and related minerals (Angove *etal* ., 1999, Chen and Lu, 2008, Hayes and Leckie, 1987). In view of abundant availability of these naturally occurring materials, a systematic study to assess the potential use of modified *Terminalia catappa* seed shell powder for the removal of Co(II) from aqueous solutions is carried out in this paper.

The adsorbent used in this study is almond tree biomass. The almond tree belongs to the family called Combretaceae (Combretum family). The botanical name of thetropical almond tree is *Terminalia catappa L* (Enemose Edith and Osakwe,

2012).So far, no reports are available for the removal of Co(II) ions using TCSS powder and thence, this paper is the first study of its kind.

2.0 Materials and Methods:

Fruit seed shells of *Terminalia catappa* werecollected from various localities in Coimbatore, washed well with distilled water and sun dried. Then the dried seedshells were ground in a laboratory blender and sorted into different particle sizesviz., 0.71 mm, 0.42 mm, 0.30 mm, 0.21 mm and 0.18 mm using standard test sieves. The varied sizes of TCSS were treated with HCl and citric acid (CA) to modify the nature of the sorbent materials. The surface characteristics of modified adsorbents were studied using FT-IR, SEM and EDAX. Experiments wereconducted to estimate the influence of varied particle sizes and dosages of the TCSS powder, prescribed agitation time intervals between the sorbent- sorbate interface, initial Co(II) concentrations, solution pH, interference of cations,anions, co-ions and effect of solution temperatures in order to establish the optimum experimental conditions for adsorption process. The stock solution of Co(II) ions were prepared from $CoCl_2.6H_2O$ salt of analaR grade using Double Distilled water. Aliquots of the adsorbate solutions of varying Co(II) ions concentrations 2-14 ppm: 2ppm intervals were prepared from the stock solution (1000 mg/L). The experiments were performed in 250ml Erlenmeyer flasks for the mentioned particle sizes and dosages of TCSS powder varying between 50- 200mg: 50mg interval for the Co(II) ion concentrations. The flasks with the contents were agitated at preset time intervals using mechanical shaker (setup given in fig 1). The other experimentally verified factors are the pH environmentsvarying between3 and 11,cationic & anionic influences of Na^+, Mg^+, and Cl^-, NO

$^-$ respectively, Cu^{2+}, Cr^{6+}, Cd^{2+}, Mn^{2+}, Ni^{2+} and Pb^{2+}ions as influential ions, temperature ranges from 20^0C to 60^0C:10^0C intervals (setup given in fig 2). The filtered samples were complexed to bluish-green colour using ammonium thiocyanate (analaR grade) and the residual concentrations were analyzed at a λ_{max} of 630 nm in UV- Visible spectrophotometer (Lab India make).The quantityof Co(II) ions adsorbed onto the adsorbent at the time of equilibrium (q_e) was calculated using the following expression:$q_e = (C_o - C_e) V/m$ --(1)

where C_o and C_e are the initial and equilibriumconcentrations (mg/L) of Co(II) ions in solution. V is volume (L) of solution and m is weight of adsorbent (g) taken for experiment.

Fig 1: Experimental setup for batch studies

Fig 2: Incubator Shaker for Temperature studies

3.0 Results and Discussion:

3. 1Adsorption mechanism of Co(II) on TCSSPowder

Scanning Electron Microscopy was used for characterizing the surface morphologyand structure of the adsorbent material. Over the adsorption period, the morphologyof the treated seed shells had undergone a remarkable physical change. The integrated and cluster arrangements for unloaded and loaded samples are presentedin figures 3a, 3c and 3b, 3d respectively which clearly indicate that the metal ion had adhered to thesurface of the adsorbent. This can be further confirmed by the less and smaller pores on the surface of the loaded sorbents that could have been occupied by metal ions (Cong Liu *etal.,* 2012).The EDAX spectra were recorded toanalyze the elemental constitution qualitatively for the modified adsorbent. Figures 4a, 4b, 4c and 4d depict the EDAX spectra for unloaded as well as Co(II) loaded samples. The presence of cobalt peaks in figures 4b and 4d confirms the adsorptionof cobalt ions by modified seed shells. Fourier Transform Infrared Spectroscopy (FTIR) has been proven and accepted to be a powerful tool for the study of biomasssamples (Fatemeh Kafshgari *etal.,* 2013). The inference from the spectra of SEM and EDAX supported the citric acid modification. Therefore, FTIR spectra were recorded for unloaded and loaded citric acid modified TCSS in the range of 400- 4000cm^{-1}andare presented in Fig 5a and 5b respectively. It is observed that the mostcharacteristic change had occurred in the range of 1700- 1737 cm^{-1}. The band centered at 1731 $cm^{-1}$is ascribed to the stretching vibrations of carboxyl group(C=Oin carboxylic acid group) on the edges of layer planes or to the conjugated carbonylgroups (Tao and Xiaoqin *etal.,*2008). Peaks at 1350-1200 cm^{-1} band interval may be due to the vibration of C-N peptidic bond of proteins present in the seed shell moiety. The strong band observed at 1041.5 cm^{-1} indicates the C-OH stretching vibrations (Stuart, 1996). Thus, it is observed that the IR-spectral analysis indicatesthe presence of ionizable functional groups and their ionization leaves behind vacant sites which might have been replaced by Co(II) ions.

Fig 3a: SEM of unloaded citric acid modifiedseed shell

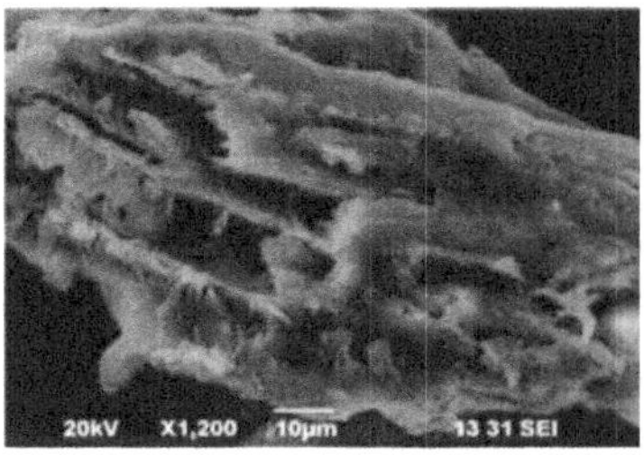

Fig 1b: SEM of loaded citric acid modified seedshell

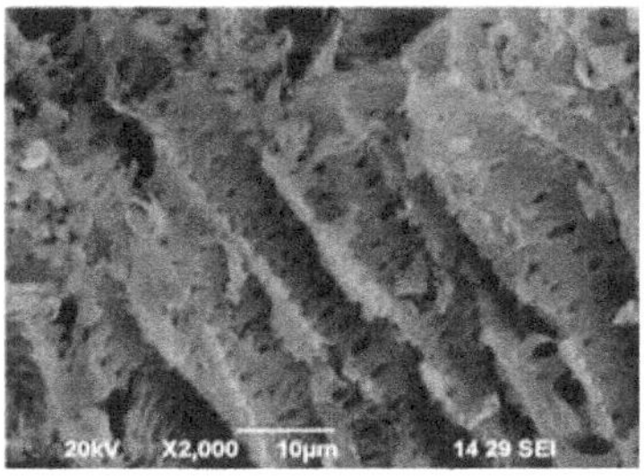

Fig 3c: SEM of unloaded HCl modified seedshell

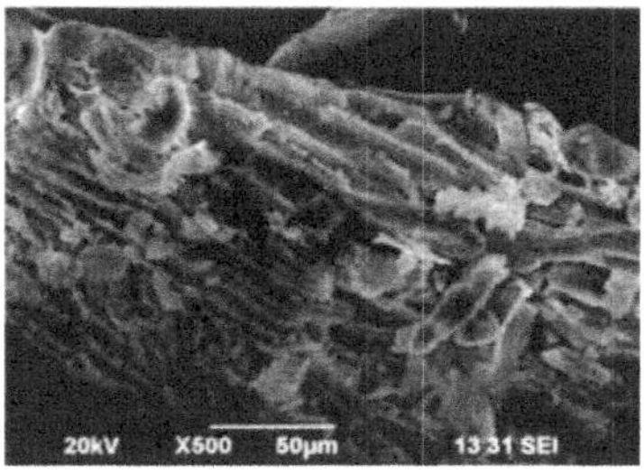

Fig 3d: SEM of loaded HCl modified seedshell

Fig 4(a): EDAX of unloaded citric acid Modified seed shell

Fig 4(b): EDAX of loaded citric acid modifiedseed shell

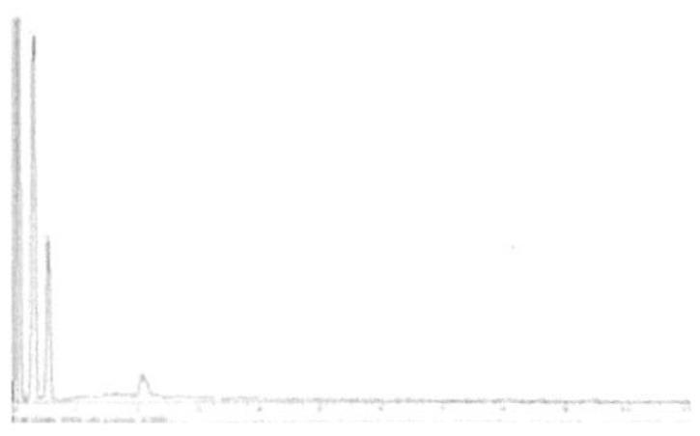

Fig 4(c): EDAX of unloaded HCl modified seed shell

Fig 4(d): EDAX of loaded HCl modified seed shell

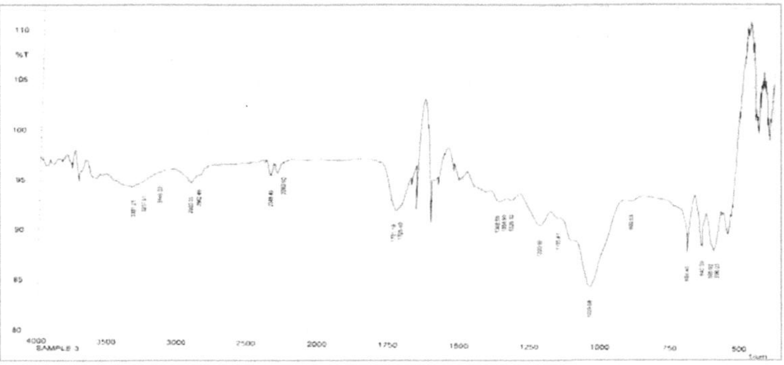

Figure 5a: IR Spectrum of unloaded citric acid modified seed shell

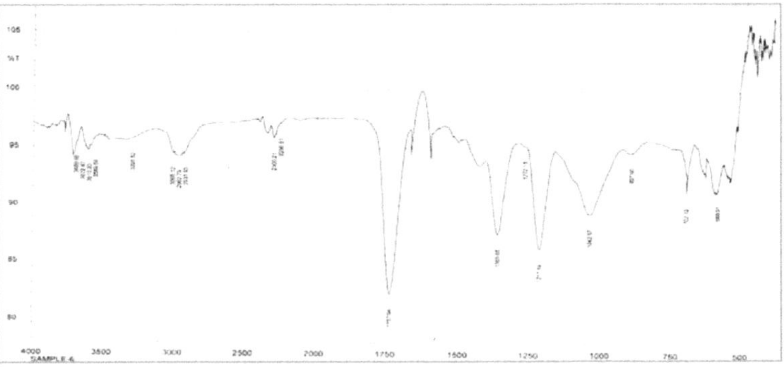

Figure 5b: IR Spectrum of loaded citric acid modified seed shell

3.2 Effect of Agitation Time

Batch experimental results for an initial concentration of 8ppm Co(II) ions at varying time intervals for the two modified adsorbent material are shown in fig 6.The smooth continuous steep rise in the curves represents increased uptake of Co(II) ions at preset time intervals (3-15 minutes: 3 minutes interval). After 9 minutes, a dip in the curve indicates that the systems tend towards saturation and had attained equilibrium at 9 minutes. Of the two modified adsorbent material, TCSS-CA registered higher amount adsorbed (q_e=

3.4 mg/g) compared to HCl modification. Hence, the study of other variable parameters were

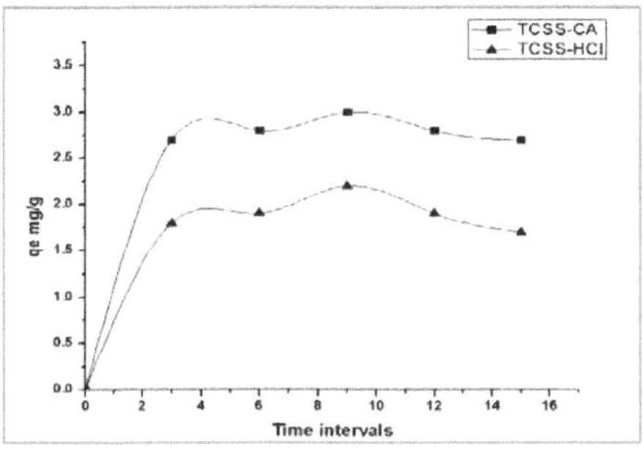

Fig 6: Effect of agitation time on the removalCo(II) by TCSS

3.2 Effect of Particle Size

Adsorbent particle size has a significant influence on the kinetics of sorption due tochange in number of adsorption sites. The removal of the Co(II) ions at different particle sizes of TCSS-CA viz., 0.71 mm, 0.42 mm, 0.30 mm, 0.21 mm and

0.18 mm is depicted in fig 5. The uptake of Co(II) ions increased with a decrease inparticle size (ie) from 0.71mm-0.18mm. Surface area of the adsorbent is an important parameter for adsorption. Smaller the particle size, greater the surface area per unit weight of the adsorbent (Montanher *etal.,* 2005). Interaction of metal

ion isfound to be more with the adsorbent possessing higher surface area and hence enhanced amount of Co(II) ions was adsorbed for the smaller particle size of 0.18mm (q_e= 3.4 mg/g) as evident from thesmooth steep maximum curve obtained at 0.18 mm particle size (Fig 7). This shows the enhancement in the amount adsorbed with respect to time compared to other particle sizes employed.

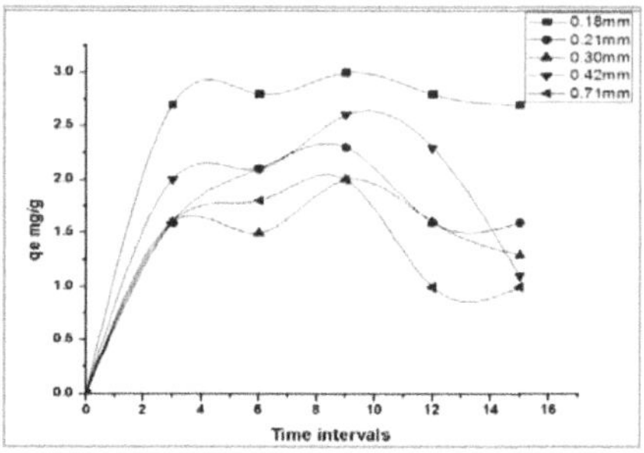

Fig 7: Effect of particle size on the removal Co(II) by TCSS-CA

3.3 *Effect of Adsorbent Dosage*

Removal of Co(II) were investigated at four different adsorbent dosages (50-200mg:50mg) the clear sharp rise in the curve for 100mg of TCSS-CA sorbent represents the maximum sorption of Co(II) ions, the other dose curves being entangled. Decreased sorption at higher dosages (150 and 200mg) can be attributedto the fact that agglomeration of the fine size (0.18mm) had occurred. This shall bedue to a decrease in available sorbent surface area per unit mass

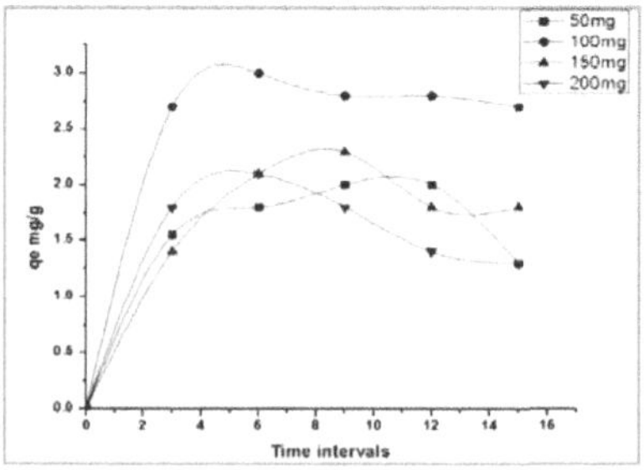

whichis caused byoverlapping and aggregation of adsorption sites (Hanim Awad and Paramalingam,2011). Hence, an optimum dose of 100 mg was fixed for all experiments.

Fig 8: Effect of adsorbent dosage on the removalCo(II) by TCSS-CA

3.4 *Effect of pH*

Metal sorption is linked with pH and the effect of pH on the solution is an importantcontrolling factor in the sorption experiment. The acidity of the medium affects thecompetition ability of hydrogen ions with metal ions towards the active sites on theadsorption surface. It also influences ionization of functional groups on the adsorbent surface and chemical speciation of metal ions. In order to optimize the pH for maximum removal efficiency, experiments were conducted at varying pH ranges from 3-11. The inverted parabolic curve with maximum removal of Co(II) (83.8%) at pH 7 is depicted in fig 9. There is absence of marked percentage removalof Co(II) at acidic and alkaline pH values. This can be due to the rapid changes in protonated and unprotonated forms of the surface sites of the sorbent. Also, it can be interpreted that the adsorption rate for pH ≥ 9 experience a decline, because of the hydroxide ion forming complexes with the Co(II) ions leading tosuppression ofthe sorption(Ajmal *etal.,* 2000).

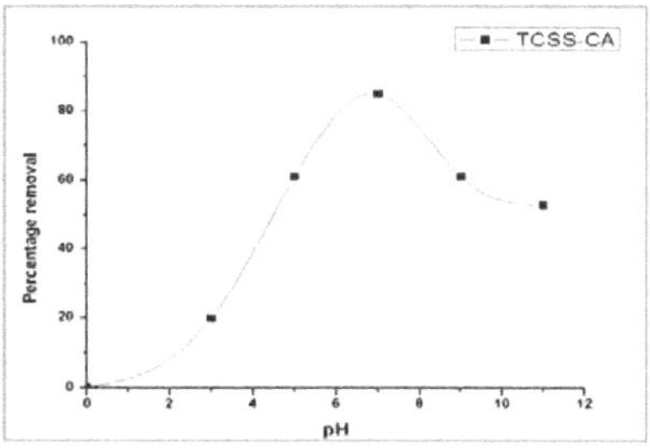

Fig 9: Effect of pH on the removal Co(II) by TCSS-CA ***Effect of cations and anions***

Industrial wastewater discharges may contain cations and anions which may interfere in the removal of heavy metal ions. Thence, the influence of these ions was studied and the results are depicted in fig 10. Of the cations($Mg^{2+}$and Na^{+}), the removal of Co(II) diminished from 83% to 35% at increasing concentrations of Mg^{2+}ions against the influence of Na^{+} ions, where the percentage removal is least inhibited ie, 85% to 77%. This is in agreement with other studies that showed stronger inhibitions with the ions of higher valence than that of lower ones (Emine Malkoc, 2006). Of the anions,(Cl^{-}andNO_3^{-}), while the concentration of Cl^{-}increased, the removal of Co(II) ions decreased from 83% to 64% against the influence of No_3^{-}ions, where the percentage removal is least inhibited (ie) 83% to 74%.

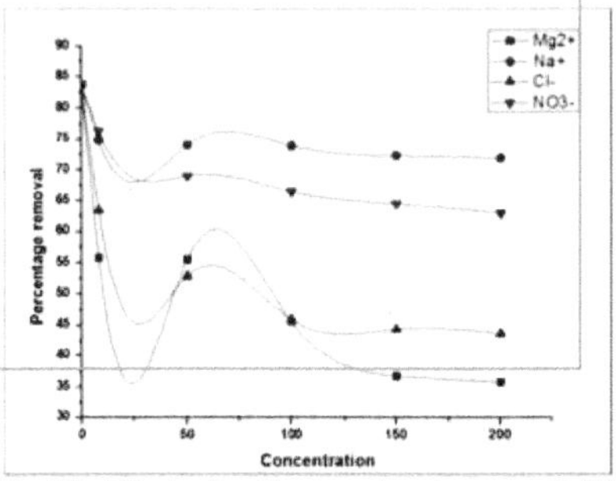

Fig 10: Effect of cations and anions on Co(II) removal by TCSS-CA

3.5 Effect of Co-ions

Study of the influence of co-ions on the removal of a specific metal ion becomes relevant when selective adsorption of metal ions occurs from the aqueous solutions. The study on the influence of co-ions in the removal of Co(II) ions gives an insight into the sorption capacity of TCSS-CA. The percentage removal values of 8ppm Co(II) in the presence of an equal concentration of influential ions is tabulated in table 1 and the data is plotted in fig.11. The inhibition characteristics was found to decrease in the order Ni^{2+}<Pb^{2+}<Cd^{2+}<Cr^{6+}< Mn^{2+} <Cu^{2+}. It is obvious from the figure, where the dip in the curve for Cu(II) ions suggests that the Cu(II) ions are preferentially adsorbed onto the adsorbent surface rather than Co(II) ions. The removal of Co(II) ions registered a decrease at higher environments of co-ions.

Table 1: Percentage removal of Co(II) on TCSS-CA in presence of co-ions

Adsorbent	Percentage Removal						
	Co^{2+} in the absence of co-ions	Ni^{2+}	Cd^{2+}	Pb^{2+}	Cu^{2+}	Mn^{2+}	Cr^{6+}
TCSS-CA	83	75	61	63	42	54	55

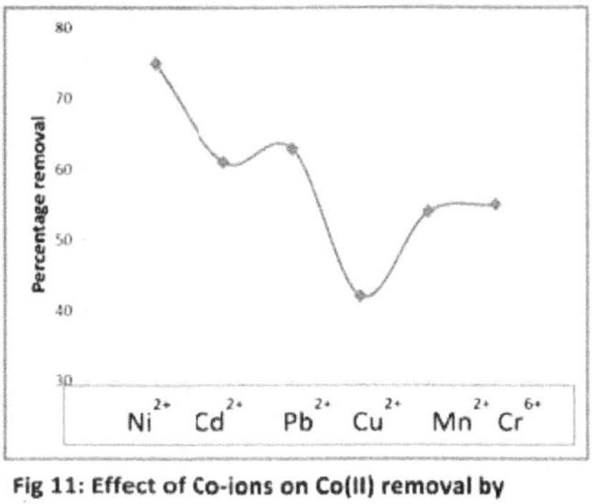

Fig 11: Effect of Co-ions on Co(II) removal by

1. ***6 Effect of Temperature*****:**

Temperature has an important effect on the process of adsorption. The percentage removal of Co(II) as a function of temperature (293K to 333K) are presented in figure 12. The smooth steep maximum increase in the curve at 303K and further decrease at high temperatures suggests that maximum percentage removal (83%) had occurred at room temperature. The decrease inpercentage values with rise in temperatures may be due to the desorption caused by an increase in the available thermal energy. Higher temperature induces higher mobility of the adsorbate molecules causing desorption (Ira N. Levine).

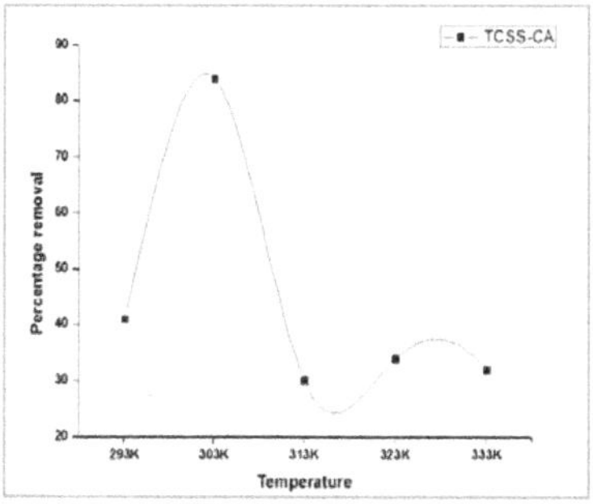

Fig 12: Effect of temperature on Co(II) removal by TCSS-CA

Adsorption Isotherm

The Langmuir and Freundlich equations are commonly used for describing the surface adsorption equilibrium. Based on the experimentalresults discussed so far,the fit in of the sorption isotherms were determined for Co(II)-TCSS-CA system ata fixed temperature (25°C) for a concentration range varying between 2 -12 mg/L.All solutions contained a fixed mass of 100mg of the adsorbent material.

Langmuir Isotherm

The mostly used Langmuir model assumes that

i. Sorption occurs at specific homogeneous adsorption sites each of which
ii. canhold only one molecule.
iii. The adsorption is limited to monolayer
iv. All adsorption sites are energetically equivalent and there is no interaction ormigration between the adsorbed ions on the surface of adsorbent

This model proposed by Langmuir suggests thatthe adsorption had occurred on homogeneous surface by monolayer sorption with no interactions between sorbed species. The linear form of Langmuir isotherm is given by the following equation:

$1/q_e = 1/q_m + 1/bq_m$

Where qe(mg/g) is the amount adsorbed at equilibrium, Ce the equilibrium concentration (mg/L)

b and q_m are Langmuir constants related to the energy and maximum adsorption capacityrespectively. Langmuir plot of Ce / q_e vs Ce depicted in fig.13 shows a straight line. The Langmuir constants, b and qm were calculated from the slope ($1/q_{max}$) and intercept ($1/q_{max}.K_c$) respectively. The R^2 value of 0.977 from the plot suggests maximum linearity and fit in of the Co(II)- TCSS-CA system on Langmuir adsorption isotherm.

Freundlich Isotherm

The Freundlich isotherm model has been widelyused for heterogeneous surface energies. The Freundlich isotherm has a general linear form. The logarithmic form of Freundlich model is given by $\text{Log } q_e = \log K_f + 1/n \log C_e$

Where qe(mg/g) is the amount adsorbed at equilibrium, Ce is the equilibrium concentration of the adsorbate (mg/L). $K_f$and 1/n are Freundlich constants relatedto adsorption capacity and adsorption intensity, respectively. Both K_f and1/n are empirical constants being indicative of sorption capacity and adsorption intensity, respectively (Hanim Awad and Paramalingam, 2011). From the slope and intercept of plots $\log q_e{}^2$vs log c (fig 12), values of K and R were determined.From the linearity coefficient value, it is obvious that the Co(II)- TCSS-CA system has a better fit for Freundlich isothermal model rather than its Langmuir counterpart.

Table 2: Langmuir and Freundlich isotherm data for adsorption of Co(II) on TCSS-CA

Adsorbents	Langmuir Constants			Freundlich Constants		
	q_m	B	R^2	K	1/n	R^2
TCSS-CA	98.04	1.8579	0.977	1.3032	0.77	0.997

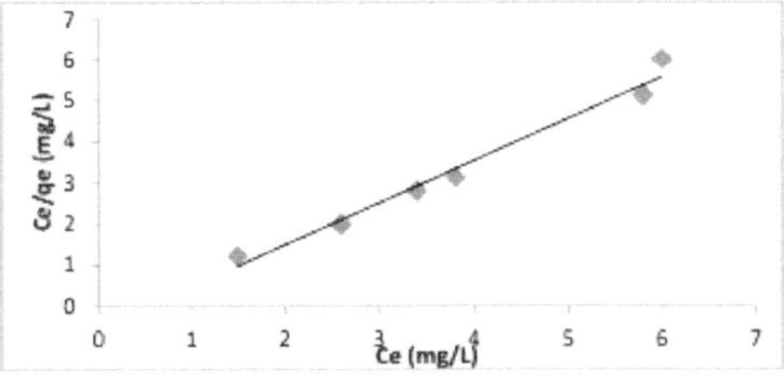

Fig 13: Langmuir Isotherm for the sorption of Co(II) by TCSS-CA

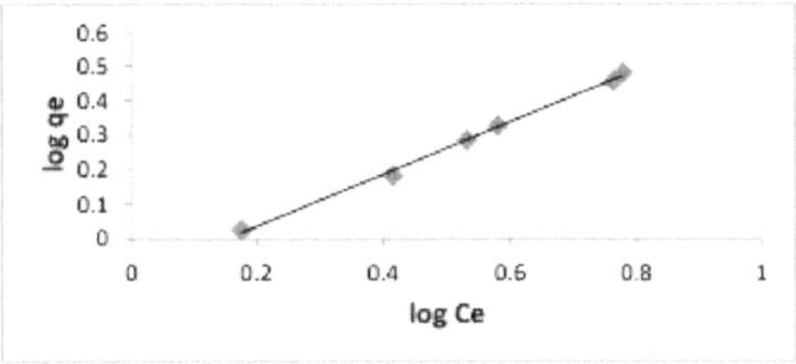

Fig 14: Freundlich Isotherm for the sorption of Co(II) by TCSS-CA

4.0 Conclusion:

The selected agricultural waste (*Terminalia catappa seed shell)* was modified usingcitric acid and HCl for the removal of Co(II) ions. Both the modification proceduresemployed show significant sorption capacity in the removal of Co(II). The studies on the characterization of both loaded and unloaded sorbent samples were carried out using FTIR, SEM and EDAX. The changes in the FTIR spectra confirmed the complexation of Co(II) ions with the functional groups present in the sorbent material. Also, the surface morphological change and appearance of sharp peak denoting the presence of Co(II) ions in SEM and EDAX respectively, supported thesorption capacity of the modified materials. However, the uptake capacity was greater for TCSS–CA. From the batch experiments conducted to study the influenceof varying parameters, 0.18mm particle size and 100 mg dosage, 9 minutes agitation time, 8ppm initial concentration of Co(II) ions, pH 7 of the solution medium, room temperature (30^0C) were fixed as the optimum conditions. The effectof cations, anions and co- ions revealed that Mg^{2+}, Cl^- and Cu^{2+} possessed higher inhibitive nature, compared to other selected ions. A decrease in the Co(II) adsorption onto TCSS-CA was observed at higher concentrations of the studied ions. The Freundlich isothermal plot registered a better linearity fit with R^2 value of0.997 than Langmuir model in describing the adsorption behavior. Thus, theexperimental results has led to the conclusion that citric acid modified *Terminalia catappa seed shell* is a promising material for Co(II) removal due to its greater adsorption capacity. Further probe of the study concentrates on the immobilizationof the trapped Co(II) ions on the surface moiety of the selected sorbent material.

5.0 Acknowledgments:

The authors would like to thank Defence Research& Development Organisation(DRDO) for their financial support to this work.

References

1) Ajmal, M., Rao, A.K.R., Ahmad, R. and Ahmad, J., (2000): Adsorption studieson Citrus reticulate (fruit peel of orange): Removal and recovery of Ni(II) fromelectroplating water. J. of Haz. Mat., (79): 117–131.

2) An, H.K., Park, B.Y. and Kim, D.S., (2001): Crab shell for the removal of heavymetals fromaqueous solutions. *Wat. Re.,*. 35: 3551–3556.

3) Angove, M.J., Wells, J.D. and Johnson, B.B.,(1999): The influence of temperature on the adsorption of cadmium (II) and cobalt (II) on goethite. *J.Coll. Inter. Sci.,* 211: 281–290.

4) Brown, P.A., Gill, S.A. and Allen, S.J., (2000): Metal removal from waste

watersusing peat. *Wat.Res,*. 34: 3907 – 3916.

5) Chen, L. and Lu, S., (2008): Sorption and desorption of radio cobalt on montmorillonite

- effects of pH, ionic strength and fulvic acid. *Appl.Radiat.Isot.,*66: 288–294.

6) Coupal, B. and Spiff, A.I., (1999): Thetreatment of waste waters with peat moss. *Wat. Res.,* 33(2): 1071 – 1076.

7) EduInam, UbongEtim, Ubong Eduok and Joseph Essien, (2012): Heavy MetalsSorption Potential of Calcareous shells of Animal Origin. Int. J. ofChe. Envir.&Phar. Res., 3(3): 184-194.

8) Ira N.Levine, Physical Chemistry, 5th edition, Tata McGraw Hill Publishing Company ,Ltd, New Delhi, India.

9) Emine Malkoc, J. of Haz. Mat., B137 (2006) 809 – 908.

10) Enemose Edith, A. and Osakwe, S.A., (2012): Studies on the effect of pH on thesorption of Al3+ and Cr6+ Ions from aqueous solutions by Almond Tree (*Terminalia catappa L.*) Biomass, *Res. J. of Che. Sci.,*2(4): 13-17.

11) Fatemeh Kafshgari, Ali Reza Keshtkar and Mohammad Ali Mousavian, (2013): Study of Mo(VI) removal from aqueous solution: application of different mathematical models to continuous biosorption data. Iranian J. of Env. Hea.Sci&Engin.,10:14.

12) Gomez-Lahoz, C., Garcia-Herruzo, F., Rodriguez-Maroto, J. M. and Rodriguez,

J.J. (1993): Cobalt(II) removal from water by chemical lreduction with borohydride. Water Res., 27:985–92.

13) Hanim Awab and Thanalechum iParamalinggam, (2011): Mussel Shell and erworks Sludge as Sorbent Materials for the Removal of Cu(II) from Aqueous Solution.J. of Fund. Sci.,7(2): 113-119.

14) Hayes, K.F. and Leckie, J.O., (1987): Modelingionic strength effects on cation adsorption at hydrous oxide/solution interfaces. *J. Coll. Interf. Sci.,*15:564–572.

15) Ho, Y.S., John Wase, D.A. and Forster, C.F., (1996): Kinetic studies of

competitiveadsorption by sphagnum moss peat. *Environ. Tech.*, 17: 71 – 77.

16) Ira N. Levine, (1995): Physical Chemistry, 5th edition, Tata McGraw HillPublishing Company Ltd, New Delhi, India

17) Montanher,S.F., Oliveira,E.A. and Rollemberg, M.C., (2005): Removal of metalions from aqueous solutions by sorption onto rice bran, *J. Hazard. Mater.*, 117:207-211.

18) Neha Gupta, Atul K. Kushwaha and Chattopadhyaya, M.C. (2011): Adsorption ofcobalt(II) from aqueous solution ontohydroxyapatite/zeolite composite. Adv.Mat. Lett.,2(4): 309-312.

19) Netzer, A. and Hughes, D.E. (1984): Adsorption of copper, lead and cobalt by activated carbon. Water Res., 18:927–33.

20) Rae, I.B. and Gibb, S.W., (2003): Removal of metals from aqueous solutions using natural chitinous materials. *Wat. Sci. Tec.l,* 47:189–196.

21) Rengaraj, S. and Seung-Hyeon Moon.(2002): Kinetics of adsorption of Co(II) removal from water and wastewater by ion exchange resins. Wat. Res., 36: 1783–1793.

22) Susan, E. B., Trudy, J. O., Mark, B. and Dean, A.D. (1999): A review of potentially low-cost sorbent for heavy metals. Water Res., 33:2469–79.

23) Stuart, B., (1996). Modern Infrared Spectroscopy , Wiley, Chister, UK.

24) Cong Liu,HuuHao Ngo and WenshanGuo., (2012): Appl. Biotech., 167:1699- 1715.

25) Tao, X.U. and Xiaoqin, L.I.U., (2008): Peanut shell Activated Carbon: Characterization, Surface Modification and Adsorption of Pb2+ from aqueous solution. Chi. J. of Che.Engin., 16(3): 401-406.

26) Volesky, B. and Holan, Z.R., (1995): Biosorption of heavy metals. *Biotel.Prog.*,11: 235–250.

Removal of Nickel from Aqueous Solution Using *Azaridachta Indica* Seed Shell Powder asAdsorbent

G Gohulavani*, and N Muthulakshmi Andal

Department of Chemistry, PSGR Krishnammal College for Women, Coimbatore-641 004,Tamil Nadu, India

Received: 28/01/2013Revised: 14/02/2013 Accepted: 15/02/2013

***For Correspondence**

Department of Chemistry, PSGR Krishnammal College for Women, Coimbatore-641 004,Tamil Nadu,India

Keywords: Nickel, Adsorption, *Azaridachta indica*, Kinetics, Isotherm

ABSTRACT

Nickel plays an important role in metallurgical, electroplating and other chemical industries. About 40% of the nickel produced is used in steel factories, nickel batteries and in theproduction of some alloys. On the other hand, it may pollute aqueous streams, arising therefore several environmental problems. Azaridachta indica seed shells were evaluated for Ni (II) adsorption. The tests were performed in batch equilibration method. The obtained maximum adsorption capacities were 66.66 mgg^{-1}. The FT-IR results showed that the sorbent have differentfunctional groups responsible for nickel ions binding, such as carboxyl, amino, peptide and hydroxide. The morphology of adsorbent surface was analyzed by scanning electron microscopy(SEM) before and after Ni(II) loading.

INTRODUCTION

Adsorption process offers the advantages of low operating costs, possibility of metal recovery, potential biosorbent regeneration, though research has been carried out for developing cost effective heavy metal removal techniques. The use of agricultural plant materials and their waste products viz., papaya wood[1],capsicum annum seeds[2] , cork biomass[3], tea industry waste[40] , grape stalk wastes[5] have been widely increased in the recent past , as these byproducts does not possess commercial value in agricultural sector. Previous work has shown that the leaf powder of Azaridachta indica has an excellent ability in removing metal ion such as Cr(VI) from aqueous solutions [6,7] . The Nickel element is present in aqueous solutions mainly in the divalent states. Nickel is released into the environment by a large number of processes such as electroplating, polishing, batteries [8] manufacturing, mining, metal finishing,paints, pigments, forging etc. Due to the toxic effects of nickel, it is necessary to eliminate it fromthe contaminated effluents. The aim of this work is to evaluate the viability of the acid- alkali pre-treated seed shell of Azaridachta indica to remove of Ni (II) from aqueous solution. Equilibrium and kinetic experiments were performed. The Langmuir isotherm model was appliedto experimental data. The pseudo-first, pseudo-second-and intra-particle diffusion models were used to fit the kinetic data. The changes in functional groups were analyzed by spectroscopy in the infrared region (FT-IR).

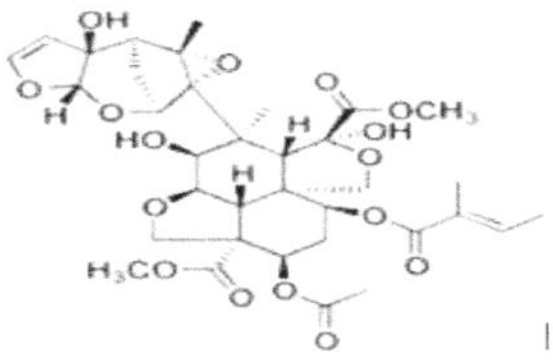

Molecular Structure of *Azaridachta indica*

Preparation of Adsorbent

MATERIALS AND METHODS

Neem tree botanically known as Azadirachta indica, belong to the family Meliaceae. The tree is tall, fast growing and is known for its drought resistance [9]. The fruits are green and yellow when ripe and are yielded around 37-55 kg per tree. It is a smooth olive- like drupe whichvaries in shape from elongate oval to nearly roundish. The white, hard inner shell of the fruit

known as the seed shells (endocarp) were collected from fertilizer manufacturing units in Coimbatore. Neem seed shell husk were thoroughly washed well with double distilled water anddried in the air. The dried shells were powdered, sieved and fractions measuring from 0.18 to
0.71 mm were treated first by boiling in 0.1 N HCl for three hours. After decanting the solution,the residue was allowed to boil with 0.1 N NaOH for three hours. Similar treatment processes were reported by A.L.Ahmad et al. and Isabel Vellaescusa et al.,[10,11] The treated sorbent of varying sieve sizes were washed well several times with double distilled water, dried and used for further experiments.

SEM/EDAX

The surface morphology of Azaridachta indica seed shell was observed using scanning electron microscope (SEM). After drying, the samples were covered with a thin layer of platinum(10 nm) using a sputter coater (SCD 0050 – Baltec, Liechenstein) and observed using the JEOL JFM-6390 scanning electron microscope (20 kV) under vacuum of 1.33 x 10-6 mBar (Jeol, Japan). To determine the chemical composition, energy dispersive X-ray spectroscopy was performed on adsorbent before and after nickel adsorption. The samples were prepared as for SEM analysis.

METAL BINDING BATCH EXPERIMENTS

Ni^{2+}solutions were prepared from sulphate salts of Nickel [Ni(NH4)2SO4.9H20]. The concentration of metal ions were determined by atomic absorption spectrometry (Perkin Elmer AA Analyst 100 with air-acetylene oxidizing flame). During the reaction process the pH was maintained constant at 8.1 throughout the experiment. Adsorption experiments were performed in continuously stirred beakers containing 50 mL of 11 mg L^{-1} heavy metals solution (150 rpm)and 150 mg adsorbent, at 30 ºC. The models of pseudo-first order, pseudo-second order and intra- particle diffusion were used to fit the experimental data. The equations are represented as follow, then Eq. (1), Eq. (2) and Eq. (3), respectively:

$$dq/dt = k1(qe-qt) \quad (1)$$

$$dq/dt = k2\ (qe-qt)^2 \quad (2)$$

$$qt = k\,p\,t^{1/2} \quad (3)$$

where: q represents the amount of adsorbate retained in the solid at equilibrium, qt represents the amount of adsorbate retained at time t, k1, k2 , kp are the rate constants of the reaction of pseudo-first order [min-1], pseudo- second order [g(mg min-1)] and intraparticle diffusion mg(g-1 min1/2)-. For equilibrium studies the concentrations of nickel solutions variedfrom 3 to 13 mg L–1 and the amount of biosorbent used was 150 mg L–1. These results were adjusted to the Langmuir model that can be expressedas:

$$qe = QobCe/1+bCe \quad (4)$$

Where: b is the nickel uuptake at equilibrium (mg g-1); Qois the maximum Langmuir uptake (mg g-1); Ce is the final chromium concentration at equilibrium (mg L-1).

Fourier transforms infrared spectroscopy (FT-IR)

FT-IR spectroscopy was used to confirm the presence of the functional groups in samplesand to observe the chemical modification after chromium biosorption. Infrared spectra were recorded in the 4000-600 cm-1 region using a Thermo Nicolet instrument, model IR-200. The ATR (attenuated total reflection) device allows getting information about the surface.

RESULTS AND DISCUSSION

SEM/EDX analysis

The morphology of *L. minor* surface was analyzed by scanning electron microscopy (SEM) adsorbent before and after nickel loading(Figure 1a &1b).

Figure 1a &1b: SEM analysis of A.indica seed shell powder before and after Ni(II) loading

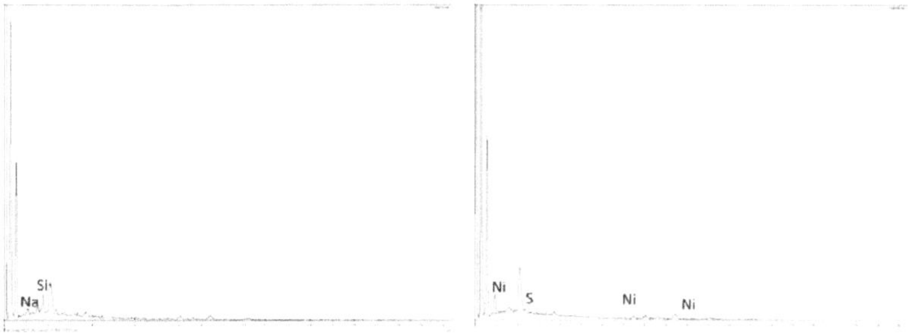

Figure 2a & 2b: EDX analysis of A.indica seed shell powder before and after Ni(II) loading

The SEM analysis of the treated adsorbent showed the rugosity surface with microstructures, as illustrate Figure 1 (a). This rugosity canbe assigned to the presence of niacin,proline, glutamic acid, aspartic acid, glutamine, tyrosine and alanine which contain polar groups like –NH2, - COOH, -OH etc. in A.indica sed shell powder (Battacharya and Sharma, 2004; Neem foundation, http://www.neemfoundation.org, 1997) [7] contribute to the negative surface charge. This rugosity is less obvious after the Ni(II) loading (Figure 1b). According to EDX analysis, the ionsinitially present in the seed shell, such as phosphorus, calcium, iron etc. [12] are removed duringthe Nickel sorption process (Figure 2b).

FT-IR Analysis

The comparison of the FT-IR spectra of *A.indica* seed shell after and before Nickel sorption, are given in Figure 3a & 3b. For unloaded A.indica, the band at 3445.94 cm−1is O-H stretching of polymeric compounds, 2895.25 cm^{-1} is symmetric vibrationof CH2, 1681.98 cm^{-1} is stretching vibration of COO, C=O, 1540.21 cm−1 band is stretching vibration of C-N of peptidic bond of proteins, 1419.66 cm^{-1} is of phenolic O-H and C=O stretching of carboxylates, 1318.39 cm^{-1} band is vibration of phosphate groups, 1231.59 cm^{-1} band is vibration of carboxylic acids. The bands < 800 cm−1 are finger print zone which is phosphate and sulphur functional groups (Saygideger et al. 2005)[13]. The FT-IR results of loaded Ni(II) A.indica seed shell powder showed that *A.indica* have different functional groups for nickel ions binding, such as carboxyl (1649.19 and 1369.50 cm-1), peptidic (~1559.50 cm−1) and hydroxide (3312.85cm-1) groups. In accordance with Sheng *et al.* (2004),[14] the peaks at 1681.98 and 1422.55 cm^{-1} is due to the carboxylate salt COO–M, where M can be Na^{+}, K^{+}, Ca^{2+}, and Mg^{2+}metals, which are naturally present in the adsorbent material. This band changes to 1649.19 and 1369.50 cm^{-1} after Nickel adsorption. The species (M) naturally present in*A.indica seed shell powder* were replaced by Nickel ions.

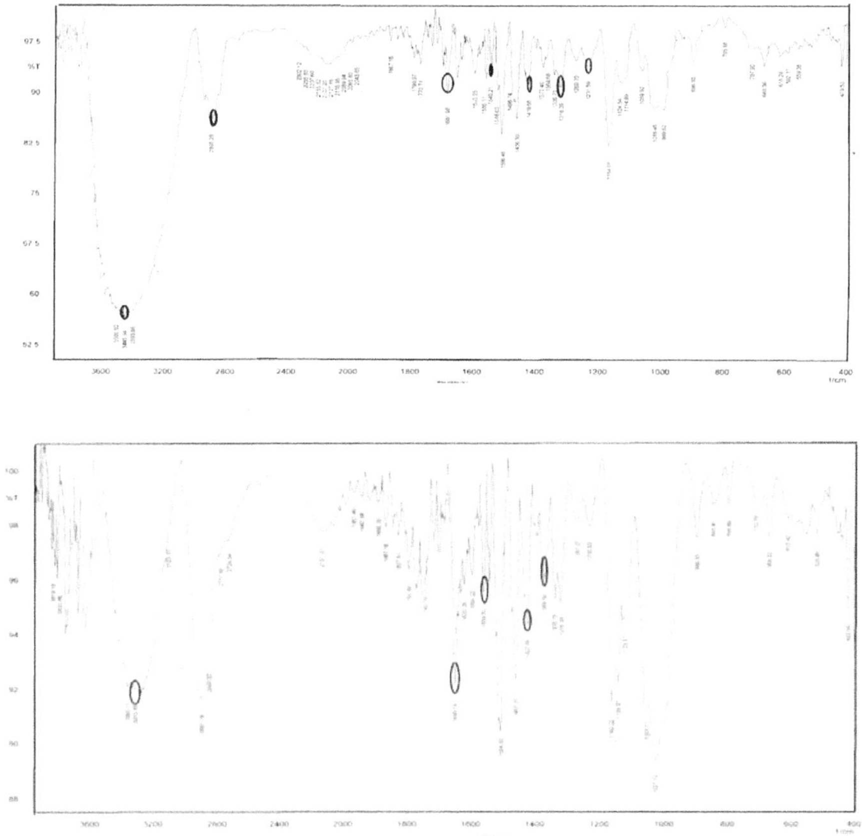

Figure 3a & 3b: FT-IR spectra of unloaded and Ni (II) loaded A.indica sed shell powder

Langmuir isotherm and Freundlich isotherm model were used to fit the experimental data. Langmuir parameters obtained were qmax of 66.66mg g^{-1}, b of 0.1666 L mg^{-1} and R^2 of 0.9967. The maximum amount adsorbed on a monolayer is represented by qmax, and b is the Langmuir association constant, which is related to the adsorption energy. The apparent Langmuir constant (b) is a measure of stability for thecomplex formed between metal ions and adsorptive surface layer of adsorbent under specified experimental conditions. For example, a small "b" value indicates that the metal ion has a high binding affinity with the adsorbent. From table 1 it couldbe understood that Langmuir model fited the experimental data as compared to Freundlich model. The values of b and qmax indicate that *A.indica* seed shell powder have affinity for Ni (II). Figure 5a and 5b presents Ni(II) ion kinetic curves adjusted with the pseudo-first-order and pseudo-second-order models. Linear correlation coefficients and pseudo-first-order (k1) and second-order (k2) constants, as well as theequilibrium capacity obtained, are shown in Table 1. The value of correlation coefficient R^2 is relatively high (>0.99) for Pseudo first order when compared to pseudo second order. Therefore, it can be concluded that the Lagregren model best describe the biosorption kinetics of metals over *A.indica* seed shell powder.

Table 1: Adsorption isotherms for Biosorption of Ni^{2+} on A.indica seed shell powder

Langmuir			Freundlich		
qmax	b	R^2	K	1/n	R^2
66.66	0.1666	0.9967	14.26	1.0055	0.9356

Table 2: The first-order and second-order kinetic constants for biosorption of Ni^{2+} ion onA.indica seed shell powder

First Pseudo-order			Second order Pseudo		
q	K1	R^2	q	K2	R^2
1.3402	0 .0999	0.9947	0.126	0.06 63	0.9615

Figure 5a & 5b: Pseudo-firstorder and pseudo-second-order kinetic models for Ni^{2+} adsorption.

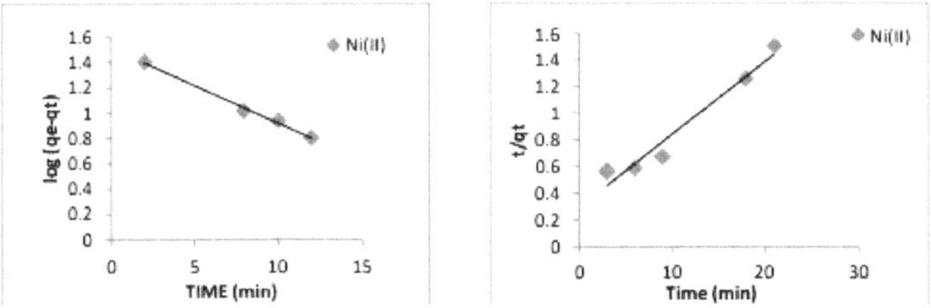

CONCLUSIONS

This study demonstrated that the *Azaridachta indica* seed shell powder (pretreated) could be used as an effective biosorbent for the treatment of wastewater containing Ni ions.The SEM and FT-IR analysis indicate the morphological changes and metal binding capacity of the employed adsorbent. The values of b and qmax indicate that *A.indica* seed shell powder have affinity for Ni(II) as obtained from Langmuir model . The value of correlation coefficient R^2 is relativelyhigh. Therefore, it can be concluded that the Lagregren model describe the biosorption kinetics of metals over *A.indica* seed shell powder.

REFERENCES

1. Saeed A, Akther M W, Iqbal M. Sep Purif Technol. 2005; 45:25-31.
2. Ozean A, SafaOzean A, Tunali S, Akar T, Kiran I. J HazardMater.

2005;124: 200-208.

3. Chubar N, Carvatho J R, Correja M J N. Colloid Surf. A ; Physicochem Eng Aspects. 2004;230:57-65.
4. Cay S, Uyanik A, Ozisik A. Sep Purif Technol. 2004;38: 273-280.
5. Villaescusa I L, Fiol N, Martinez M, Miralles N, Poch J, SerarolsJ. Water Res. 2004;38:992-1002
6. Venkateswarlu P, Venkata Ratnam M, Subba Rao D, Venkateswara Rao M. Int J Phys Sci. 2007;2(8):188-195.
7. Bhattacharya KG, Sarma A. J Hazard Mater. 2004; B113:97-109.
8. Bailey S E, Olin T J, Bricka RM, Adrian DD. Water Res. 1999;33:2469.
9. Vijayalaxmi K, Radha K S, Vandana Shiva. Neem - A User's Manual, New Delhi 1995 .
10. Ahmad A L, Sumathi S, Hameed B H. Wat Res. 2005; 39:2483-2494.
11. Isabel Villaescusa, Nuria Fiol, Maria Mortinez, Naria Miralles, JordiPoch, Joan Serarols. Wat Res. 2004;38: 992-1002
12. Udhayasekara Rao P. J Am Oil Chem Soc. 1987; 64 (9): 1348-1351.
13. Saygideger S, Gulnaz O, Istifli ES, Yucel N. J Hazard Mater. 2005; B126:96–104.
14. Sheng P. X, Ting Y,Chen J P, Hong L. Sorption of lead,copper,cadmium,zinc and Nickel by marine algal biomass; characterization of biosorptive capacity and investigation of mechanisms.J Colloid Interface Sci. 2004;275:131-141.

A Comparative Study on the Sorption Characteristicsof Pb(II)and Hg(II) onto Activated Carbon

N. MUTHULAKSHMI ANDAL* and V. SAKTHI

*Department of Chemistry, PSGR Krishnammal College for Women,Coimbatore - 641 004, India.

Department of Chemistry, N.G.M.College, Pollachi- 642001,India.

drandalcbe@yahoo.co.in

Received 22 October 2009; Accepted 20 December 2009

Abstract: Biosorption equilibrium and kinetics of Pb(II) and Hg(II)on coconut shell carbon (CSC) were investigated by batch equilibration method. The effects of pH, adsorbent dosage, contact time, temperature and initial concentration of Pb(II) and Hg(II) on the activated carbon of coconut shell wastes were studied. Maximumadsorption of Pb(II) occurred at pH 4.5 and Hg(II) at pH 6. The sorptive mechanism followed the pseudo second order kinetics. The equilibriumdata were analysed by Langmuir, Freundlich and Dubinin-Radushkevich isotherm models. The equilibration data fitted well with both Langmuir and Freundlich isotherm model. The Langmuir adsorption capacity for Pb(II) was greater than Hg(II). The mean freeenergy of adsorption calculated from Dubinin- Radushkevich (D-R) isotherm model indicated that the adsorption of metal ions was foundto be by chemical ion exchange. Thermodynamic parameter showed thatthe sorption process of Pb(II) onto SDC was feasible, spontaneous and endothermicunder studied conditions. A comparison was evaluatedfor the two metals.

Keywords: Hg(II) ions, Pb(II) ions, CSC, Isotherms, Kinetics.

Introduction

The presence of heavy metals in water is of great concern becauseof their toxicity to life forms even in small concentration. Mercuryand lead are two important toxic elements found in industrialeffluents causing water pollution. Pollution by mercury usually comes from industries such as chloroalkali, paint, pharmaceutical,pulp and paper[1]. Lead pollution is caused by storage batteries, leadsmelting and plating industries[2]. The presence of lead in drinking water above the permissible limit of 0.05 mg/L[3] can delay the physical and mental development of children and develop high blood pressure in adults[4] whereas the presence of mercury in water in above quantity will cause damage to the central nervous system and chromosomes and will cause chest pain[5]. Because of increased environmental awareness, the development of cost effective methods for the removal of lead and mercury from aqueous solutions has gained momentum.

Adsorption by activated carbon from cheap and readily available sources like ricehusk ash[1], Ceiba pentandra hulls[6], Polymerised sawdust[7], Carica papaya[8], tea waste[9] and hazelnut husks[10] have been reported for the removal of lead and mercury from aqueous solution. The current investigation discusses the characterization of a low cost adsorbent for the removalof leadand mercury.

Coconut shell, a hard and thick bony endocarp material, is an abundantly available agricultural waste from the local coconut industry. Conversion of coconut shell into activated carbon, whichcan be used as an adsorbent for water purification, would add its economic value, help reduce the cost of waste disposal, and most importantly, provide a potentially inexpensive alternative to the existing commercial activated carbon.

The present study aims at investigating the feasibility of Pb(II) and Hg(II) removal from aqueous solutions by an indigenous,economically viable, coconut shell carbon as adsorbent under varied conditions.

Experimental

Coconut shells were crushed, washed with deionised water and sundried. The activation was done by treating with concentrated sulphuric acid and held in an oven at 150 ^{0}C for 24 h. Theobtainedacid carbonized coconut shell was washed several times withdeionised water and 5% sodium bicarbonate solution followed by deionised water again. The adsorbent was dried at 90 ^{0}C for 10 hours, powdered and then sieved into different sizes and a particlesize of 90 μm was used in this study.

Stock solutions of 1000 mg/L Pb(II) and Hg(II) were prepared by dissolving lead nitrate and mercuric chloride salts (AR grade) indouble distilled water. All aliquots of varying concentrations wereprepared from the stock solution. 50 mL of the metal solutions of different concentrations were taken in iodine flasks of 100 mL capacity. Known amounts of the adsorbent were added into the flasks and the solutions were agitated in a mechanical shaker at preset time intervals and at desired pH environments. The spent CSC materials were separated by centrifugation and the supernatant adsorbate solutions were analyzed for the residual Pb(II) and Hg(II) concentrations spectrophotometrically by using UV Visible spectrophotometer. (Genesus 2 Thermospectronic make) employing PAR reagent (2, 4- Pyridyl azo resorcinol) and Rhodamine 6G as a complexing agent respectively.

Specifications of various studies

Isothermal, pH and kinetic studies were carried out at different initial concentrations of metal ions from 10 to 50 mg/L while maintaining the adsorbent dosage at 0.8 g/L. Temperature effect was studied with a concentration of 50 mg/L of the adsorbate at temperatures 20^{0}, 30°, 40° and 50 ^{o}C.

Results and Discussion

Effect of pH

Knowledge of the optimum pH is very important since pH affects not only the surface charge of adsorbent, but also the degree of ionization and speciation of adsorbate during reaction. Thebindingof metal ions to the adsorbent as a function of pH is represented inFigures 1 and 2. It is clear that the metal binding is strongly pH dependent and maximum adsorption occurs 0.at pH 5 for lead andpH 6 for mercury. The increase in metal removal with increase in pH is due to a decrease in competition between proton and metal ions for the surface sites and by the decrease in positive surface charge. The decrease in metal removal at low pH values is due to increasing competition of protons for the same binding sites that metal can use.

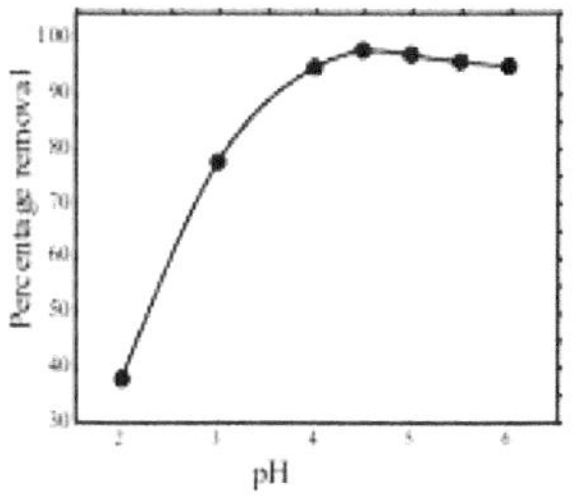

Figure 1. Effect of pH on the removal of Pb(II) ions by CSC (Initial Concentration-10 mg L^{-1}).

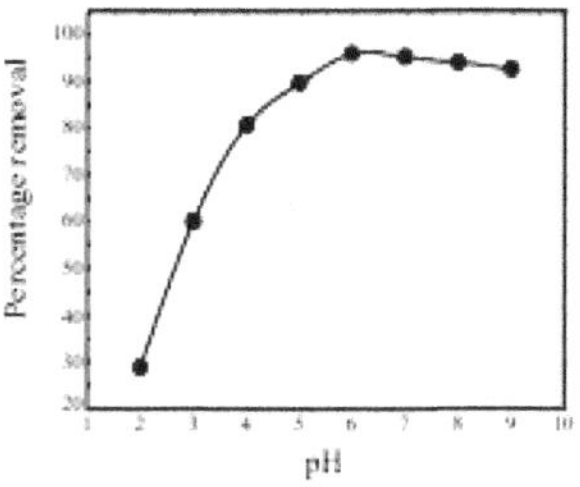

Figure 2. Effect of pH on the removal of Hg(II) ions on CSC (Initial Concentration-10 mg L^{-1}).

The Pb(II) and Hg(II) speciation diagram indicates that in thehighest sorption range, thedominant species of adsorption are Pb^{2+}, $Pb(OH)^{+}$, Hg^{2+}, $Hg(OH)^{+}$[7,11]. The exchange /sorption properties of coconut shell are due to the presence of some functional groups, such as carboxylic, hydroxyl and lactone, which have high affinity for metal ions[12]. The ion exchangemechanism between H^{+} ions at theCSC surface and metal ion may be illustrated as follows[13].

$$2S\text{-}COH + M^{2+} \leftrightarrow (R\text{-}CO)_2M + 2H^+$$
$$S\text{-}COH + MOH^+ \leftrightarrow RCOMOH + H^+$$

Where S is the CSC surface.

A slight decrease in adsorption was observed above pH 4.5 for lead and at pH 6, Pb(II) started precipitating as $Pb(OH)_2$. Hence pH experiments at alkaline ranges were restricted. Hg(II) adsorption athigher pH is due to electrostatic attraction between $Hg(OH)^+$ ions and the negatively charged adsorbent surface[14].

Effect of agitation time and initial concentration of metal ions on adsorption

Effect of agitation time and initial metal concentration on the removal of metal by CSC is represented in Figure 3. The uptake ofmetal (mg/g) increased with time and high amount of metal ions was adsorbed in the first 15-30 minutes. The attainment of equilibrium was observed at 120 minutes for Pb(II) and 90 minutesfor Hg(II). The removal curves are single, smooth and continuous leading to saturation suggesting the possible monolayer coverage of metal ions on the surface of the adsorbent. The rate of metal removal is higher in the beginning due to a larger surface of the adsorbent being available for the adsorption of metals. The percentmetal removal decreased from 99.08 to 62.76 for Pb(II) and 96 to53.64 as the metal concentration was increased from 10 to 50 mg/L. This trend can be attributed to the progressive increase in electrostatic interactions relative to covalent interactions, of the sites with lower affinity for metal ions with an increase in the initialconcentrations[15].

Effect of adsorbent dose on adsorption

Figure 4 is indicative of the removal of metal as a function of different adsorbent dosages foran initial concentration of 10 mg/Lsolution. It is obvious from the Figure 4 that, the increase in the dose of adsorbent, decreases the adsorption per unit mass of adsorbent. At an initial concentration of 10 mg/L metal ions, the unit adsorption decreased from 62.41 mg/g to 12.383 mg/g for Pb(II) and 59 mg /g to 12 mg/g at an adsorbent concentration of 0.1 mg/L to 0.8 g/L. This may be due to the aggregation/agglomeration of sorbent particles at higher sorbent concentrations, which would lead to a decrease in the total surfacearea of the adsorbent particles available to the metal ions. As a result the adsorption capacity of the adsorbent is not fully utilized[16].

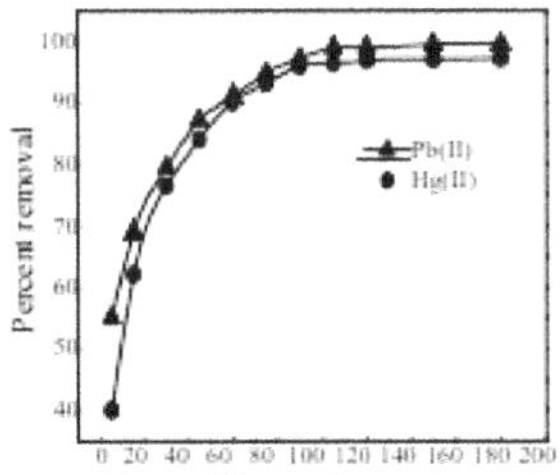

Figure 3. Effect of contact time on the removal of Pb(II) and Hg(II) ions on CSC (Initial Concentration- 10 mg L^{-1})

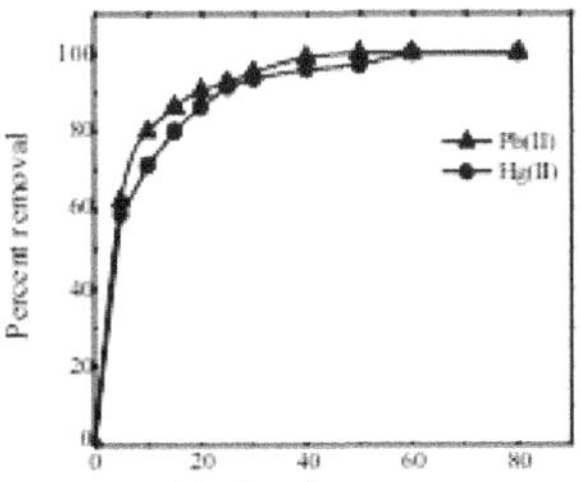

Figure 4. Effect of adsorbent dose on the removal of Pb(II) and Hg(II) ions on CSC (Initial Concentration- 10 mg L^{-1}).

Adsorption kinetics

Pseudo-second order kinetic model

The linearised form of pseudo second order equation can be written as[17]

$$t / q_t = 1 / k_2 q_e^2 + t / q_e \qquad (1)$$

where, k_2 is the rate constant of second order adsorption (g/mg/min). k_2 and q_e can be obtained from the intercept and slopeof plotting t/q_t *vs.* t (Figure 5). The correlationcoefficient was foundto be 0.9995 for all the concentrations studied and the calculated q_e values showed that pseudo second order model describes the adsorption kinetics of both the metals. The k_2 values are shown in Table 1.

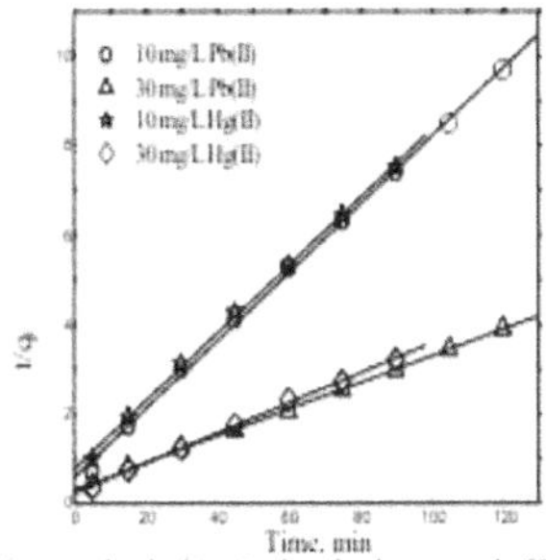

Figure 5. Pseudo second order kinetic plots for the removal of Pb(II) and Hg(II) ions on CSC.

Table 1. Pseudo second-order rate constants and thermodynamic parameters for Pb(II) and Hg(II).

	Pb(II)	Hg(II)		Pb(II)		Hg(II)	
Initial Concentration, mg/L	k_2 x10^{-3} g/mg/min	k_2 x10^{-3} g/mg/min	Temperature K	k_2 x10^{-3} g/mg/min	$-\Delta G$ kJ mol^{-1}	k_2 x10^{-3} g/mg/min	$-\Delta G$ kJ mol^{-1}
10	9.84	7.43	293	1.11	0.19	3.66	0.10
20	4.12	5.21	303	1.33	1.32	3.95	0.37
30	2.89	5.09	313	1.65	2.29	4.05	1.26
40	1.78	4.99	323	2.16	4.03	4.26	2.15
50	1.33	3.95					

Adsorption Isotherms

Langmuir isotherm

The Langmuir equation assumes that the solid surface presents a finite number of identical sites which are energetically uniform anda monolayer is formed when the solid surface reaches saturation. Langmuir equation can be described by the linearised form[18]

$$C_e / q_e \quad = 1 / Q_0 b + Ce / Q_0 \qquad (2)$$

Where C_e is the equilibrium concentration of adsorbate (mg/L), q_e is the amount adsorbed at equilibrium (mg /g), Q_o and b are Langmuir constants related to adsorption capacity and energy of adsorption respectively. The linear plot of C_e/q_e versus C_e showed that the adsorption obeys Langmuir isotherm. The essential characteristics of Langmuir isotherm can be expressedin terms of adimensionless constant separation factor R_L which is defined by

$$R_{L} = 1/1+b\ C_0$$

Where, C_o is the initial concentration of adsorbate. R_L values between 0.02 to 0.09 for Pb(II), 0.03 to 0.133 for Hg(II) indicate favourable adsorption of metals on CSC at the concentration studied[19].

Freundlich isotherm

The Freundlich expression is an empirical equation based on sorption on a heterogeneous surface suggesting that binding sites are not equivalent and/or independent. The linearised logarithmic form of Freundlich equation is[18]:

$$qe = \log K_F + 1/n \log C_e \qquad (3)$$

Where K_F and n are Freundlich constants related to the sorption capacity and sorption intensity of the sorbent respectively. Linear plots of log q_e *versus* log C_e showed that the adsorption obeys Freundlich isotherm. The regression values indicate that the adsorptiondata for metal removal fitted well within the Langmuir and the Freundlich isothermal plots for the concentration studied.

Dubinin- Radushkevich (D-R) isotherm model

The equilibrium data were also applied to the D-R isotherm modelto determine the nature of adsorption processes as physical or chemical The linearized form of the the D-R isotherm equation[20] is:

(4)

$$\ln q_e = \ln q_m - \beta \varepsilon^2$$

where q_e is the amount of metal ions adsorbed on per unit weight of adsorbent (mol/g), q_m the theoretical saturation sorption capacity (mol/g) , β is a constant related to mean free energy of sorption per mole of the sorbate (mol^2/J^2),and the Polanyi Potential, which is equal to RT ln $(1+1/C_e)$,where R is the gas constant ($J\ mol^{-1}\ K^{-1}$) andT(K) is the absolute temperature. A plot of ln q_e should give straightline (Figure 9 and 10).

The constant β gives an idea about the mean free energy E (kJ/mol) of sorption per molecule of the sorbate when it is transferred to the surface of the solid from infinity in the solution and can be calculated using the relationship.

$$E = 1/(2\beta)^{1/2} \quad (5)$$

This parameter gives information about sorption mechanism as chemical ion exchange or physical sorption. The magnitude of E is between 8 and 16 kJ/mol, the sorption process follows chemical ion exchange, while for the values of $E < 8$ kJ/mol, the sorption processis of a physical nature[21]. In the present study, the mean free energyof sorption was found to be

12.23 and 10.19 kJ/mol for Pb(II) and Hg(II), indicating support to chemical ion exchange mechanism or chemisorption.

The values supportive of Langmuir, Freundlich and D-R parameters for the adsorptionof Pb(II) onto CSC are listed in Table2.

Table 2. Isotherm constants for the metal ions.

Metal ion	Langmuir Constants			Freundlich Constants			D-R isotherm constants		
	Q_0 mg/g	b L/g	R^2	K_F mg/g	n	R^2	q_e mg/g	E kJ/mol	R^2
Pb(II)	40.49	1.01	0.9969	20.99	4.52	0.9993	42.27	12.29	0.9896
Hg(II)	35.09	0.652	0.9982	15.65	3.89	0.9947	38.60	10.19	0.9980

Effect of temperature on adsorption rate

It was observed that Pb(II) and Hg(II) biosorption followed pseudosecond order kinetics at different temperatures. Linear plots of t/q_t *versus* t corresponding to the pseudo second order kinetic model were obtained. The values of kinetic parameters obtained are presented in Table 1. The rise in temperature increased the values of q_e and k_2.

In the present investigation, it was observed that the adsorptionof Pb(II) and Hg(II) increase with rise in temperature. The ratio ofincrease in the rate constant was 1.2 for Pb(II) and 1.05 for Hg(II)for every 10 °C rise in temperature.

The enhancement of adsorption capacity at higher temperature may be attributed to the enlargement of pore size and/or activation of adsorbent surface[22].

Adsorption dynamics

The thermodynamic equilibrium constant K_c for the adsorption was determined by using the equation $K_c = C_A/C_e$, where C_A is the amount of Pb(II) adsorbed on the adsorbent of the solution at equilibrium (mol /L) and C_e is the equilibrium concentration of Pb(II) in the solution (mol/L).The change in standard free energy, enthalpy and entropy of adsorption were calculated using thefollowing equation:

$$\Delta G^{o} = -RT \ln K_c \qquad (6)$$

where R is the gas constant and T the temperature (K).According to

Vant Hoff's equation:

$$\ln K_c = \Delta S^{o}/R - \Delta H^{o}/RT \qquad (7)$$

A plot of ln K_c *versus* 1 / T was found to be linear. The values were evaluated from the slope and intercept of Vant Hoff's plot .The positive value 38.45 kJ mol^{-1} for Pb(II) and 23.82 kJ mol^{-1} forHg(II) confirms the endothermic nature of adsorption. When the temperature was increased from 20°C to 50°C, the magnitude of free energy change shifted to a high negative value suggesting thatthe adsorption was rapid and spontaneous (Table1). The positive value of S 131.1 and 80.3 J $mol^{-1}K^{-1}$ for Pb(II)and Hg(II) suggested an increase in randomness at the solid – solution interface during the adsorption of Pb(II) on SDC[23]..

Comparative study

The experimental results showed that the adsorption capacity (Q_o) and the adsorption intensity (K_F) obtained from the Langmuir and Freundlich model is greater for Pb(II)compared to Hg(II). The Q_0 andq_m obtained from Langmuir and D-R isotherm model are almost in the same order. For ions of same valence, the sorbent prefers the metal with higheratomic number. Moreover, the observed order wasthe same as that of their increasing ionic radii or their decreasing hydrated ionic radii. Smaller the hydrated ionic radius, the greater isits efficiency to activate groups of the adsorbent. This suggests that the energy required inthe dehydration of metal ions, in order that they could occupy a site in the adsorbent, playsan important role in determining the selectivityseries for the metal ions[24].

The higher binding capacity of Pb(II) ions over Hg(II) ions on the CSC could be explained by comparing the covalent index of Pb(II) (7.18) with Hg(II) (6.75) ions[15]. The covalent index of metal ions can be calculated by Eq. (8), taking into accounts of electro negativity (X) and ionic radius (r) of the metal ions. The value 0.85in Eq. (8) represents an appropriate constant assumed to reflect theradius of oxygen and nitrogen donor atoms. The high value of the covalent index shows the high degree of binding capacity of metal ions to the functional groups of the adsorbent.

$$X_m^2(r+0.85) \qquad (8)$$

Working with crop milling waste as an adsorbent Asma Saeed *et al*[25] concluded that higher is the value of b and n, the higher is the affinity of adsorbent towards the metal being adsorbed. It is evident from the values of b and n for Pb(II) and Hg(II), that the affinity of CSC to adsorb is highest for Pb(II). Increase in temperature stimulates more adsorption for both the metals especially for lead. The thermodynamic parameters H and

S are greater for Pb(II) compared to Hg(II).

Conclusion

The study indicated that activated carbon prepared from coconut shell carbon can be used as an effective adsorbent material for the treatment of lead and mercury bearing aqueous wastewaters. The adsorption of lead and mercury on activated carbon is found to be time, concentration, pH and dose dependent. The Langmuir and Freundlich equation agrees very well with the equilibrium isothermfor the entire concentration range studied. The pseudo second order kinetic model agrees very well with the dynamical behaviour for theadsorptionof lead and mercury on CSC under different initial metalion concentrations. This suggests thatthe rate limiting step may be chemical sorption rather than diffusion. The thermodynamic calculations indicated the feasibility, endothermic and spontaneousnature of the adsorption of Pb(II) and Hg(II) onto CSC. The present study enlighten that Pb(II) is preferentially adsorbed on the surfaceof the coconut shell carbon than Hg(II). Hence, acid activated coconut shell carbon is an effective and alternative adsorbent for theremoval of lead and mercury from wastewater in terms of its high sorption capacity, abundant availability and inexpensive nature.

References

1. Qingge Feng, Qinggu Lin, Fuzhong Gong, Shuichi Sugita and Masami Shiya, *JColloid Interface Sci.*, 2004, **278**, 1-8.
2. Shekinah P, Kadirvelu K, Kanmani P, Senthilkumar P and Subburam V, *J ChemTechnol Biotechnol.*, 2002, **77,** 458 – 464.
3. Tarun Kumar Naiya, Ashim Kumar Bhattacharya, Sailendranath Mandal and Sudip Kumar Das, *J Hazard Mater.*, 2009, **163**, 1254- 1264.
4. Vandana Singh, Stuti Tiwari, Ajit Kumar Sharma and Rashmi Sanghi, *J ColloidInterface Sci.*, 2007, **316**, 224-232.
5. Namasivayam C and Kadirvelu K, *Carbon*, 1999, **37**, 79-84.
6. Madhava Rao M, Kumar Reddy D H K, Padala Venkateswarulu and Seshaiah K, *JEnviron Mang.*, 2008, **90**, 634-643.
7. Raji C and Anirudhan T S, *Indian J Chem Technol.*, 1996, **3**, 49-54.

8. Shaik Basha., Murthy Z V P and Jha B, *Chem Eng J.*, 2009, **147**, 226-234.
9. Amarasinghe B M W P K and Williams R A, *Chem Eng .J.*, 2007, **132**,299-309.
10. Mustafa Imamoglu and Oktay Tekir, *Desalination*, 2008, 228, 108-113.
11. Manju G N, Anoop Krishnan K, Vinod V P and Anirudhan TS, *J Hazard Mater.*,2002, **91**, 221-238.
12. Tan W T, Ooi S T and Lee C K, *Environ Technol.*, 1993, **14**, 277-282.
13. Alka shukla, Yu Hui Zhang, Dubey P, Margrave J L and Shyam S. Shukla, *J HazardMater.*, 2002, B95, 137-152.
14. Anirudhan T S, Priya Senan and Unnithan M R, *Sep PurifTechnol.*, 2007, 52, 512-519.
15. Ismail Kiran, Tamer Akar and Sibel Tunali, *Process Biochem.*,2005, **40**, 3550-3558.
16. Shubha K P, Raji C and Anirudhan, *Water Res.*, 2001, **35**, 300-310.
17. Nathalie Chiron, Richard Guilet and Eric Deydier, *Water Res.*,2003, **37**, 3079 – 3086.
18. Dinesh Mohan, GuptaV K Srivastava S K and Chander S, *Colloid Surf.*, A2001, **177**,169-181.
19. Ahmet Sari, Mustafa Tuzen, Ozgur Dogan Uluozlu and Mustafa Soylak, *BiochemEng J.*, 2007, **37**, 151-158.
20. Ozgur Dogan Uluozlu, Ahmet Sari, Mustafa Tuzen and Mustafa Soylak, *BioresTechnol.*, 2008, **99**, 2972-2980.
21. Ting Fan, Yunguo Liu, Baoying Feng, Guangming Zeng, Chunping Yang, Ming Zhou,Haizhou Zhou, Zhenfeng Tan, Xin Wang, 2008. *J Hazard Mater.,* 160, 655-661.
22. Arzu Y. Dursun,. *Biochem Eng J.,* 2006, **28**, 187-195.
23. Anoop Krishnan K and Anirudhan T S, *Proces Biochem.*, 2003, **29**, 147-156.
24. Asma Saeed, Muhammed Iqbal and Waheed Akhtar, *J Hazard Mater.*, 2005,

Removal of Pb(II) ions from aqueous solutions using *Bombax ceiba* saw dust activated carbon

N. Muthulakshmi Andal[a] and V. Sakthi[b]

[b]*Department of Chemistry, N.G.M.College, Pollachi, India*

[a]*Department of Chemistry, PSGR Krishnammal College for Women, Coimbatore–641 004, India email: drandalcbe@yahoo.co.in*

Received 4 October 2009; Accepted 6 December 2009

AB S T R AC T

The isotherms, kinetics and thermodynamics of Pb(II) ions from aqueous solution by activated carbon prepared from *Bombax ceiba* sawdust (SDC) were carried out in a batch adsorption system. The effects of pH, adsorbent dosage, contact time, initial concentration ofPb(II) and temperature on the adsorption were studied. Maximum adsorption of Pb(II) occurred at pH 5. Pseudo first order, pseudo second order and intraparticle diffusion modelswere applied to kinetic data. The sorptive mechanism followed the pseudo second-order kinetics and intra particle diffusion model. The equilibration data fitted well with both Langmuir and Freundlich isotherm model with maximum sorption capacity of 209 mg g^{-1}. The mean free energy of adsorption calculated from Dubinin–Radushkevich (D-R) isotherm model indicated that the adsorption of metal ions was found to be by chemical ion exchange. Thermodynamic parameters and Tempkin constant showed that the sorption process of Pb(II)onto SDC was feasible, spontaneous and endothermic under studied conditions.

Keywords: Adsorption; Isotherms; Kinetics; Pb(II); Saw dust carbon; Thermodynamics

1. Introduction

Heavy metals are major pollutants in marine, ground and industrial wastewaters. Lead is attracting wideattention of environmentalists as one of the most toxic heavy metals. The major sources of lead released into the environment by waste streams are storagebatter- ies; lead smelting, mining, plating, ammunition, ceramic and glass industries [1]. The permissible limit of lead in drinking water is 0.05 mg L^{-1} [2]. The presence of lead in drinking water above the permissible limit can result in delays in physical andmental development, along with slight deficits in attention span and learningabilities in babies and children. In adults, it can cause blood pressure [3]. Removal of toxic Pb(II) ions from wastewaters before being discharged into the sewage system or into the aquatic environment is considered essential and the need of the hour.

Research has been carried out for developing cost effective heavy metal removal techniques. Physicochemical methods, such as chemical precipitation, chemicaloxidation or reduction, filtration, electrochemical treatment, application of membrane technology, solvent extraction and ion exchange process have beentradition- ally employed for the heavy metal removal from industrial waste water [4].The adsorption method is preferred to other methods in view of its efficiency and theease with which it can apply in the treatment of wastewatercontaining heavy metals.

Recently agricultural materials such as waste wool,nut wastes, tree barks, modified cotton and sawdust have

been used to remove heavy metals from wastewater. Mostof the agricultural byproductsare not having any commercial value which is largely discarded in agriculturalsector.Agricultural wastes such as bark and saw dust, which are in large quantities in timber mills, often present a disposal problem. Several researchers have reported the use of sawdust and modified sawdust for the removalof heavy metals from aqueous solutions [5–15].

The present study is to evaluate and observe the adsorption capacity of chemicallyactivated *Bombax ceiba* sawdust carbon for Pb(II) ion removal from aqueous solution. This sawdust is a by-product of wood industry waste and available in large quantities.Further more no reports are available on the use of this material for the treatment of Pb(II) ion from aqueous stream. The batchadsorption technique reported here are of the influenceof temperature on the rate constant for the removal of Pb(II) ion. The activation parameters calculated from the adsorption measurements in the present work, are to be very useful in elucidating the nature of adsorption for environmental engineers.

2. Materials and methods

2.1. *Chemicals*

All the chemicals used in this study were of analytical grade obtained from Ranbaxy, Qualigens and S.D Fine Chemicals Ltd, India. Stock solution of lead was pre- pared using lead nitrate in double distilled water. Pb(II) solutions of different concentrations were obtained by diluting the stock solution. Standard acid and base solutions (0.1 M HCland 0.1 M NaOH) were used for pH adjustments.

2.2. *Equipment*

pH measurements were made using a pH meter (model Elico-LI-120). The concentration of lead solutions was analyzed using a UV-visible spectrophotometer (model – Genesis 2). A temperature controlled mechanical shaker (model – TECHNO)was used for agitating the solutions. Carbon, hydrohen, nitrogen, sulphurs and oxygencontent of the activated carbon was analyzed by elemental analyzer model Vario EL III(Vario EL, Ele- mental Analyzer system, Germany).

2.3. *Preparation of Bombax ceiba saw dust activated carbon*

Sawdust of *Bombax ceiba* plant was collected from wood industries in Coimbatore,India and washed with deionized water four to five times for removing dirt and earthy

impurities. The washed saw dusts were dried at 90 for 10 h. The dried saw dusts

were used for carbon preparation by mixing with concentrated sulphuric acid in the ratio

of 1:1 and then it was kept atC150 for 12 h. This carbonized material was washed

with distilled water several times to remove the free acid and soaked in 5% sodium bicarbonate solution to remove any residual acid. This material was then again washed withdistilled water and dried at 105 C in a hot air oven for 10 h andthen sieved into differentparticle sizes ranges from 90 to 600 m. The material was placed in airtight container forfurther study

2.4. *Characterization of adsorbent*

The physico-chemical characteristics of *Bombax ceiba*saw dust activated carbon was characterized by means of pH, ash content, decolorizing power, water-soluble mat- ter, acid soluble matter, as per standard procedures [16]. The Zero point charge of the adsorbent was determinedby the solid addition method [17]. The surface area of the adsorbent was determined by acetic acid adsorption method [18]. The physico-chemical properties along with the percentage of carbon, hydrogen, nitrogen andsulphur in the activated carbon made from sawdust of *Bombax ceiba* were presented in Table 1.

2.5. *Batch adsorption experiments*

The kinetics of adsorption of Pb(II) ions on the SDCwas carried out using 10 to 50

mg L initial concentra- tion of the metal ions. 50 mL of Pb(II) solution of required

concentration were adjusted to desired pH and taken in 100 mL reaction bottle for adsorption studies. Differentdosages of the adsorbent were added into this solution and agitated for a predetermined period in a mechanical shaker. The adsorbents wereseparated by centrifugation

Table 1
Characteristic properties of activated carbon.

Parameter	Value
pH (1%solution)	7.2
Moisture (%)	5.75
Ash (%)	2.11
Surface area ($m^2 g^{-1}$)	369.4
Decolorizing power ($mg\ g^{-1}$)	15.0
Bulk density ($g\ L^{-1}$)	0.740
Water-soluble matter (%)	1.76
Acid-soluble matter 0.25 M HCl (%)	4.01
pH_{ZPC}	5.2
Carbon (%)	48.01
Hydrogen (%)	3.869
Nitrogen (%)	0.124
Sulphur (%)	2.243
Oxygen (%)	45.754

and the supernatant was analyzed for the residual Pb(II) concentration spectrophotometrically at a wavelength of 520 nm using PAR reagent (2,4-Pyridyl azo resorcinol)

[19] as a complexing agent. The same method was used in the subsequent experiments.

Langmuir adsorption isotherm and kinetic studies were carried out with differentinitial concentrations of Pb(II) from 10 to 50 mg L^{-1} at constant adsorbent dosages of

0.2 g L^{-1}. Freundlich adsorption isotherm studies were carried out with different adsorbent dosages from 0.1 to

0.8 g L^{-1} for different initial concentration of Pb(II). The effects of pH and particle sizewere studied by employ- ing 10 and 30 mg L^{-1} of the Pb(II) solutions and an adsor- bent dosage of 0.2 g L^{-1}. Temperature effect was studiedwith a concentration of 50 mg L^{-1} of the adsorbate andan adsorbent dose of 0.2 g L^{-1} at temperatures 30°C, 40°C and 50°C. In order to correct for any adsorption of Pb(II) on the container surface, control experiments were carried out without adsorbent. It was found that no adsorption occurred by the container walls. In addition, all mixing vessels were kept sealed throughout theduration of each isotherm test to minimize dissolution ofgaseous species in the atmosphere.

3. **Results and discussions**

3.1. *Effect of pH*

One of the most important factors affecting sorption of metal ions is the pH of thesolution [20]. The pH of the solution affects the charge on the surface of the adsorbents, so the change in pH affects the adsorption process.This is due to two

factors (i) H are competingwith metal ions for adsorption. (ii) Solution pH influ- ences the ionization of functional groups on the adsorbent surfaces and chemical speciation of metal ions. The influence of pH on the sorption of Pb(II) onto sawdust was studied by varying the pH of Pb(II) solution from 2 to 6 and the results were shown in Fig. 1. The maximum sorption was found to be at pH 5.0. So all the experiments were carried out at pH 5.0. The removal of Pb(II)by sawdust adsorption increased with increasing pH from 2.0 to 5.0. A slight decrease in the adsorption wasobserved above pH 5 and at pH 6 Pb(II) started precipitating as $Pb(OH)_2$ and hencepH experiments at alkaline ranges were restricted. Similar results were reported byYu *et al.* [21], Ayyappan *et al.* [22] and Jayaram *et al.* [23] for adsorption of Pb(II).

At lower pH, the H^+ions compete with metal cations

for the exchange sites on the SDC. The perusal of Pb(II) speciation diagram indicates that the dominant speciesof adsorption are Pb^{2+} and PbOH at pH less than six and $Pb(OH)_2$ at pH greater than six. [8]. It was observed that at

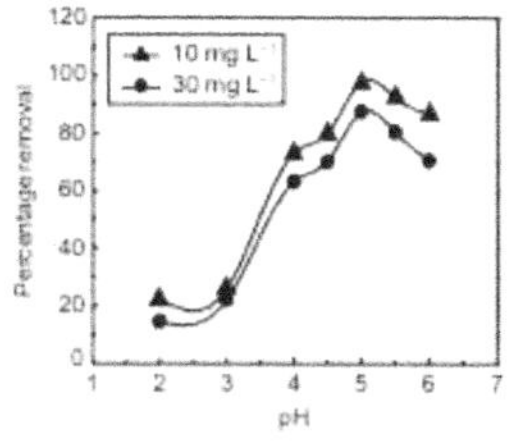

Fig. 1. Effect of pH on the sorption of Pb(II) onto SDC (metal ion concentration = 10 and 30 mg L^{-1}; adsorbent dose = 0.2 g L^{-1}; contact time = 120 min; temperature = 303 K).

low pH the adsorption wasvery low. This can be

explained that at low pH values the H^+ ion concentration is high and therefore protons compete with lead ions for surface sites, because at low pH the lead are present as Pb^{2+}ions. When the pH is increased deprotonation ofthe surface functional groups ofthe adsorbent results a decrease in the surface positive charge [24]. As a conse-quence the electrostatic repulsion between the positively charged metal ions and the surface is lowered favoringadsorption.

The cell walls of sawdust mainly consist of cellulose, lignin and many hydroxyl groups, such as tannins or other phenolic compounds. All these components are active ion exchange compounds. Based on the electron donating nature of the O, S, N containing groups in sawdust materials and the e^- accepting nature of Pb(II) ion, the ion exchange mechanism is proposed. The ionexchange mechanism is shown in the following equations [25]:

$$2R\text{-}COH + M^{2+} \leftrightarrow (R\text{-}CO)_2M + 2H^+$$
$$R\text{-}COH + MOH^+ \leftrightarrow RCOMOH + H^+,$$

where R is the matrix of sawdust.

3.2. *Adsorption isotherms*

3.2.1. *Langmuir isotherm*

The adsorption data has been analyzed in the light of Langmuir and Freundlichadsorption models. The Langmuir equation may be described as [26]

$$C_e/q_e = 1/Q_o b + C_e/Q_o \quad (1)$$

where C_e is the equilibrium concentration of metal (mg L^{-1}), Q_o and b are Langmuir constants relating to adsorption capacity and adsorption energy respec- tively. The Langmuir adsorption isotherm is valid for

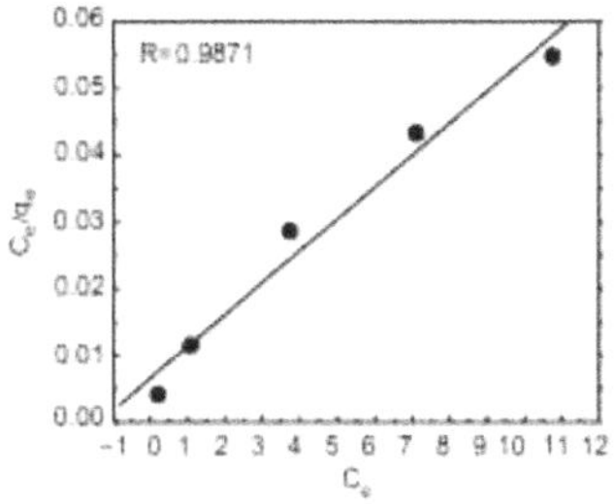

Fig. 2. Langmuir isotherm plot for the adsorption of Pb(II) onto SDC (adsorbent dose = 0.2 g L^{-1}; contact time = 120 min; pH = 5; temperature = 303 K).

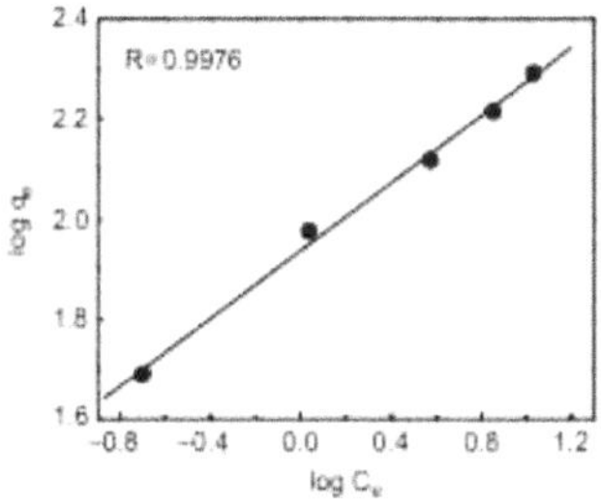

Fig. 3. Freundlich isotherm plot for the sorption of Pb(II) onto SDC (adsorbent dose = 0.2 g L^{-1}; contact time = 120 min; pH = 5; temperature = 303 K).

monolayer adsorption onto the surface of the adsorbentcontaining a finite number of identical sites. The plot of Ce/qe against Ce gives a straight line (Fig. 2) showing the applicability of Langmuir isotherm. The adsorption capacity Qo and energy of adsorption b were found to be 209 mg g^{-1} and 0.7253 L g^{-1} The essential features of the Langmuir Isotherm can be expressed in terms of dimensionless constant referred to as separation factorfor equilibrium parameter (R_L) which is defined by thefollowing relation [27]:

$$R_L = 1/1 + bC_o, \quad (2)$$

processes as physical or chemical [32].

R_L values lies between 0 and 1, the adsorption process is favorable. Further, thecalculated R_L values are 0.121, 0.065, 0.044, 0.033 and 0.027 for the initial

concentrations 10 to 50 mg L respectively. Therefore, sorptionwas favorable.

The adsorption capacity of other saw dusts for Pb(II) obtained by few other researchers □1 □1

chemically modified saw dust [8]; (iii) 21.05 mg^{-1} g for sawdust of poplar tree [29];

(iv) 34.246 mg^{-1} g for meranti sawdust [30]. A comparison of these values with the one obtained in this study (209 mg^{-1} g) showed that saw dustof *Bombax ceiba* used in this work exhibited a reasonablecapacity for Pb(II) adsorption from aqueoussolution.

3.2.2. Freundlich isotherm

The Freundlich isotherm, which is an empirical model used to describe adsorption inaqueous systemsis shown as [31] of logq_e against logC_e (Fig. 3) for different initial metal ion concentrations was found to be linear indicating the applicability of the Freundlich model. The intercept of the line is roughly an indicator of the adsorption capacity, and the slope is the indication of adsorption intensity.Adsorption of Pb(II) ontoSDC fitted both the Langmuirand Freundlich isotherms.

3.2.3. Dubinin–Radushkevich (D-R) isotherm model

The equilibrium data were also applied to the D-Risotherm model to determine the nature of adsorption $\ln q_e = \ln q_m - \beta\varepsilon^2,$

where q_e is the amount of metal ions adsorbed on per unit weight of adsorbent (mol g^{-1}), q_m the theoretical saturation sorption capacity (mol g^{-1}), constant related to mean free energy of sorption per mole of the sorbate (mol^2 J^{-2}), and Polanyi Potential, which is equal to RT ln(1+1/C_e), where *R* is the gas constant (Jmol^{-1} K^{-1}) and T(K) is the absolute temperature. A plot of lnq_e should give straightline (Fig 4).

The constant gives an idea about the mean free energy *E* (kJ mol^{-1}) of sorption per molecule of the sor-bate when it is transferred to the surface of the solid from infinity in the solution and can be calculated using the relationship:

$$E = 1/(2\beta)^{1/2}.$$

This parameter gives information about sorption mechanism as chemical ion exchange or physical sorption. The magnitude of E is between 8 and 16 kJ mol^{-1}, the where k_F and n are constants related to adsorption capacity and the adsorptionintensity, respectively.

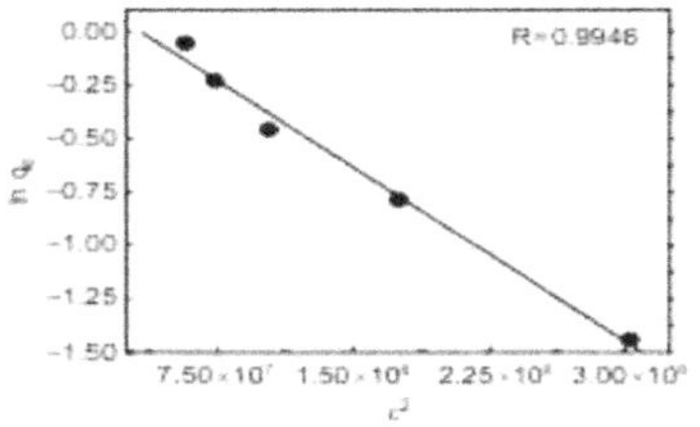

Fig. 4. Dubinin–Redushkevich isotherm plot for the sorption of Pb(II) onto SDC (adsorbent dose = 0.2 g L^{-1}; contact time = 120 min, pH = 5; temperature = 303 K).

3.2.4. Tempkin isotherm model

Heat of adsorption and the adsorbate–adsorbateinteraction on adsorption isotherms were studied by Tempkin and Pyshev, who suggested that because of these interactions the energy of adsorption of all the molecules decreases linearly with coverage. Tempkin isotherm is given as [35]

$$q_e = B\ln(K_T C_e). \qquad (6)$$

It can be expressed in the linear form as

$$q_e = B\ln K_T + B\ln C_e, \qquad (7)$$

where B = *RT*/*b*, *R* is the universal gas constant and *T* isthe absolute temperature

3.3. *Adsorption kinetics*

The efficiency of the adsorbent was evaluated by studying adsorption kinetics. Therate constant of adsorption is determined from the pseudo-first-

$$dq_t/dt = k_1(q_e - q_t), \quad (8)$$

where q_e and q_t are the adsorption capacity at equilibrium and at time t, respectively (mg g^{-1}) and k_1 is the rate constant for pseudo-first-order adsorption (min^{-1}). After integration and applying the boundary conditions $t = 0$ to t and $q_t = 0$ to q_e, the integrated form of equation (8) becomes

$$\log(q_e - q_t) = \log q_e - k_1t/2.303. \quad (9)$$

The slopes and intercepts of plots of $\log(q_e - q_t)$ versus t were used to determine the first-order rate constant k_1 and equilibrium adsorption capacity q_e. In the present study, although the correlation coefficients were reasonably high the model did not predict q_e values equal or even reasonably close, to experimental q_e values (Table. 3). This suggests that the adsorption of Pb(II) onto SDC is not a pseudo-first-order reaction. The pseudo-second-order adsorption kinetic rate equation is expressed as [38]:

$$dq_t/dt = k_2(q_e - q_t)^2, \quad (10)$$

where k_2 is the rate constant for pseudo-second-order adsorption g mg^{-1} min^{-1}. For the boundary conditions $t = 0$ to $t = t$ and $q_t = 0$ to $q_t = q_e$, the integrated form of equation (10) becomes

$$1/(q_e - q_t) = 1/q_e + k_2t, \quad (11)$$

order equation

Table 2
Isotherm constants for Pb(II) ion adsorption onto SDC.

Langmuir		Freundlich		D-R isotherm	
q_s (mg g^{-1})	209.205	K_f (mg g^{-1})	86.50	q_m (mg g^{-1})	247.592
b (L g^{-1})	0.7253	n	2.95	E (kJ mol^{-1})	9.602
R^2	0.9871	R^2	0.9976	R^2	0.9946

(Initial metal ion concentration–10 to 50 mg L^{-1}; pH–5; contact time–120 min; temperature–303 K; adsorbent dose–0.2 g L^{-1}).

Table 3
Comparison of the pseudo-first-order, pseudo-second-order and Intra particle diffusion rate Constants for different Initial Pb(II) concentrations.

Initial concentration (mg L^{-1})	exp q_e (mg g^{-1})	Pseudo first order			Pseudo second order			Intra particle diffusion		
		$K_1 \times 10^{-2}$ (min^{-1})	Cal q_e (mg g^{-1})	R^2	$K_2 \times 10^{-3}$ g mg^{-1} min^{-1}	Cal q_e mg g^{-1}	R^2	K_p mg g^{-1} $min^{1/2}$	q_e mg g^{-1}	R^2
10	49.00	4.339	32.66	0.9612	4.257	50.51	0.9989	2.033	49.64	0.9948
20	94.52	3.109	36.73	0.9974	1.916	98.04	0.9984	4.138	95.19	0.9971
30	131.25	3.961	58.61	0.9942	1.429	136.99	0.9993	5.329	135.31	0.9875
40	164.50	3.132	71.12	0.9717	0.962	169.49	0.9989	6.229	168.64	0.9850
50	196.25	3.017	87.30	0.9959	0.748	204.08	0.9990	8.164	202.04	0.985

(Initial metal ion concentration–10 to 50 mg L^{-1}; pH–5; contact time–120 min; temperature–303 K; adsorbent dose–0.2 g L^{-1}).

The slopes and intercepts of plots of t/q_t versus t were used to calculate the q_e and second-order rate constant k_2. The straight lines in plot of t/q_t versus t (Fig. 5) shows a good agreement of experimental data with thesecond-order kinetic model for differentinitial metal ionconcentrations. The correlation coefficients for this modelare greater than 0.995 for all the cases (Table. 3). The bestcorrelation between predicted and experimental q_e values indicates that the adsorption system studied belongs to second-order kinetic model. Further the values of k_2, how- ever, were found to decrease with increase of initial concentration [39]. Increasing the metal ion concentration in solution seems to reduce the diffusion of metal ions in the boundary layer and enhance the diffusion in the solid.

3.4. *Intra particle diffusion model*

The first- and second-order kinetic models cannot identify the diffusion mechanism and the kinetic results were then analyzed using intra particle diffusion model. The equation can be written as [40]

$$q_t = k_P\, t^{1/2} + C, \qquad (13)$$

where C is the intercept and k_p is the intra particle diffusion constant (mg g^{-1} $min^{1/2}$). According to this model, the plot of q_t versus $t^{1/2}$ should be linear if intra particu lar diffusion is involved in the adsorption process and if these lines pass through the origin the intra particle diffusion is the rate controlling step. In many cases, a linear portion follows an initial steep sloped portion indicating external mass transfer to the intra particle diffusion andplateau to the equilibrium.

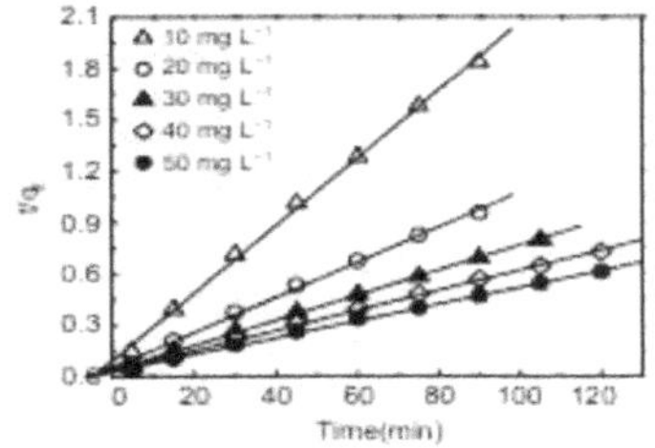

Fig. 5. Pseudo-second order kinetic plots for different initial concentrations (adsorbent dose = 0.2 g L^{-1}; contact time = 120 min; pH = 5; temperature = 303 K).

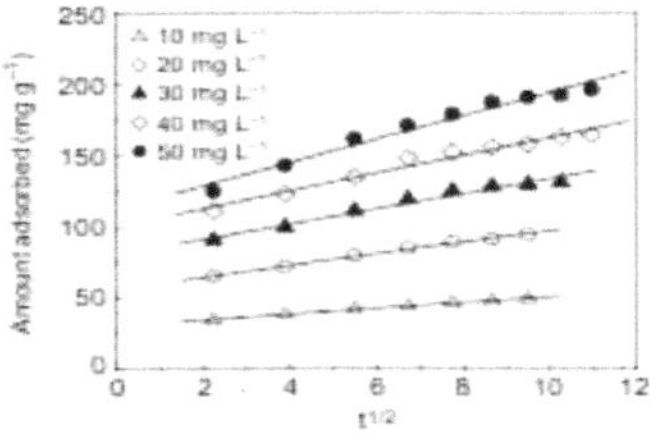

Fig. 6. Intra particle diffusion plots for different initial concentrations (adsorbent dose = 0.2 g L^{-1}; contact time = 120 min; pH = 5; temperature = 303 K).

This model was applied to the adsorption of Pb(II) on SDC as a function of different initial concentrationsof metal ions and the variation of q versus $t^{1/2}$ was givenin Fig.

6. The linear portions of the curves did not pass through the origin indicating that themechanism of Pb(II) removal on SDC was by some degree of boundary layer controlfurther implying that the intra particle dif-fusion was not the only rate limiting step but also other kinetic models may control the rate of adsorption, all of which may be operating simultaneously [41,42]. The intra particle rate constant values

calculated from the slopes of the linear portions of the plots of q versus t $^{1/2}$ are given in Table3.

3.5. *Effect of temperature on adsorption rate*

It was observed that Pb(II) biosorption followed pseudo-second-order kinetics at different temperatures. Linear plots of t/q_t versus t corresponding to the pseudo-second- order kinetic model were obtained (Fig. 7). Thevalues of kinetic parameters obtainedare presented in Table 4. The rise in temperature increased the values of q_e and k_2. In thepresent investigation it was observed that the adsorption of Pb(II) increase with increase in temperature. The adsorption capacity of the metal is found to increasefrom196 mg g^{-1} to 235 mg g^{-1} and the rate of the reaction was almost doubled for every 10°C rise in temperature. The enhancement of adsorption capacity at higher temperature may be attributed to either increasein the number of active sites available for adsorption on the adsorbent or due to decrease in the boundary layer thickness surrounding the sorbent, so that the mass transfer resistance of adsorbate in the boundary layer decreased [43].

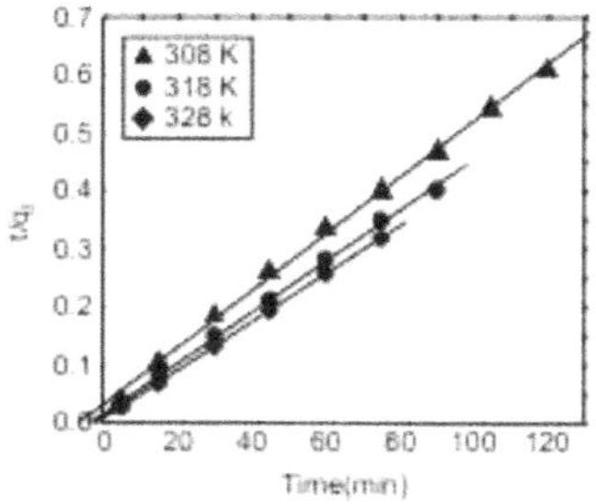

Fig. 7. Pseudo-second order kinetic plots for different temperatures (adsorbent dose = 0.2 g L^{-1}; Pb(II) concentration = 50 mg L^{-1}; pH = 5).

Adsorption dynamics

The thermodynamic equilibrium constant K_C for theadsorption was determined byusing the equation $K_C = C_A/C_e$, where C_A is the amount of Pb(II) adsorbed on the adsorbent of the solution at equilibrium (mol L^{-1}) and C_e is the equilibrium concentration of Pb(II) in the solution(mol L^{-1}). The change in standard free energy,enthalpy and entropy of adsorption were calculated using the following equation:

$$\Delta G^\circ = -RT \ln K_c$$

where R is the gas constant and T the temperature (K). 0.7 0.6 0.5 0.4 0.3 0.2 0.1 0.0

0 20, 40, 60, 80, 100, 120 Time(min)

According to Vant Hoff's equation:

$\ln K_c = \Delta S^\circ/R - \Delta H^\circ/RT$ (15)

where ΔS° and ΔH° , the changes in entropy and enthalpy of adsorption respectively.A plot of ln Kc versus 1/T is linear (Fig. 8). Values of ΔH° and ΔS° were evaluated fromthe slope and intercept of Vant Hoff's plot (Table 5). The positive value of ΔH° confirmsthe endothermic nature of adsorption [44]. When the temperature was increased from 30°C to 50°C, the magnitude of free energy change shifted to a high negative value suggesting that the adsorption was rapid Fig. 7. Pseudo-second order kinetic plots for different temperatures (adsorbent dose = 0.2 g L−1 ; Pb(II) concentration = 50 mg L −1 ;pH = 5). and spontaneous [45]. The positive value of ΔS° suggested an increase in randomness at the solid–solution interface during the adsorption of Pb(II) on SDC [46].

Table 4

Comparison of the pseudo-first-order and pseudo-second-order adsorption rate constants for different temperatures.

Temperature K	Exp q_e (mg g^{-1})	Pseudo-first order			Pseudo-second order		
		$K_1 \times 10^{-2}$ (min^{-1})	Cal q_e (mg g^{-1})	R^2	$K_2 \times 10^{-3}$ g mg^{-1} min^{-1}	Cal q_e mg g^{-1}	R^2
303	196.25	3.017	87.30	0.9959	0.748	204.08	0.9990
313	223.77	2.764	59.16	0.9582	1.291	227.27	0.9993
323	234.99	5.804	57.28	0.9675	2.248	240.39	1.000

(Initial metal ion concentration–50 mg L^{-1}; pH–5; contact time–120 min; adsorbent dose 0.2 g L^{-1}).

Table 5

Thermodynamic parameters and Tempkin constant for the adsorption of Pb(II) onto SDC.

Temperature K	$-\Delta G$ kJ mol^{-1}	ΔH kJ mol^{-1}	ΔS J mol^{-1}	Tempkin isotherm constant	
				B	K_t (L g^{-1})
303	3.262			-82.253	121.049
313	5.579	60.484	210.702	-51.096	464.331
323	7.384			-38.291	1575.792

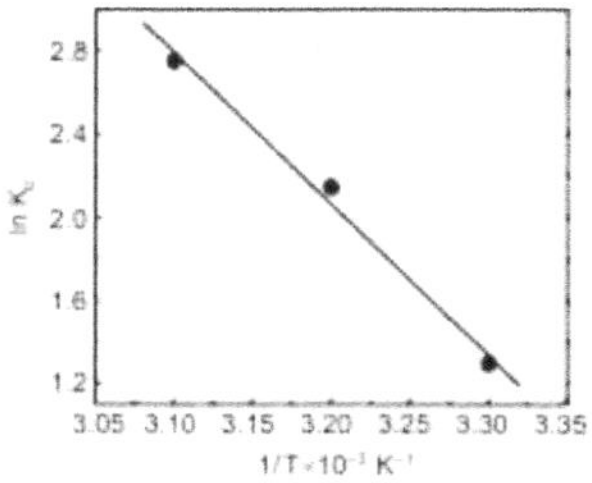

Fig. 8. Vant Hoff plot for the calculation of thermodynamic parameters for adsorption of Pb(II) onto SDC.

Conclusion

The present study clearly established that the saw- dust of *Bombax ceiba* is an effective adsorbent for Pb(II) removal from aqueous solutions. The adsorption of Pb(II) ions was heavily dependent on the amount of adsorbent, agitation time and initial pH of thesolution. Both Freundlich and Langmuir isotherms modeled the adsorptionequilibrium. The mean free energy of adsorption E was determined in the D-R equation. The adsorption capacity of the sawdust is high 209 mg g^{-1} compared to other reported in the literature. The kinetics of sorption followed pseudo-second-order mechanism and also intraparticle diffusion model, where diffusion is not the onlyrate-controlling step. Thethermodynamic calculations indicated the feasibility, endothermic and spontaneous nature of the adsorption of Pb(II) onto SDC from 303 K to 323 K. Kinetic and thermodynamic data obtained in the present study are useful for the fabrication and designing of wastewater treatment. It can be concluded that thehigh adsorption capacity and the endothermic nature of*Bombax ceiba* sawdust appears to be a promising adsor-bent for the removal of Pb(II) ions from aqueous solution and it is an inexpensive material for treating the industrialwastewater. At these adsorption levels a process usingsawdust

for the removal and recovery of a heavy metal is more economical than current processtechnology.

Acknowledgement

The authors would like to thank Secretary, Principaland the Management of N.G.M.College, Pollachi, Indiafor providing all the facilities to carry out this researchwork.

References

[1] P. Shekinah, K. Kadirvelu, P. Kanmani, P. Senthilkumar and

[2] V. Subburam, Adsorption of Lead from aqueous solution by activated carbonprepared from *Eichhornia*, J. Chem. Technol. Biot., 77 (2002) 458–464.

[3] T.K. Naiya, A.K. Bhattacharya, S. Mandal and S.K. Das, The sorption of Lead (II)ions on rice husk ash, J. Hazard. Mater., 163 (2009) 1254–1264.

[4] V. Singh, S. Tiwari, A.K. Sharma and R. Sanghi, Removal of lead from aqueous solutions using *Cassia grandis seed* gum- graft-poly (methymethacrylate), J. Colloid. Interf. Sci., 316 (2007) 224–232.

[5] R. Donat, A. Akdogan, E. Erdem and H. Cetisli, Thermody- namics of

Pb^2 and Ni^2 adsorption onto natural bentonite from aqueous solutions, J. Colloid. Interface. Sci., 286 (2005) 43–52.

[6] S.Q. Memon, N. Memon, S.W. Shah, M.Y. Khuhawar and

[7] M.I. Bhanger, Sawdust—A green and economical sorbent for the removal of Cd(II)ions, J. Hazard. Mater., 139 (2007) 116–121.

[8] M. Sciban, M. Klasnja and B. Skrbic, Modified softwood saw- dust as adsorbent ofheavy metal ions from water, J. Hazard.

[9] Mater., 136 (2006) 266–271

[10] K. Selvi, S. Pattabhi and K. Kadirvelu, Removal of Cr(VI) from aqueous solution by adsorption onto activated carbon, Biore- source Technol., 80 (2001) 87–89

[11] C. Raji and T.S. Anirudhan, Use of chemically modified saw- dust in the removal of Pb(II) ions from aqueous media, Indian J. Environ. Health, 39 (1997) 230–238.

[12] L.J. Yu, S.S. Shukla, K.L. Dorris, A. Shukla and J.L. Margrave, Adsorption of Chromium from aqueous solutions by maple saw dust, J. Hazard. Mater., 100 (2003) 53–63.

[13] M.E. Argun, S. Dursun, C. Ozdemir and M. Karatas, Heavy metal adsorptionby modified oak sawdust: Thermodynamics and Kinetics, J. Hazard. Mater., 141 (2007) 77–85.

[14] T. Karthikeyan, S. Rajagopal and L.R. Miranda, 2005. Chro- mium (VI) adsorption from aqueous solution by *Hevea Brasi- linesis* sawdust activated carbon,J. Hazard. Mater., B124 (2005) 192–199.

[15] B. Yasemin and T. Zeki, Removal of heavy metals from aque- ous solution by sawdust adsorption, J. Environ. Sci., 19 (2007) 160–166.

[16] L.G. Gutierrez and J.C. Moreno- Pirajan, Pb(II) and Cr(VI) adsorption from aqueous solution on activated carbons obtained from sugar cane husk and sawdust,J. Anal. Appl. Pyrol., 81 (2008) 278–284.

[17] S. Gupta and B.V. Babu, Removal of toxic metal Cr(IV) from

aqueous solutions using sawdust as adsorbent: Equilibrium, Kinetics and regeneration studies, Chem. Eng. J., 150 (2009) 352–365.

[18] A.K. Bhattacharya, T.K. Naiya, S.W.Mandal and S.K. Das, Adsorpion, Kinetics and equilibrium studies on removal of Cr(VI) from aqueous solutions usingdifferent low-cost adsor- bents, Chem. Eng. J., 137 (2008) 529–541

[19] K. Kadirvelu, Preparation and characterization of coirpith carbon and its utilization treatment of metal bearing waste, Ph.D., Thesis, Bharathiar University,
Coimbatore. Tamil Nadu, India. 1998, 52–57

[20] V.C. Srivastava, I.D. Mall and I.M. Mishra, Characterization of mesoporous rice husk ash (RHA) and adsorption kinetics of metal ions from aqueous solution onto RHA, J. Hazard. Mater., 134 (2006) 257–267.

[21] M.M. Rao, G.P. Chandra Rao, K. Seshaiah, N.V. Choudary and

[22] M.C. Wang, Activated carbon from *Ceiba pentandra hulls,* an agriculturalwaste, as an adsorbent in the removal of lead and zinc from aqueous solutions, Waste Manage., 28 (2008) 849–858

[23] APHA, Standard methods for the examination of water and waste water, 18th Edn, Washington DC American Public Health Association, 1992.

[24] O.D. Uluozlu, A. Sari, M. Tuzen and M. Soylak, Biosorption of Pb(II) and Cr
(II) from aqueous solution by lichen (*Parmelina tillaceae*) biomass, Bioresource Technol., 99 (2008) 2972–2980.

[25] B. Yu, Y. Zhang, A. Shukla, S.S. Shukla, K..L. Doris, The removal of heavy metals from aqueous solution by sawdust adsorp- tion –Removal of lead and comparison of its adsorption with copper, J. Hazard. Mater., B84 (2001) 83–94.

[26] R. Ayyappan, A. Carmalin Sophia, K. Swaminathan and

[27] S. Sandhya, Removal of Pb(II) from aqueous solution using carbon derived from agricultural wastes, Process Biochem., 40 (2005) 1293–1299.

[28] K. Jayaram, I.Y.L.N. Murthy, H. Lalhruaitluanga and M.N.V. Prasad, Biosorption of lead from aqueous solutions by seed powder of *Strychnos potatorum*,L. Colloid Surf., B 71 (2009) 248–254.

[29] K. Li and X. Wang, Adsorptive removal of Pb(II) by activated carbon preparedfrom *Spartina alterniflora*: Equilibrium, kinet- ics and thermodynamics, BioresourceTechnol., 100 (2009) 2810–2815.

[30] A. Shukla, Y.H. Zhang, P. Dubey, J.L. Margrave and S.S. Shukla, The role of sawdust in the removal of unwanted materials fromwater, J. Hazard. Mater., 95 (2002)137–152.

[31] K.P. Shubha, C. Raji and T.S. Anirudhan, Immobilization of heavy metals fromaqueous solutions using polyacrylamide grafted hydrous Tin (IV) oxide gel having carboxylate func- tional groups, Water Res., 35 (2001) 300–310.

[32] K. Jayaram and M.N.V. Prasad, Removal of Pb(II) from aque- ous solution byseed powder of *Prosopis julifloraDC*, J. Hazard. Mater., 169 (2009) 991–997.

[33] V.C. Taty-Costodes, H. Fauduet, C. Porte and A. Delacroix, Removal of Cd(II)and Pb(II) ions, from aqueous solutions, by adsorption onto Sawdust of PinusSylvestris, J. Hazard. Mater., 105 (2003) 121–142.

[34] Q. Li, J. Zhai, W. Zhang, M. Wang and J. Zhou, Kinetic studies of adsorptionof Pb(II), Cr(III) and Cu(II) from aqueous solu- tion by sawdust and modified peanuthusk, J. Hazard. Mater., 141 (2007) 163–167.

[35] M. Rafatullah, O. Sulaiman, R. Hashim, and A. Ahmad, Adsorp- tion of copper(II), Chromium (III), nickel(II) and lead(II) ions

[36] from aqueous solutions by meranti sawdust, J. Hazard. Mater.,

170 (2009) 969–977.

[37] V.K. Gupta and Imran Ali, Removal of lead and chromium from wastewater using bagasse fly ash–A sugar industry waste, J. Colloid. Interf. Sci., 271 (2004) 321–328

[38] A. Sari, M. Tuzen, O.D. Uluozlu and M. Soylak, Biosorption of Pb(II) and Ni

(II) from aqueous solution by *lichen* (Caledonia furcata) biomass, Biochem. Eng. J.,37 (2007) 151–158.

[39] A. Sari and M. Tuzen, Biosorption of Pb(II) and Cd(II) from aqueous solutionusing green alga (Ulva lactuca) biomass, J. Hazard. Mater., 152 (2008) 302–30.

[40] S. Tunali, T. Akar, S. Ozcan, I. Kiran and A. Ozcan, Equilibriumand kineticsof biosorption of lead (II) from aqueous solutions by *Cephalosporium aphidicola*, Sep. Purif. Technol., 147 (2006) 105–112.

[41] J.U. Kennedy, Oubagaranadin, N. Sathyamurthy and Z.V.P. Murthy, Evalution of Fuller's earth for the adsorption of mer- cury from aqueous solutions: A comparative study with acti- vated carbon, J. Hazard. Mater., 142 (2007) 165–174.

[42] A. Seker, T. Shahwan, A.E. Eroglu, S. Yilmaz, Z. Demirel and

[43] M.C. Dalay, Equilibrium, thermodynamic and kinetic studies for the biosorption of aqueous lead(II), Cd(II) and Ni(II) ions on Spirulina platensis,J. Hazard. Mater., 154 (2008) 973–980.

[44] N. Chiron, Richard Guilet and Eric Deydier, Adsorption of Cu(II) and Pb(II) onto a grafted silica: Isotherms and kinetic models, Water Res., 37 (2003) 3079– 3086.

[45] S.H. Hasan, K.K. Singh, O. Prakash, M. Talat and Y.S. Ho, Removal of

Cr(VI) from aqueous solutions using agricultural waste maize bran, J. Hazard. Mater., 152 (2008) 356–365.

[46] B.M.W.P.K. Amarasinghe and R.A. Williams, Tea waste as a low cost adsorbent for the removal of Cu and Pb from waste water, Chem. Eng. J., 132 (2007)299–309

[47] F. Kaczala, M. Margues and W. Hogland, Lead and Vanadium removal fromindustrial wastewater by gravitational setting/ sedimentation and sorption onto PinusSylvestris Sawdust, Bioresource Technol., 100 (2008) 235–243.

[48] A. Ozer, G. Akkaya and M. Turabik, Biosorption of Acid red 274 on *Enteromorpha prolifera* in a batch system, J. Hazard. Mater., 126 (2005) 119–127.

Adsorption studies of iron(III) on chitin

G KARTHIKEYAN*, N MUTHULAKSHMI ANDAL and K ANBALAGAN

Department of Chemistry, Gandhigram Rural Institute – Deemed University,Gandhigram 624 302, India e-mail: drg_karthikeyan@rediffmail.com

MS received 18 December 2003; revised 4 June 2005

Abstract. Adsorption of ferric ions by chitin was studied by the batch equilibration method. The influence of particle size and dosage of the adsorbant, contact time, initial concentration of the adsorbate and temperature were experimentally verified.The effect of anions like chloride, nitrate and sulphate and also of cations like zinc, chromium and copper on the adsorption of iron(III) was determined. The time depend- ence of fraction of adsorption, Y_t, at varying particle sizes and doses of chitin and the intraparticle diffusion rate constants, k_p, of the adsorption process were calculated. Thermodynamic and equilibrium parameters of the reaction were determined to understand the sorption behaviour of chitin. The results revealed thatthe adsorption of iron(III) by chitin is spontaneous, endothermic and favourable.

Keywords. Adsorption; chitin; variable parameters; fraction of adsorption; temperature effect.

1. Introduction

Iron is one of the major constituents of the litho- sphere and comprises approximately 5%of it. It is routinely detected in municipal waste effluent, particularly in cities where ironand steel are manufactured. Iron readily complexes with sulphates in the sediments of many surface waters. The primary concernabout the presence of iron in drinking water isits objectionable taste. The taste of iron in drinking water canbe easily detected even at low concentrations of about 1–8 mg/l.[1] There are many problems that result from iron toxicity. These include anorexia, oligura, diarrhoea, hypothermia, diphasic shock, metabolic acidosis and even death. In addition to these, the patient may experience vascular congestion of the gastrointestinal tract, liver, kidneys, heart, brain, spleen, adrenals and thymus. With acute iron poisoning, much of the damage to the gastrointestinal tract and liver may result from the high localized iron concentration and free radical production leading to heptatoxicity vialipid peroxidation and destruction of the hepatic mitochondria. As a result of iron storage disease, the liver becomes cirrhotic. Hepatoma, a primary cancerof the liver, has become the most common cause of death among patients with hemochromatosis.[2] This is an iron storage disease that results from the inabilityof the intestine to keep out unwanted iron. Instead, this iron accumulates in the liver causing siderosis[3] (accumulation of storage iron in tissues) and causes damage to the storage organs. Also, when siderosis becomes severe in young people, it leads to myocardialdisease which is a common cause of death. Impotence mayalso occur in young men and amenorrhea in youngwomen. Both these problems relatingto reproduction are due to iron loading in the anterior pituitary.

This study deals with the removal of ferric ionsfrom aqueous systems using chitin as an adsorbent. Chitin is a polysaccharide and is a major componentof arthropod and crustacean shells, like those of lobsters, shrimps or crabs. Chitin is a non-toxic and biodegradable polymer of N-acetylglucosamine and glucosamine residues.[4] The constituent monosaccharideunits in chitin are linked together by *b* - 1,4-glycosi- dic bonds and hence have a relatively rigid configuration. It is easily filmable and spinnableand the materials thus processed have very good mechanical properties. It is insoluble inwater and alkali but solublein strong mineral acids and in anhydrous formic acid. It forms complexes with transition metal ions. A systematic study of the adsorption of ferricion by chitin under different experimental conditions has beencarried out to understand the nature of the reaction.

2. Experimental

Commercially available chitin (Otto Kemi, C1904)was obtained in the form of flakes from M/s BR corporation, Mumbai. The raw sample was pulverized using mortar and pestle and sieved into three sizes, viz., 0·21, 0·50 and 0·71 mm, using scientific test molecular sieves manufactured by Elite Scientific Instruments Co., Mumbai. Chitin samples of definite particle sizes were further purified by first boiling in 0·1 N HCl for three hours. The solution was decanted and the residue was treated further by boiling in 0·1 N NaOH for another three hours. The purified samples were washed several times and were soaked in double distilled water till neutral pH. Soaking of chitin resulted in

swelling which results in more reaction sites in the adsorbent material. Later the samples were dried in an oven. Only the chemically treated samples were used for the adsorptionexperiments. All experimental solutions were prepared using only double distilled water.The adsorption experiments were carried out by batch equilibration method. All the chemicals employed for the analysis were AnalaR grade. A stock solution of the adsorbatecontaining 1000 mg/l of iron(III) ions was prepared using ammonium ferric sulphate [NH4Fe(SO4)2.12H2O]. From the stock solution, aliquots containing different concentrations viz., 2, 4, 6, 8, 10, 12 and 14 mg/l of iron(III) ions were measured out into separate iodine flasks. A sample adsorbate of each concentration (50 ml) was mixed with10 mg of chitin and the mixtures were agitated in a mechanical shaker. Adsorption experiments with chitin samples of particle sizes 0·21, 0·50 and 0·71 mm were separatelycarried out. The solutions were filtered at regular intervals of 2 minutes using Whatman No. 41 filter paper and the concentration of iron(III) ions in the filtrate at each stage wasdetermined using an atomic absorption spectrometer (Model AA 100, Perkin-Elmer). Adsorption experiments were conducted at pH 3. This was because of the formation of ferric hydroxide at higher pH range with sodium hydroxide which was added to adjust the pH of the acidic ferric solution. The experiments were repeated at four different temperatures viz., 293, 303, 313 and 323 K to determine the effect of temperature on thesorptive behaviour of chitin. All experiments were done in a thermostat with a shaker attached (Techno) which maintained temperature within ±0·1°C. Adsorption studies of iron(III) ions by chitin were repeated in the presence of copper, zinc and chromium ionsto examine the influence of co-ions on the sorptive nature of the adsorbent. The concentrations of residual iron(III) ions from the mixture containing equal amounts of copper, zinc and chromium ions were determined. Duplicate measurements were made such that the residual concentration values were reproducible within ±2%. 3. Results and discussion 3.1 Effect of particle size Preliminary adsorption studies of metal ions by chitinindicated that sorption capacity is dependent on its particle size. Figure 1 represents the uptake of iron(III) by chitin of three different particle sizes, viz., 0·21, 0·50 and 0·71 mm.The sample with particle size 0·21 mm exhibits maximum adsorption efficiency and thisis attributed to the larger surface area of the adsorbent material. Diffusional resistance tomass transport in the case of adsorbent with larger particle sizes is higher and most of the

internal surface of these particles may not be utilized for adsorption. Consequently, the amount of metal ions adsorbed is less in such cases.5 Since the extent of adsorption is considerably higher in the case of chitin samples of 0·21 mm particle sizes, it was decidedto limit the discussions of the adsorption experiments to that with chitin of 0·21 mm particle size. 3.2 Effect of contact time Contact time required for the maximum removal of iron(III) by chitin is represented by figure 2.

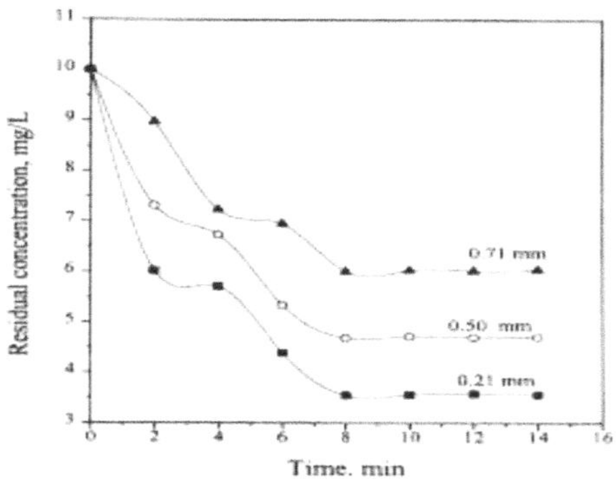

Figure 1. Effect of particle size.

There is a steep increase in the adsorption in the initial stages of the reaction as seen in the curve.6 Maximum adsorption occurs at the 8th minute after which the adsorption remains uniform, represented by the attainment of equilibrium. The amount of iron(III) adsorbed at this point is the maximum under the particular operating conditions. The time variation curveis smooth and continuous and this indicates the formation of a monolayer coverage on the outer interface of the adsorbent.7–9

3.3 Effect of initial concentration

The adsorption curves of iron(III) from solutions containing different initial concentrationsare shown in figure 3. The dependence of the system on the concentration factor is explainedby low iron(III) to chitin ratio at lower concentrations. Subsequently, the fractional adsorption becomes independent of the initial concentration.10 However, at higher concentrations, the available sites of adsorption become fewer and hence the percentage removal of iron(III) is dependent upon the initial concentration. It is evident from the figurethat the adsorption rate increased with an increase in the concentration. The actual amount of iron(III) adsorbed is more for higher equilibrium concentration value, Ce at increasing concentrations of the adsorbate.11 Similar findings were reported by Rao et al12 while working on Blue MR using activated carbon.

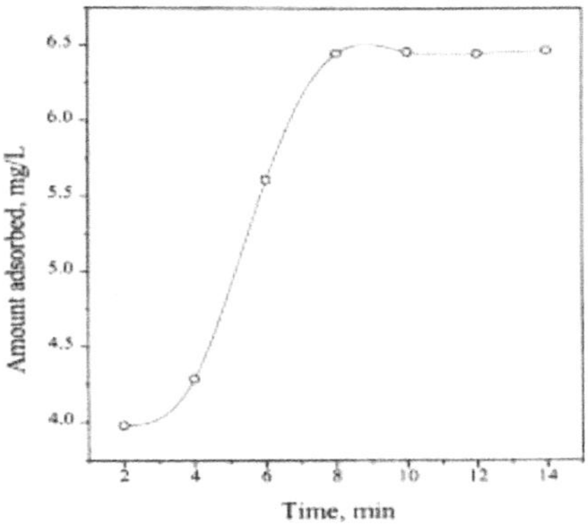

Figure 2. Effect of contact time.

In the present study the initial concentration of the iron(III) solutions was maintained at 10 mg/l for all the adsorption experiments. 3.4 Effect of dosage Results of the adsorptionexperiments using three different dosages, viz. 10, 20 and 30 mg of chitin and 10 mg/l ofiron(III) are given in table 2. The values indicate an increase in the adsorption capacity ofchitin with dosage. This is because of an increase in the number of available adsorption sites of the adsorbent which results in enhanced removal of iron(III).13 Rate of adsorptionof iron(III) by chitin is appreciable even at an initial dose of 10 mg although at higher doses like 20 and 30 mg it records an increase. Further, the rate of adsorption does not register a proportionate increase as per the experimental results. In view of this observation, it was decided to conveniently fix the dose of chitin as 10 mg for the remaining experiments. Since the particle size and contact time of the adsorbent material,determined as above, accounts for the maximum removal of adsorbate, this was taken asthe optimum condition for further adsorption studies using chitin

Table 2. Effect of dosage.

	Amount adsorbed (mg/l) at concentration (in mg)		
Time (min)	10	20	30
2	3.98	5.69	7.01
4	4.29	6.38	7.87
6	5.61	7.41	8.41
8	6.45	7.85	8.73
10	6.46	7.82	8.71
12	6.45	7.84	8.70
14	6.47	7.83	8.72

Initial concentration of iron(III) = 10 mg/l; pH = 3.0; temperature = 303 K; particle size = 0.21 mm

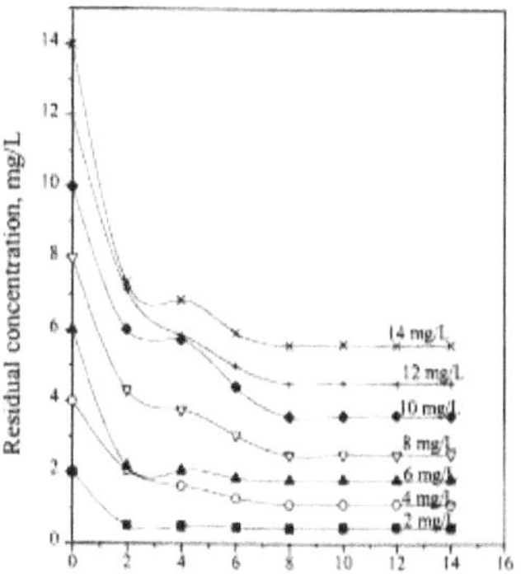

Figure 3. Effect of initial concentration.

Table 1. Effect of contact time.

Time (min)	Amount adsorbed (mg/l)
2	3.98
4	4.29
6	5.61
8	6.45
10	6.46
12	6.45
14	6.47

Initial concentration of iron(III) = 10 mg/l; pH = 3.0; temperature = 303 K; adsorbant dose = 10 mg; particle size = 0.21 mm

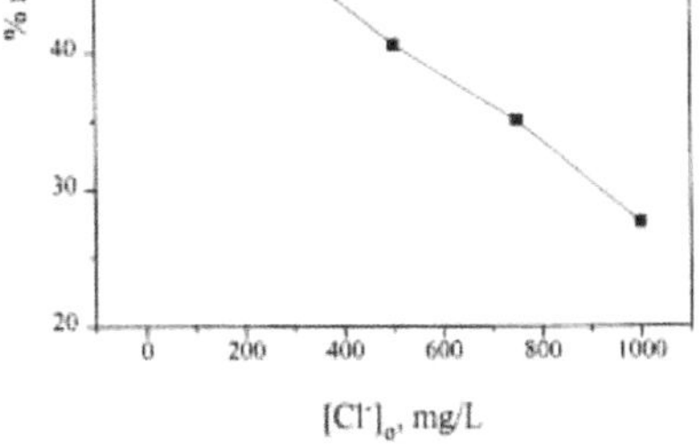

Figure 4. Effect of chloride ion concentration.

3.5 Effect of chloride, nitrate and sulphate

Results of the adsorption experiments showing the effects of chloride, nitrate and sulphate ions on the amount of Fe(III) ions removed by chitin are provided in table **3.** The influence of chloride ions on the adsorption of iron(III) by chitin is illustrated in figure 4. The percentage of iron(III) removed decreases from 64·5% to 27·65% as thechloride ion concentration increases from 0–1000 mg/l. Formation of soluble chloro complexes at higher concentrations of chloride is responsible for this observedinhibition rate.14 This also results in lowering of free iron(III). The observed decreasein the adsorption rate indicates that these chloro complexes have lesser affinity than free metal ions towards chitin. Results indicate simi lar reduction in the chelating ability of chitin with iron(III) at higher nitrate environments also. This is due to the propensity of iron(III) to complex preferentially with nitrate rather than gettingadsorbed onto the chitin surface.15 Adsorption studies of iron(III) by chitin in the presence of excess sulphate ion sindicate a two-fold decrease in the percentage of iron(III) removed. The results reveal that 42·39% of iron(III) is reduced in 500 mg/l of the sulphate environment whereas the amount reduced is 21·43% at a sulphate concentration of 1000 mg/L

3.6 Effect of ionic strength

Figure 5 shows the influence of ionic strength on the adsorption of iron(III) by chitin.Perchlorate solutions ranging from 10–1000 mg/l were prepared and mixed with ferric solution of 10 mg/l concentration. The absence of any significant decrease in the uptake of iron(III) in the presence of sodium perchlolrate up to 750 mg/l is evident from the absence of any curve as indicated in the figure. However, at 1000 mg/l of perchlorate concentration, considerable reduction in the amounts of iron(III) adsorbedis evident, as implied by the bend in the curve. In other words, the increased amount of perchlorate (1000 mg/l) helps to swamp the surface of chitin, which decreases the access of Fe(III) ions to the chitin surface for adsorption

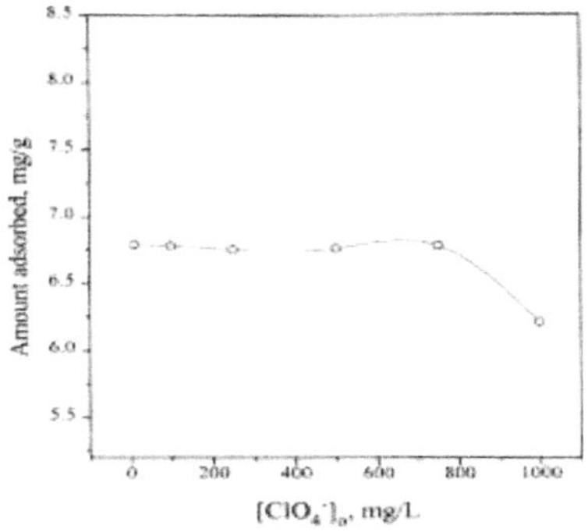

Figure 5. Effect of ionic strength.

Table 3. Effect of chloride, nitrate and sulphate.

$[Cl^-]_0$	% removal	$[NO_3^-]_0$	% removal	$[SO_4^{2-}]_0$	% removal
0	64.50	0	64.50	0	64.50
10	58.09	10	46.63	10	58.80
100	51.46	100	41.52	100	51.22
250	47.78	250	38.56	250	45.33
500	40.56	500	36.42	500	42.39
750	35.03	750	35.94	750	35.59
1000	26.65	1000	32.48	1000	21.43

Initial concentration of iron(III) = 10 mg/l; pH = 3.0; temperature = 303 K; adsorbant dose = 10 mg; contact time = 8 min; particle size = 0.21 mm

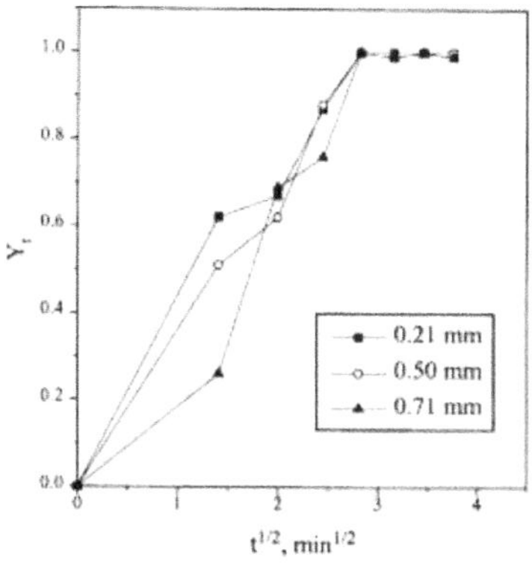

Figure 6. Time dependence of the fraction of adsorption at different particle sizes of chitin.

Adsorption is sensitive to changes in the concentration of the supporting electrolytes, if electrostatic attraction is the significant mechanism for sorbate removal. It is therefore concluded that electrostatic attraction plays a negligible role in the removal of iron(III) by chitin.16 3.7 Time-dependence of fraction of adsorption at different particle sizes and dosesof chitin Mechanism of sorption of iron(III) by chitin is explained by the fraction of adsorption Yt. The plots of Yt (figures 6 and 7) for different particle sizes and dosages have three-stage shapes indicating external surface adsorption, intraparticle diffusion and final equilibrium. The values given in tables 4 and 5 show enhanced rates of sorption at increaseddosage levels and reduced particle sizes. The first, sharp portion of the curve corresponds tothe external surface adsorption stage or instantaneous adsorption stage. The second, less- sloping linear portion indicates the gradual adsorption stage and the final portion representsthe equilibrium stage. In this model, the chitin sample is treated as being surrounded by a boundary layer film through which the metal ions must diffuse prior to adsorption on the chitin surface. The second portion of the figures indicates the intraparticle diffusion, whichis predominant in the rate-controlling step where diffusive transport occurs through the internal pores of the adsorbent. As the bulk and surface metal concentrations start to decrease, the third section of the figures shows decrease in the rate of diffusion reach ing a point of equilibrium.17 Both increasing dose and decreasing particle size increase the driving force of metal ions from the bulk phase on and into the solid chitin. An examinationof the values of t 1/2 and Yt leads to the following facts. The fraction of adsorption in termsof retention time, t 1/2 shows that the first stage corresponds to a retention time of 1·41 min1/2 . The second stage corresponds to a retention time for t 1/2 between 1·41 min1/2 and2·45 min1/2 . Values of t 1/2 , above 2·45 min1/2 represent the equilibrium stage reached and this corresponds to the third stage.

3.7 Intraparticle diffusion

Plots of time squared and amount adsorbed for particle sizes and dosages of chitin as variables are shown in figures 8 and 9. The initial curved portions represent the boundary- layer diffusion effect. The intraparticle diffusion rate constants, kp, calculated from the slopes of the linear portion of curves which represent the intraparticle diffusion effects are given in tables 6 and 7.7,18 The kp values increase with decrease in size. Otherwise, increasing the particle size of the adsorbent requires more time to reach equilibrium. This results in the lowering of intraparticle diffusion rate constants at larger particle sizes and viseversa smaller particles sizes. Values of kp increase with the dosage of chitin, due to increasein the number of chitin molecules as more active binding sites are available for iron(III) to diffuse into.

3.8 Effect of co-ions

Experimental results of the co-ionic effect on the adsorption of iron(III) ions from an aqueous solution containing copper(II), zinc(II) and chromium(VI) in equal concentrationsreveal that the presence of foreign ions diminishes the adsorption of iron(III) ions. However,the presence of an equal amount of copper(II) ions reduces the adsorption rate considerably,

Table 4. Time dependence of the fraction of adsorption at different particle sizes of chitin.

$t^{1/2}$ (min$^{1/2}$)	Y_t (mm) 0.21	0.50	0.71
0	0	0	0
1.41	0.62	0.51	0.26
2.00	0.67	0.62	0.69
2.45	0.87	0.88	0.76
2.83	1.00	1.00	1.00
3.46	1.00	0.99	0.99
3.46	1.00	1.00	1.00
3.76	1.00	1.00	0.99

Table 5. Time dependence of the fraction of adsorption at different doses of chitin.

$t^{1/2}$ (min$^{1/2}$)	Y_t (mg) 10	20	30
0	0	0	0
1.41	0.62	0.72	0.81
2.00	0.67	0.81	0.90
2.45	0.87	0.91	0.93
2.83	1.00	1.00	1.00
3.46	1.00	1.00	1.00
3.46	1.00	1.00	1.00
3.76	1.00	1.00	1.00

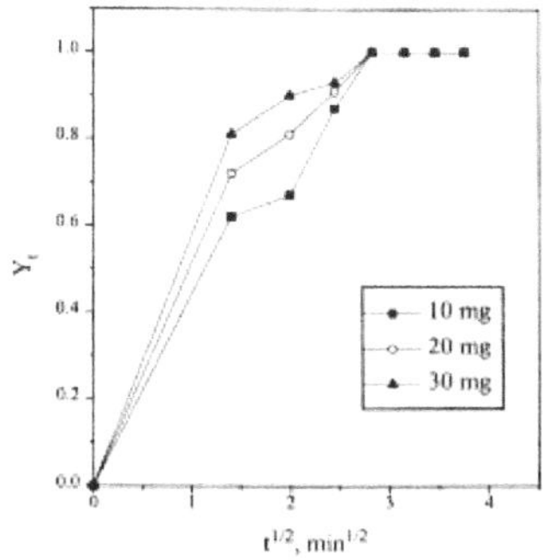

Figure 7. Time dependence of the fraction of adsorption at different doses of chitin.

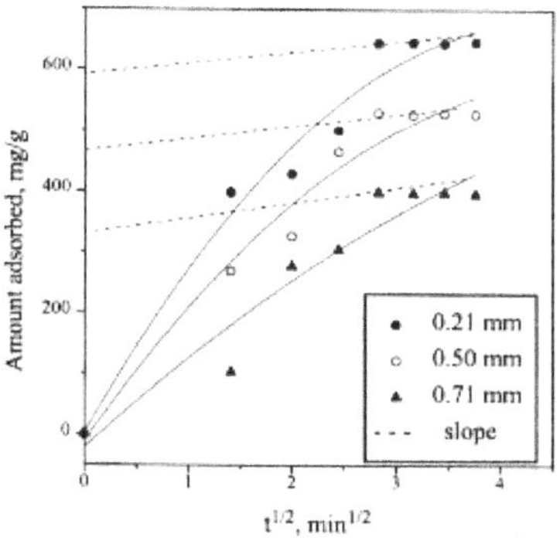

Figure 8. Intraparticle diffusion plot for the adsorption of Fe(III) on chitin at different particle sizes.

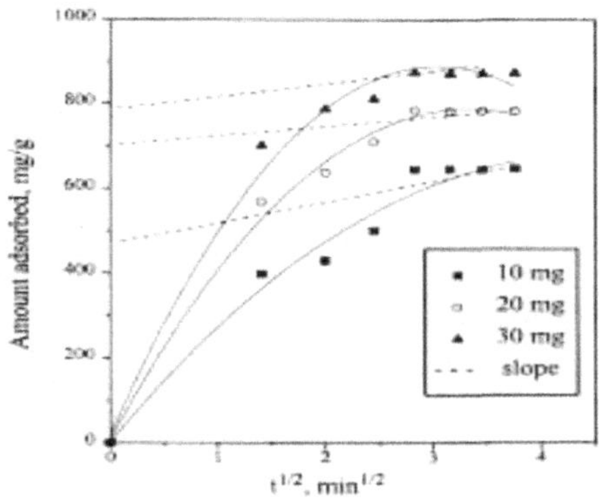

Figure 9. Intraparticle diffusion plot for the adsorption of Fe(III) on chitin at different doses.

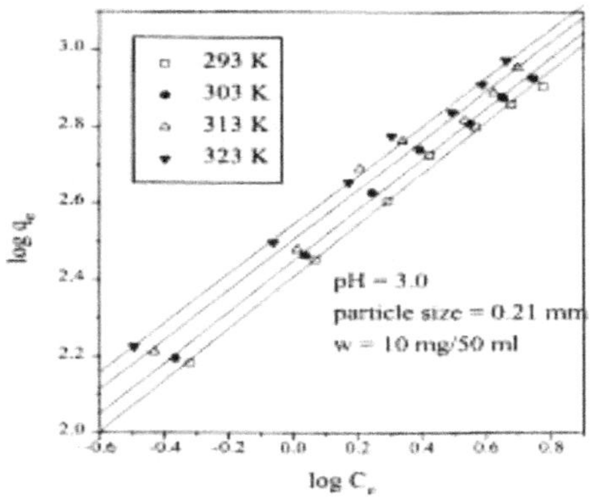

Figure 10. Freundlich isotherm.

while the presence of chromium(VI) and zinc(II), has a lesser effect. Jha *et al*[16] havereported similar co-ionic effects in the removal of specific metal ionsin their studies.

Table 6. Intraparticle diffusion rate constants depending on particle size

Particle size (mm)	0.21	0.60	0.71
k_p	2.09	1.67	1.61

Table 7. Intraparticle diffusion rate constants depending on dosage.

Dosage (mg)	10	20	30
k_p	2.09	2.61	3.23

Table 8. Freundlich parameters.

Temperature (K)	K	$1/n$
293	2.41	0.68
303	2.45	0.67
313	2.60	0.65
323	2.64	0.64

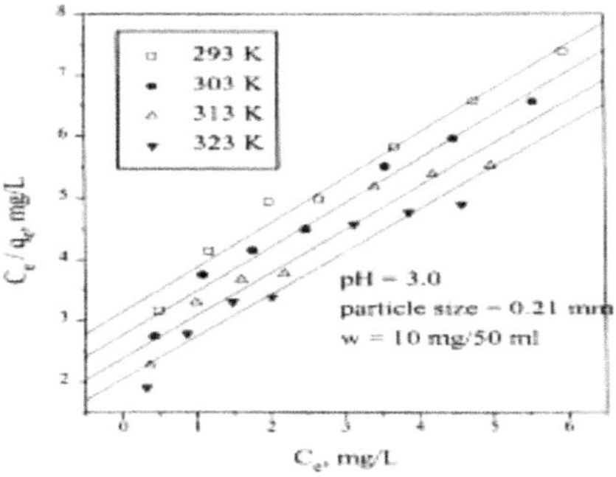

Figure 11. Langmuir isotherm.

Table 9. Effect of temperature on equilibrium parameters.

Concentration of Fe(III) (mg/l)	C_e (mg/l) at temp. (K)				C_e/q_e (mg/l) at temp. (K)			
	293	303	313	323	293	303	313	323
2	0.48	0.43	0.37	0.32	3.16	2.74	2.27	1.90
4	1.17	1.09	0.99	0.87	4.13	3.75	3.29	2.78
6	1.98	1.76	1.61	1.49	4.93	4.15	3.67	3.30
8	2.66	2.48	2.19	2.02	4.98	4.49	3.77	3.38
10	3.68	3.65	3.41	3.13	5.82	5.50	5.17	4.56
12	4.76	4.48	4.20	3.87	6.57	5.96	5.38	4.76
14	5.95	5.65	4.98	4.69	7.39	6.57	5.52	4.88

Table 10. Langmuir constants and thermodynamic parameters.

Temp. (K)	Langmuir constants		Thermodynamic parameters			
	b (l mg^{-1})	Q (mg g^{-1})	K_o	G° (kJ mol^{-1})	H° (kJ mol^{-1})	S° (J mol^{-1} K^{-1})
293	0.2311	1.3778	5.88 ±0.06	−4.32 ±0.03	2.16	22.06
303	0.2591	1.3982	6.02 ±0.06	−4.52 ±0.03		
313	0.2934	1.4323	6.17 ±0.11	−4.72 ±0.04		
323	0.3356	1.4537	6.36 ±0.10	−4.97 ±0.04		

3.9 Adsorption isotherms

Freundlich and Langmuir adsorption isotherm models were applied for adsorption equilibrium at all four temperatures. 3.10a Freundlich isotherm: Batch isothermal datafitted to the linear form of the Freundlich isotherm are shown in figure 10. The isothermal equation is as follows: log qe = log K + 1/n log Ce, where qe is the amountof copper sorbed per unit weight of the sorbent (mg/g) and Ce is the equilibrium concentration (mg/l) of Fe(III) in solution. The fit of the data imply that the intercept,K, is roughly an indicator of the sorption capacity and the slope, 1/n, of the sorption intensity, the constants incorporating all the factors affecting the adsorption process. Values of K and 1/n are calculated from the graph and the data are provided in table

8. The linearity of the plots ($0{\cdot}9911 < r < 0{\cdot}9979$ and $0{\cdot}0185 < sd < 0{\cdot}0378$) at differenttemperatures is evident from figure 10. This supports the applicability of the Freundlich adsorption isotherm indicating that the adsorption by chitin may begoverned by physisorption.19 The values of n for this system were calculated from the slope of the curve and are found to be between 1 and 2. Our findings are in good agreement with the findings of Treybal20 who mathematically evaluated values of n for a number of mass transfer operations of systems and reported that values of n between 1 and 10 would represent beneficial adsorption. 3.10 b Langmuir isotherm: The linear equation of the Langmuir adsorption isotherm was applied in the form, Ce/qe = 1/Qb + Ce/Q, where Q and b are the Langmuir constants, indicating adsorptioncapacity and energy of adsorption respectively. Ce and qe have their usual significance. The values for the plot Ce/qe versus Ce for the adsorption by chitin at different temperatures are calculated and presented in table 9. The linear plots ($0{\cdot}9702 < r < 0{\cdot}9893$ and $0{\cdot}2175 < sd < 0{\cdot}3212$) of Ce/qe versus Ce at different temperatures are given in figure 11 and they indicate the applicability of the Langmuir adsorption isotherm to the adsorption process. The linear form of the regression equations are given below.

Ce/qe = 3·14 ±0·17 + 0·73 ±0·05 Ce (at 293 K),Ce/qe = 2·76 ±0·16 + 0·72 ±0·05 Ce (at 303 K),

Ce/qe = 2·38 ±0·23 + 0·70 ±0·08 Ce (at 313 K)Ce/qe = 2·05 ±0·21 + 0·69 ±0·08 Ce (at 323 K).

Values of Q and b at different temperatures were determined from the above plots and are given in table 10. An increase in Ko value with temperature indicates the endothermic character of the adsorption. The results reveal that the energy of adsorption, b, for trapping of iron(III) by chitin increases with temperature and hence leads to greater of Fe(III) adsorption at higher temperatures.21 Annadurai et al. 22 who studied the sorptive character of chitin on basic dyes have also reported similar observations. The influence of temperature on chitin and its enhanced adsorption behaviour further emphasises to the phenomenon of swelling in the internal structureof the chitin

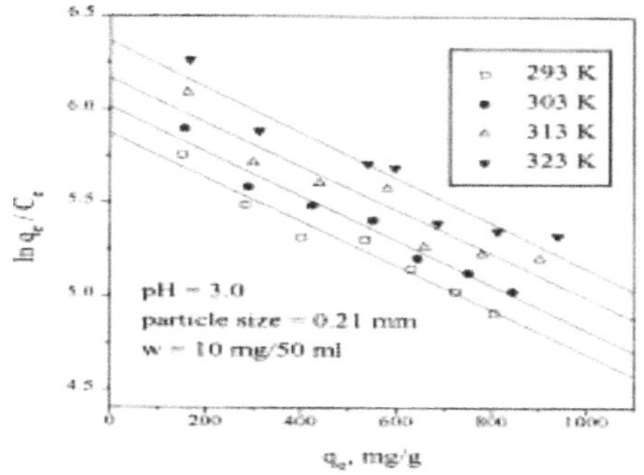

Figure 12. Plot of ln q_e/C_e versus q_e.

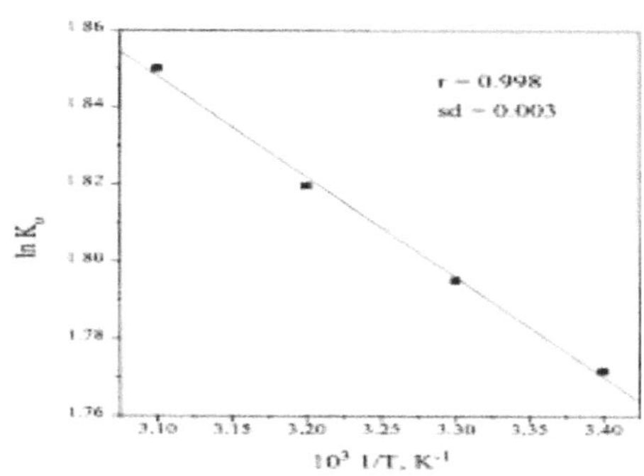

Figure 13. Plot of ln K_o versus 1/T.

molecules to accommodate more iron(III) on its surface

Equilibrium parameter

Values of the dimensionless equilibrium parameter, RL, of different temperatures arecalculated for the initial concentrations of iron(III) and are given in table 11. The parameters explain the differences in the shapes of the isotherm. The RL values lyingbetween 0 and 1 indicate favourable adsorption at the four temperatures.7 Theobserved RL values for iron(III) in the solution are 0·18–0·68 and correspond to the concentration range 2–14 mg/l, indicative of the favourable adsorption. The plots of ln qe/Ce versus qe of the reactions are given in figure 12. The equilibrium constant, Ko for the sorption reaction was determined from the plot ln qe/Ce versus qe and extrapolating to zero qe. An increase in Ko value with temperature indicates that the adsorption process of iron(III) by chitin is endothermic. The effect of increase in temperature on the adsorption of iron(III) at the chitin–solution inter face has been studied in the temperature range of 293– 313 K. A plot of ln Ko vs 1/T is found to be linear as shown in figure 13. ΔHo and ΔSo are determined from the slopes and intercepts of the plot23 . The negative values of the standard free energy change (ΔGo

) for this reaction suggests that the sorptive nature is spontaneous. The positive value of ΔHo (2·16 kJ/ mo1) indicates that the process of adsorption is endothermic and irreversible, probably due to nonpolar interactions.24 The positive value of entropy change ΔS o (22·06 J/K/mo1) suggests a high degree of disorderliness at the solid– solution interface during the adsorption process of iron(III) by chitin.25 It also reflectsthe affinity of the adsorbent for iron(III)

Table 11. Equilibrium parameter (R_L).

Concentration of Fe(III) (mg/l)	Temperature (K)			
	293	303	313	323
2	0·68	0·66	0·63	0·60
4	0·52	0·49	0·46	0·43
6	0·42	0·39	0·36	0·33
8	0·35	0·33	0·30	0·27
10	0·30	0·28	0·25	0·23
12	0·27	0·24	0·22	0·20
14	0·24	0·22	0·20	0·18

and suggests some structural changes in adsorbate and adsorbent. Enhancement of adsorption capacity of chitin at higher temperatures may be attributed to the enlargement of the pore size and/or activation of the adsorbent surface and increase in the mobility of the metal ions.

4. Conclusion

Based on the experimental results, it is observed that the adsorption of Fe^{3+} ions by chitinis maximum with the least particle size and minimum contact time of eight minutes. Sorption of Fe^{3+} ions increases at higher initial concentrations and doses. The influence ofpH on the adsorption of iron(III) by chitin could not be studied as its solution itself is acidic. The inhibitive effects of chloride, nitrate and sulphate and the absence of any significant influence of ionic strength over the sorption rate are observed. However, marked reduction in the sorption processis noticed under the influence of co-ions. Y_t plots depict the three stages in the sorption process; the calculated k_p values increase withdosages and de-crease with particle sizes. The Fe(III)–chitin system follows Freundlich and Langmuir isotherms which indicates the governance of physisorption and/or chemisorption. Values of thermodynamic functions show that the reaction is endothermic,random and beneficial.

References

1. Arms K 1990 *Environmental science* (San Francisco,CA: Saunders College)
2. Lauffer R 1992 *Iron and human diseases* (London: CRC Press)
3. Emery T 1991 *Iron and your health: Facts and falla- cies* (London: CRC Press)
4. Gonzalez-Davila M, Santana-Casiano M J and Mil- lero F J 1990 *J. Colloid. Inter.Sci.* **137** 102
5. Slejko F 1985 *Adsorption technology: A step by step approach to process calculationand application* (New York: Marcel Decker)
6. Singh D K and Bhavna Srivastava 2001 *Indian J. Chem. Technol.* **8** 133
7. Singh D K, Garg S K and Bharadwaj R K 2001 *In-dian J. Environ. Protection* **21 604**

8. Khattri S D and Singh M K 1999 *Indian J. Chem.Technol.* **6** 112

9. Desai M, Dogra A, Vora S, Bahadur P and Ram R N 1997 *Indian J. Chem.* **A36** 938

10. Parekh D C, Patel J B, Sudhakar P and Koshy V J2002 *Indian J. Chem. Technol.* **9** 540

11. De D S and Basu J K 1999 *Indian J. Environ. Protec- tion* **19** 416

12. Rao L N, Krishnaiah K C K and Ashutosh A 1994 *Indian J. Chem. Technol.* **1** 13

13. Siddiqui B A, Sharama P P and Sultan M 1999 *Indian J. Environ. Protection* **19** 846

14. Anirudhan T S and Sreedhar M K 1998 *Poll. Res.* **17** 381

15. Viswanadham M, Sriramulu N and Chary A M 2000*Indian J. Environ. Protection* **20** 515

16. Jha I N, Iyengar L and Prabhakara Rao A V S 1998 *J.Environ. Engg.* **114** 962

17. Ruey-Shin Juang, Ru-Ling Tseng, Feng-Chin Wu,Shwu-Hwa Lee 1997 *J. Chem. Tech.Biotechnol.* **70** 391

18. Rengaraj S, Sivabalan R, Arabindo B and MurugesanV 2000 *J. Indian Chem. Soc.* **7** 127

19. Rama Mohan K, Ramesh A and Seshaiah K 2001 *In-dian J. Chem. Technol.* **8** 278

20. Treybal R E 1980 *Mass transfer operations* (NewYork: McGraw Hill)

21. Karthikeyan G, Pius A and Alagumuthu G 2002 *In-dian J. Chem. Technol.* **9** 397

22. Annadurai G, Chellapandian M and Krishnan M R V 1997 *Indian J. Environ. Protection* **17** 95

23. Manju G N, Gigi M C and Anirudhan T S 1999 *In-dian J. Chem. Technol.* **6** 134

24. Stephan Inbaraj B and Sulochana N 2002 *Indian J.Chem. Technol.* **9** 201

25. Viswakarma P P, Yadava K P and Singh V N 1989

Adsorption dynamics and equilibrium studies of Zn (II) onto chitosan

G KARTHIKEYAN*, K ANBALAGAN and N MUTHULAKSHMI ANDAL

Department of Chemistry, Gandhigram Rural Institute – Deemed University, Gandhigram 624 302, India e-mail: drg_karthikeyan@rediffmail.com

MS received 3 June 2003; revised 12 November 2003

Abstract. Batch equilibration studies are conducted to determine the nature of adsorption of zinc (II) over chitosan. The factors affecting the adsorption process like particle size, contact time, dosage, pH, effects of chloride and nitrate are identified. The influence of temperature and co-ions on the adsorption process is verified. The fraction of adsorption, Y_t and the intraparticle diffusion rate constant, k_p are calculated at different environments and the results are discussed. The natureof adsorption of the zinc (II) – chitosan system is explained using Freundlich, Langmuir isotherms and thermodynamic parameters.

Keywords. Adsorption dynamics; chitosan; pH; fraction of adsorption; intraparticle diffusion; co-ions.

1. Introduction

Zinc occurring in the earth's crust at an average concentration of 70 mg/kg is classified asa border- line metal. Under aerobic conditions, Zn^{2+} is the predominant species at acidic pH, but it is replacedby $Zn(OH)_2$ at pH 8–11.[1] Ingestion of > 2 g zinc produces toxic symptoms (fever, diarrhoea and gastrointestinal tract irritation) in humans. There are also many incidents of extensive mortality and/or contamination of fish and other aquatic species from the discharge of zinc-containing wastewater.

The present study deals with the sorption of zinc (II) by chitosan, a deacetylated product of chi-tin as an adsorbent. Chitosan is obtained by deacetylation of chitin, whichis extracted from the shells of shrimps, crabs and other crustaceans and somefungi.[2]

Chitosan chelates five to six times greater amounts of metals than chitin.[3] This is attributed to the deacetylation process. Chitosan has many useful features e.g. hydrophilicity, biocompatibility, bio- degradability, anti-bacterial property and remark- able affinity for many proteins.[4] This polymer is a well known sorbent, effective in the uptake of metal ions, since the amine groups on the chitosan chain can serve as chelation sites for metals. The mechanism by which metal ions are bound by chitosan probably involves attachment of these ions to $-NH_2$ groups. Because of this difference, it may be suit able for scavenging important heavy metal ions and complexes that cannotbe adequately treated by other natural polymers.[5,6] Guibal *et al*[7] studied the enhancementof metal ion sorption performances of chitosan. Chitosan displayed a surface controlled sorption mechanism, indicating a monolayer sorption with interactions between the sorbed molecules and heterogeneous distribution of sorption energies. Chitosan interaction with metal ions has also been studied by Schmuhl *et al*[8].

2. Experimental

The raw sample of chitosan was supplied by M/s BR Corporation, Mumbai, in the form of coarse grains. Chitosan was pulverized and sieved into two grades, 0·21 mmand 0·71 mm. The materials with different dimensions were treated with 0·1 N hydrochloric acid and 0·1 N sodium hydroxide for three hours each. The purified samples were washed several times with double distilled water till it registered neutralpH, as swelling would make more sites available. Later the samples were dried in anoven. All the chemicals employed for the analysis were either AnalaR, GR or other high purity grade and used without further purification. Zinc sulphate, sodium chloride, sodium nitrate, hydrochloric acid and sodium hydroxide were used for the experiments. The experiments were performed using double distilled water From thestock solution containing 1000 mg/L of zinc ions, aliquots of the adsorbate containingvarying concentrations from 2–14 mg/L of zinc ions were prepared. 50 ml of the samewas mixed with 10 mg each of two grades of chitosan and the mixtures were agitatedin a mechanical shaker. The solution was filtered at preset time intervals (2 min) and the residual metal ion concentrations were measured using Atomic Absorption Spectrometer (Model AA 100, Perkin Elmer make). The experiments were conductedat four different temperatures, viz., 293 K, 303 K, 313 K and 323 K using a thermostat attached with a shaker, TECHNO make whose accuracy is within ±0·5°C. Adsorptionstudies of zinc (II) ions in the presence of three metal ions viz., iron (III), copper (II)and chromium (VI) were experimentally verified to study the effect of co-ions. This involved the determination of the residual zinc (II) ions in the mixture containing equal amounts of iron (III), copper (II) and chromium (VI) ions. The pH was measured using Elico L1 120 pH meter. Duplicate measurements were also conducted to ensurethe reproducibility of the values of residual concentration within ±2%.

3. Results and discussion

3.1 Effect of particle size

The adsorption rate of zinc ions by chitosan with two particle sizes (0·21 and 0·50 mm) is illustrated in figure 1 and the values are given in table 1.

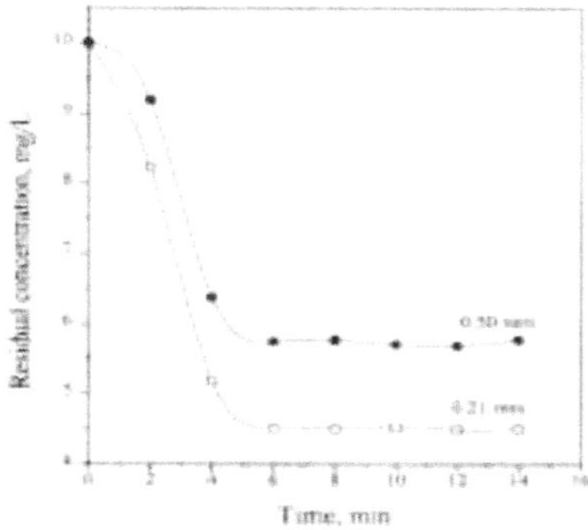

Figure 1. Effect of particle size.

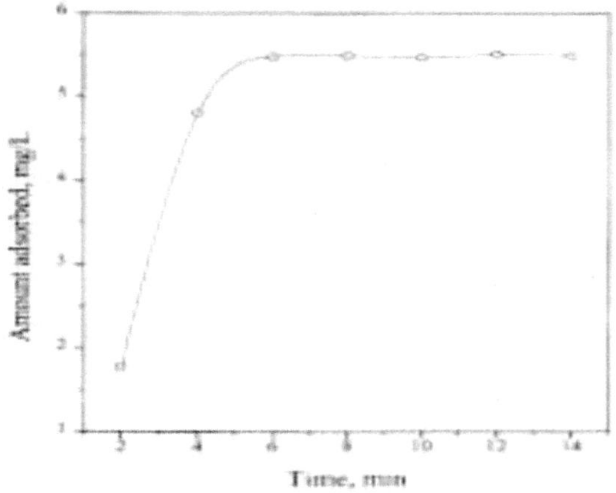

Figure 2. Effect of contact time.

Table 1. Effect of particle size.

Time (min)	Concentration (mg/L)	
	0.21 mm	0.60 mm
0	10.00	10.00
2	8.22	9.79
4	5.49	6.38
6	4.62	5.75
8	4.61	5.77
10	4.63	5.71
12	4.60	5.69
14	4.61	5.78

Initial concentration of zinc ions $[Zn]_o$ = 10 mg/L, pH = 7; temperature = 303 K; adsorbant dose = 10 mg

Table 2. Effect of contact time.

Time (min)	Amount adsorbed (mg/L)
2	1.78
4	4.81
6	5.48
8	5.49
10	5.47
12	5.50
14	5.49

Initial concentration of zinc ions $[Zn]_o$ = 10 mg/L; pH = 7; temperature = 303 K; adsorbant dose = 10 mg

evident from the table that the concentration of zinc (II) adsorbed decreases with an increase in the particle size of chitosan, explaining the diminishing sorption rate at large dimensions. High rate of ad sorption by chitosan with smaller particle sizes is due to the availability of more specific surface area on the adsorbent. According to Weber and Morris,9 the breaking of larger particles tends to open tiny cracks and channels on the particle surface of the material resulting in more accessibility to betterdiffusion, owing to the smaller particle size.10

3.2 Effect of contact time

The nature of the adsorption of zinc (II) on chitosan is illustrated in figure 2 and the experimental results are given in table 2. An initial steep increase in the adsorption curve followed by a bend at the sixth minute was observed. No further adsorption of zinc (II) was evident and this indicated the formation of an equilibrium at the sixth minute.

3.3 Effect of dose The experimental results using three doses viz. 10, 20 and 30 mg of chitosan and 10 mg/L of zinc (II) revealed a definite increase in the adsorption capacity of chitosan with dosage. This is due to the larger number of available adsorption sites favouring the enhanced uptake of the metal ion. The dependence of adsorption on doses can be explained in terms of the concentration factor involving low metal-to adsorbent ratio at low metal concentrations. Subsequently, fractional adsorption becomes independent of initial concentration. However, at higher concentrations, the available sites of adsorption become fewer and hence the percentage of metal ion re moved is dependent upon the initial concentration.11 The rate of adsorption was appreciable at an initial dose of 10 mg of chitosan although at its higher doses the rate registered an increase. However, a proportionate increase in the rate was absent at higher doses. In view of this, 10 mg of chitosan was fixed as the dose for the remaining experiments along with 0·21 mm particle size and six minutescontact time as the optimum conditions for the further experiments.

3.4 Effect of pH

The sorption characteristics of chitosan determined at pH 3, 5, 7 and 8 against time isillustrated in figure 3. The system is strongly pH dependent, because the properties ofboth the chitosan surface (charge and potential) and the solution composition (metal ion speciation) change with pH.12 The number of active sites on the surface of the adsorbent may change with varying pH. The rate of adsorption is maximum at pH 7. Table 3 explains the details of the residual concentrations of zinc (II) in solution at definite time intervals at four pH levels. The sorption rate is lower at acidic and alkaline ranges. Low pH would favour protonation of the amino sites on the surface, resulting in a reversal of charge leading to a reduction in the chelating ability of chitosan.8 At alkaline range, the formation of Zn (OH)2 complexes prevent further adsorption of zinc (II). 3.5 Effect of chloride and nitrate The adsorption of zinc (II) by chitosan in presence of excess chloride is shown in figure 4. The percentage of zinc

(II) removed by chitosan decreased at higher concentrations of chloride ions. The inhibition in the sorption rate of chitosan is due to the affinity between Zn2+ ions andCl– ions. The low adsorption rate of zinc ions viz., 16·6% at a concentration of 1000 mg/L of Cl– in the solution is due to the formation of zinc chloride. That is, the amount of free zinc ions remaining in solution for adsorption is reduced considerably and hence the low adsorption rate of zinc ions. Our findings are in good agreement with the observations of earlier workers.13,14 A similar inhibitory trend in the adsorptionof zinc was observed in nitrate environment also. A gradual reduction in the adsorption rate resulted at higher concentrations of nitrate ions. The adsorption rate of zinc (II) is reduced to half of its original value, as observed at 500 mg/L of NO– amounting to a reduction from 54·80% to 26·35%.

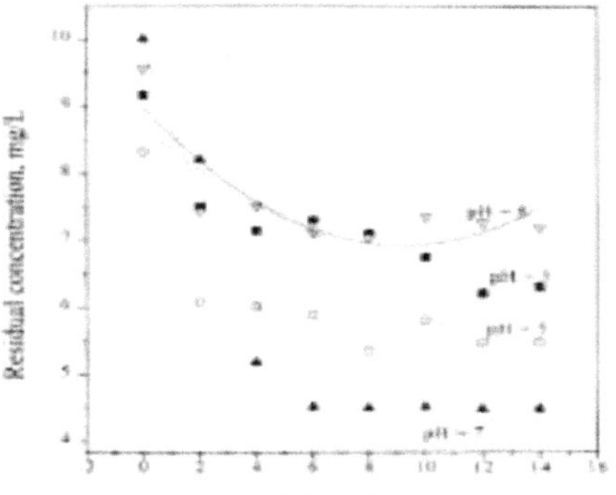

Figure 3. Effect of pH.

Table 3. Effect of pH.

Time (min)	Concentration (mg/L)			
	pH = 3	pH = 5	pH = 7	pH = 8
0	9.47	8.33	10.00	9.65
2	7.60	6.09	8.22	7.43
4	7.45	6.02	5.49	7.61
6	7.30	5.89	4.62	7.42
8	7.40	5.36	4.61	7.05
10	6.75	5.82	4.63	7.35
12	6.22	5.48	4.60	7.26
14	6.31	5.60	4.61	7.28

Initial concentration of zinc ions $[Zn]_0$ = 10 mg/L; temperature = 303 K; adsorbant dose = 10 mg

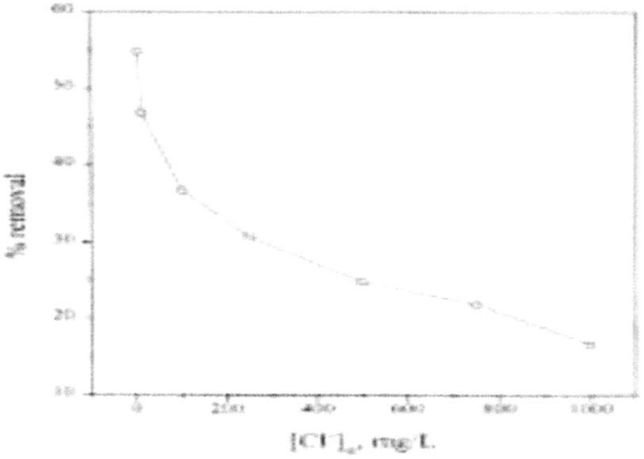

Figure 4. Effect of chloride ion concentration.

Table 4. Time dependence of the fraction of adsorption at different particle sizes.

$T^{1/2}$ ($min^{1/2}$)	F	
	0.21 mm	0.60 mm
0	0	0
1.41	0.32	0.49
2.00	0.87	0.85
2.45	1.00	1.00
2.83	0.95	1.00
3.16	0.94	1.01
3.46	0.95	1.01
3.76	0.94	0.99

Table 5. Time dependence of the fraction of adsorption at different amounts.

$t^{1/2}$ ($min^{1/2}$)	Y_t		
	10 mg	20 mg	30 mg
0	0	0	0
1.41	0.32	0.63	0.89
2.00	0.87	0.94	0.94
2.45	1.00	1.00	1.00
2.83	0.95	1.04	1.01
3.16	0.94	1.03	1.01
3.46	0.95	0.99	1.00
3.76	0.94	1.03	1.00

3.6 Time dependence of the fraction of adsorption at different particle sizes and doses of chitosan

Time dependence (tables 4 and 5) is determined by plotting the fraction of adsorption, Yt =C0 – Ct/C0 – Ce against the retention time, t 1/2 , where C0 denotes the initial concentrationof metal ions, Ct, the con centration of metal ions at the preset time intervals and Ce, the equilibrium concentration. The rate of fraction of adsorption, Yt, expressed as square root of time,7 is estimated from the three-staged curves illustrated for different particle sizes anddoses of chitosan as shown in figures 5 and 6. The initial stage of the curves relate to the transfer of metal ions from the bulk of solution to the boundary film of the adsorbent and later to its surface. The second stage, corresponds to the transfer of the metal ions from the surface to the intraparticular active sites of chitosan. This stage is the rate-determining stepof the reaction. The final stage shows the completion of sorption reaching an equilibrium.15 At equilibrium, there is a defined distribution of the solute particle between the solution andthe solid chitosan. Ruey–Shin Juang et al6 reported similar observations while studying the adsorption of reactive dyes over chitosan. 3.7 Intraparticle diffusion Sorption performancesof chitosan are determined by the influence of diffusion mechanisms on zinc (II) uptake. Weber and Morris9 pointed out that a functional relationship common to the majority of the intraparticle diffusion treatments is that the up take varies almost proportionately with the half power of time. The influence of intraparticle diffusion is estimated by evaluation of initial sorption rate. Figures 7 and 8 describe the plots of the sorption capacity, qe against time squared. The kp values given in tables 6 and 7 for the two different particle sizes are 3·04 and 0·75, and those for different doses are 3·04, 9·74 and 13·9 respectively. This showsthat the initial slope of the curves indicating the sorption rate increases with doses and decreases with particle size of the material.16

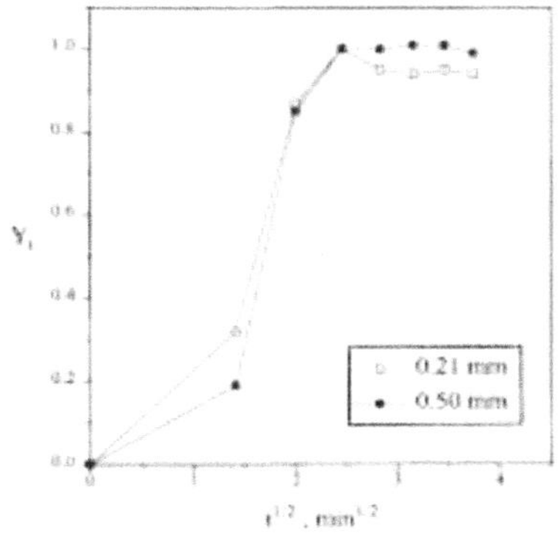

Figure 5. Time-dependence of the fraction of adsorption at different particle sizes of chitosan.

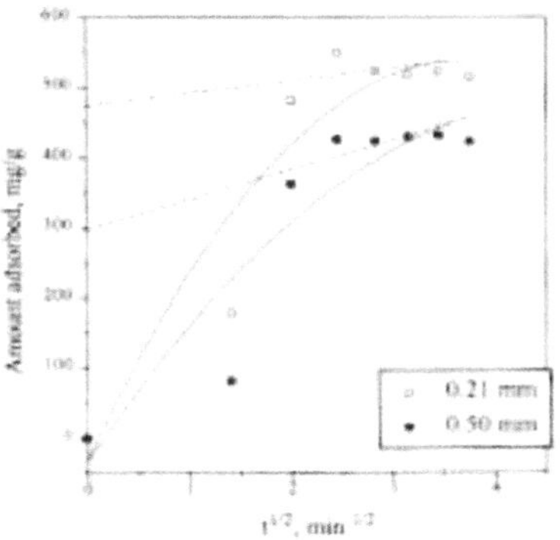

Figure 7. Intraparticle diffusion plot for the adsorption of zinc on chitosan at different particle sizes.

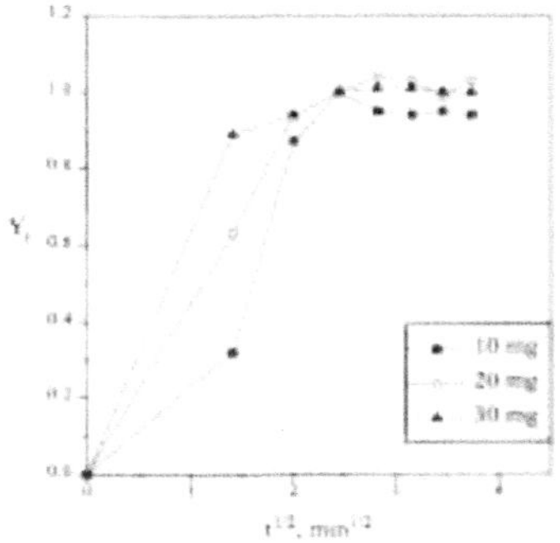

Figure 6. Time-dependence of the fraction of adsorption at different amounts of chitosan.

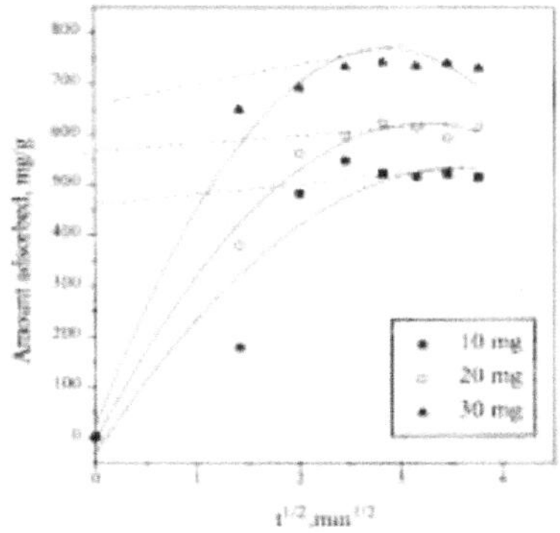

Figure 8. Intraparticle diffusion plot for the adsorption of zinc on chitosan at different dosages.

3.8 Effect of temperature

The temperature dependence of zinc (II) by chitosan was studiedover the range of 293–323 K. The considerable increase in the amounts of zinc (II) adsorbed with temperature confirms the endothermic nature of the process. The enhancement of adsorption capacity of chitosan, with temperature is attributed to the possible increase in the number of active sites available for adsorption on the surface.17

Table 6. Intraparticle diffusion rate constant.

Particle size (mm)	k_p value
0·21	3·04
0·60	0·75

Table 7.

Dosage (mg)	k_p value
10	3.04
20	9.74
30	13.90

Effect of co-ions

The study of the influence of co-ions on the removal of zinc (II) ions becomes relevant when the selective adsorption of the metal ion occurs from the aqueous system. The experimental results revealed that the presence of foreign ions such as Fe (III), Cu (II) andCr (VI) diminished the adsorption of zinc (II) ions. This is attributed to the fact that the hydrated copper ions have smaller radii (0·419 nm) than hydrated zinc ions (0·430 nm) andthus have greater access to the surface of the adsorbent.18 Jha et al13 have also reported asimilar co-ionic effect in the removal of specific metal ions in their studies using chitosan

3.11 Adsorption isotherms

Freundlich19 and Langmuir20 adsorption isotherm models were applied to the system, studied at four temperatures. 3.10a Freundlich isotherm: Batch adsorption isothermal data, fitted into the linear form of the Freundlich isotherm is shown in figure 9 (logqe versus log Ce). The adsorption capacity, K and the ad sorption intensity, 1/n are directly obtained from the slopes and the intercepts of the linear plot,respectively and the data are provided in table 8. The K, which is a measure of adsorption capacity, increased with temperature. The magnitude of the exponent, n gives an indication of the favourability and capacity of the adsorbent/adsorbate system. Treybal21 has reported that 'n' values between 1 and 10 represent favourableadsorption conditions.22 In all cases reported here, the exponent is $1 < n < 2$ showingbeneficial adsorption for the system. The plot of logqe versus logCe for various initial concentrations is lin ear indicating the applicability of Freundlich adsorption isotherm ($0{\cdot}9894 < r < 0{\cdot}9982$ and 0·0163

< sd < 0·0408). The fit of the data to the Freundlich model indicate that the forces of adsorption by chitosan are governed by physisorption.

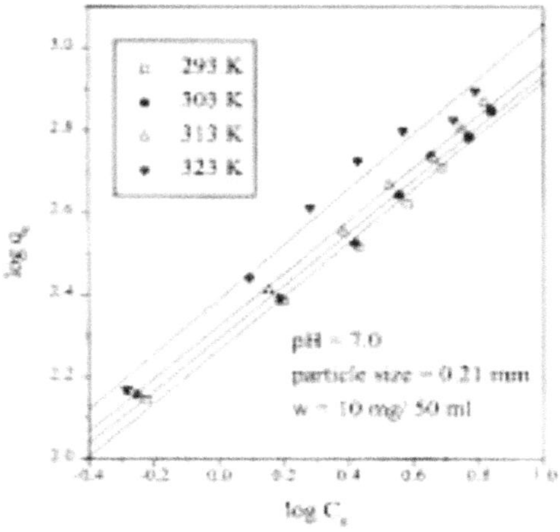

Figure 9. Freundlich isotherm.

Table 8. Freundlich parameters.

Temperature (K)	*K*	1/*n*
293	2·2658	0·6529
303	2·2937	0·6422
313	2·3236	0·6428
323	2·3893	0·6707

3.9 b Langmuir isotherm:

Zinc(II)–chitosan system obeys the Langmuir adsorption isotherm. The Ce and Ce/qe values and the Langmuir constants, Q and b for different temperatures calculated are presented in tables 9 and 10 respectively. The linear plots of Ce/qe versus Ce (0·8977 < r <0·9845 and 0·3298 < sd < 1·0437) at the four temperatures indicate the applicability of Langmuir adsorption isotherm on the adsorption process occurring through a monolayer coverage. The Langmuir plots showing the increase in the rate of adsorption at higher temperatures are presented in figure 10. The Ce values of adsorbate in solution are higher atlower temperatures and they decrease with increasing temperature indicating greater adsorption at higher temperatures. The increasing trends observed in Q and b provided

Table 9. Effect of temperature on equilibrium parameters.

Concentration of zinc ion (mg/L)	C_e (mg/L) Temperature (K) 293	303	313	323	C_e/q_e (mg/L) Temperature (K) 293	303	313	323
2	0.60	0.56	0.54	0.52	3.49	3.89	3.70	3.51
4	1.58	1.54	1.42	1.24	6.63	6.26	5.60	4.49
6	2.71	2.64	2.42	1.91	8.24	7.86	6.76	4.66
8	3.82	3.60	3.34	2.68	9.44	8.48	7.47	5.04
10	4.89	4.52	4.60	3.70	9.57	8.24	8.62	5.87
12	5.91	5.88	5.60	5.30	9.70	9.61	8.75	7.91
14	6.93	6.90	6.58	6.49	9.80	9.71	8.86	7.93

Table 10. Langmuir constants and thermodynamic parameters.

Temp. (K)	Langmuir constants b (L mg^{-1})	Q (mg g^{-1})	Thermodynamic parameters K_o	$\Delta G°$ (kJ mol^{-1})	$\Delta H°$ (kJ mol^{-1})	$\Delta S°$ (J mol^{-1} K^{-1})
293	0.4703	1.4857	5.39	−4.09		
303	0.4778	1.2095	5.61	−4.30	2.2090	21.50
313	0.4985	1.2145	5.65	−4.52		
323	0.2546	1.2442	5.84	−4.73		

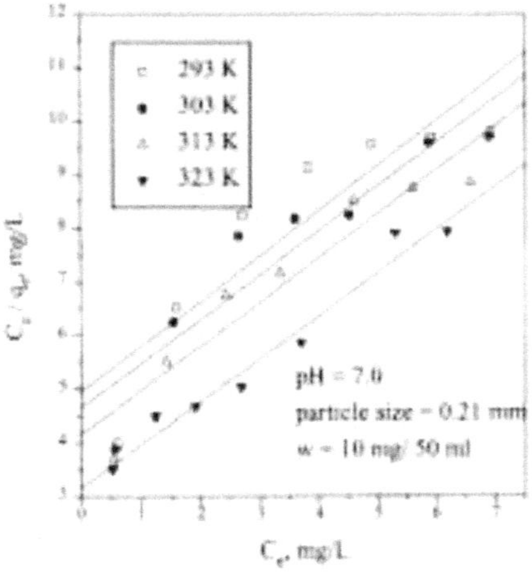

Figure 10. Langmuir isotherm.

in the table confirm the enhanced adsorption at higher temperature. The enhanced sorptivenature of chitosan at higher temperature is due to the swelling in the internal pores of the adsorbent to trap in more zinc ions on its surface. The above results are in good agreementwith the earlier reports from this laboratory, pertaining to fluoride adsorption studies by dolomite.23

3.10 Equilibrium parameter

The essential characteristics of Langmuir equation can be expressed in terms of adimensionless separation factor or equilibrium parameter, RL. 24 These values indicate the shape of the isotherm to be either unfavourable (RL > 1), linear (RL = 1),favourable (0 < RL < 1) or irreversible (RL = 0). The values of RL obtained for the zinc (II) – chitosan system are provided in table 11. The RL values, for 2–14 mg/L ofzinc ions in the solution, lying between 0 and 1, indicate favourable adsorption at fourtemperatures.

3.11 Adsorption dynamics

The thermodynamic equilibrium constants, Ko for the sorption reactions are determined by the method suggested by Khan and Singh25 by plotting ln(qe/Ce)versus qe (figure 11) and extrapolating to zero. The endothermic nature of adsorption is indicated by an increase in the Ko values with a rise in temperature.Thermodynamic parameters are calculated from the variation of the thermodynamic equilibrium constants Ko and the values are given in table

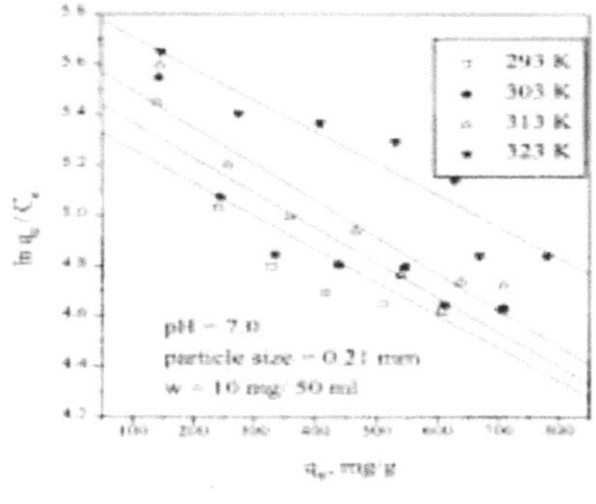

Figure 11. Plot of $\ln q_e/C_e$ versus q_e.

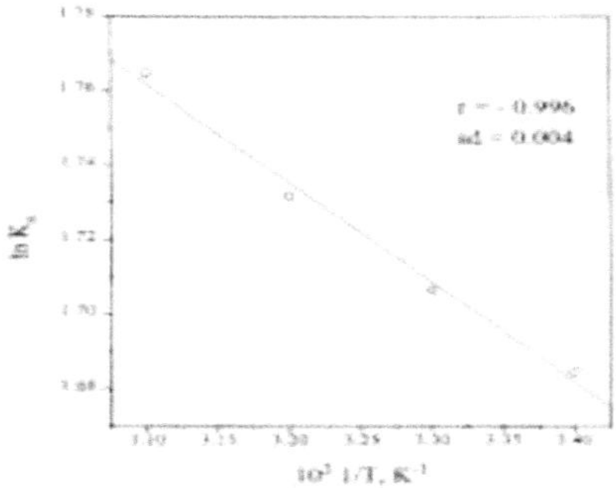

Figure 12. Plot of $\ln K_o$ versus $1/T$.

10. Aplot of lnKo versus 1/T is linear as indicated in figure 12. ΔH°and ΔS°are determined from the slopes and intercepts of the plot and the values are given intable

10. The negative values of ΔG°indicate the feasibility of the process and also the spontaneity of adsorption reaction. The amount of zinc (II) adsorbed at equilibrium must increase with increasing temperature, because ΔG° decreases with the rise in temperature of the solution. This explains why the value of ΔG°is negative with the rise in temperature. The value of ΔH°(2·2090 kJ/mol) is positive and it confirms the endothermic character of the reaction. The positive value of ΔS°(21·50 J/K/mol) shows the increased randomness at the solid-solution interface during the adsorption of Zn (II) on chitosan.26

Table 11. Equilibrium parameter.

Concentration of zinc ion (mg/L)	Temperature (K)			
	293	303	313	323
2	0·746	0·738	0·716	0·663
4	0·695	0·684	0·657	0·495
6	0·495	0·484	0·456	0·396
8	0·423	0·413	0·386	0·329
10	0·370	0·360	0·335	0·282
12	0·329	0·319	0·296	0·247
14	0·295	0·286	0·265	0·219

4. Conclusion

It is observed from the present study that an optimum contact period of six minutes is required forthe maximum removal of zinc (II) by chitosan. The adsorption mechanismalso depends upon the particle size of the adsorbent, pH of the medium and the presenceof other anions like chlorides and nitrates. Also, the co-ions like Fe (III), Cu (II) and Cr

(VI)were observed to inhibit the adsorption rate significantly. The fraction of adsorptionincreases with de- crease in particle size and increase in the dosage of chitosan. The intraparticle diffusion rate constant, k_p is found to be maximum for smaller particle size and higher dose of the adsorbent. The adsorption mechanism obeys Freundlich and Langmuir equations indicating beneficial adsorption occurring through a monolayer mechanism involving physisorption and/or chemisorption. Thermodynamic and equilibrium parameters show that the adsorption process is endothermic and spontaneous, which im- plies increased sorption at higher temperatures.

References

1. Cunningham W P 1999 *Environmental encyclopedia* 2nd edn (Chennai: Jaia)
2. McKay G, Blair H S and Findon A 1989 *Indian J.Chem.* **A28** 356
3. Yang T C and Zall R R 1984 *Indian Eng. Chem.Prod. Res. Dev.* **23** 168
4. Ravi Kumar M N V 1999 *Bull. Mater. Sci.* **22** 905
5. Randall J M, Randall V G, McDonald G M and Young R N 1979 *J. Appl. Polym. Sci*

   23 727
6. Ruey-Shin Juang, Ru-Ling Tseng, Feng-Chin Wu and Shwu-Hwa Lee

   1997 *J.Chem. Technol. Biotechnol.* **70** 391
7. Guibal E, Jansson-Charrier M, Saucedo I and Le Clo- riec P 1995 *Langmuir* **11** 591
8. Schmuhl R, Krieg H M and Keizer K 2001 *Water SA* **27** 1
9. Weber W J and Morris J C 1963 *J. Sanit. Engg. Div.ASCE* **89** (SA2) 31
10. Selvaraj K, Chandramohan V and Pattabhi S 1998 *Indian J. Environ. Protect.*

    18 641
11. Parekh D C, Patel J B, Sudhakar P and Koshy V J2002 *Indian J. Chem. Technol.*

    9 540
12. Gonzalez M, Magdalena Santana-Casiano JDand Millero F J

    1990 *J.Colloid Interface Sci.* **137**102
13. Jha I N, Leela Iyengar and Prabhakara Rao A V S1988 *J. Environ. Engg.* **114** 962
14. Anirudhan T S and Sreedhar M K 1988 *Poll. Res.* **17**381
15. Peniche-Covas C, Alvarez L W and Arguelles-MonalW 1992 *J. Appl. Polym. Sci.*

46 1147

16. Raji C and Anirudhan T S 1998 *Ecol. Environ. Conserv.* **4** 33

17. Anoop Krishnan K and Anirudhan T S 2002 *Indian J.Chem. Technol.* **9** 32

18. Nightmighle K R 1959 *J. Phys. Chem.* **63** 1381

19. Freundlich H 1926 *Colloid capillary chemistry* (Lon-don: Metheun)

20. Langmuir I J 1918 *J. Am. Chem. Soc.* **40** 1361

21. Treybal R E 1980 *Mass transfer operations* (New York: McGraw Hill)

22. Stephan Inbaraj B and Sulochana N 2002 *Indian J. Chem. Technol.* **9** 201

23. Karthikeyan G, Pius A and Alagumuthu G 2002 In-dian J. Chem. Technol. 9 397

24. Singh D K and Srivastava B 2001 Indian J. em. Technol. 8 133

25. Khan A A and Singh R P 1987 *Colloid Surf.* **24** 33

26. Thomas R and Anirudhan T S 1999 *Indian J. Chem. Technol.* **6** 268

Publisher: Eliva Press Global Ltd

Email: info@elivapress.com

Made in the USA
Middletown, DE
04 November 2026

36553125R00245